ISBN 978-0-265-87593-3
PIBN 10903001

SEMINAR REPORTS

I Semester 1942-43

CLEAVAGE OF ARALKYL ETHERS (1)

A considerable proportion of the natural products for which syntheses have been and are being sought contain phenolic hydroxyl groups. In nearly all cases it is necessary to start such a synthesis with a compound containing this group in a modified form, so that its active hydrogen will not interfere with the reactions to be carried out in the course of the synthesis. The choice of a modifying group must be made, bearing in mind first, the conditions under which it must hold up during the course of the reactions to follow, and second, the conditions necessary to remove it at the end of the synthesis and the effect of these conditions upon the rest of the molecule.

In general, esterification of the phenolic hydroxyl affords insufficient protection, since most phenolic esters are highly susceptible to hydrolytic conditions. As a consequence, it is most often necessary to etherify the phenol in order to have assurance that the protecting group will stay in place until it is desirable to remove it. In recent years, therefore, the problem of cleavage of these ethers at the end of a synthesis has come up repeatedly. Since, in most cases, the molecules are relatively complex, the choice of a reagent for the cleavage must be carefully made. It is valuable, then, to the organic chemist to have available to him a knowledge of the various conditions used to effect these ether cleavages. This review is presented for that reason.

<u>Thermal Cleavage</u>

While aliphatic ethers usually are stable at 500° and above, most aralkyl ethers are cleaved at temperatures below 400°. The products are olefins and the phenol.

$$C_6H_5OC_6H_{11} \rightarrow C_6H_5OH + C_6H_{10}$$

$$C_6H_5OCH_3 \rightarrow C_6H_5OH + C_2H_4$$

Although phenol-alkyl ethers require drastic conditions for thermal cleavage, the β-alkenyl ethers are split at relatively low temperatures. This is particularly true when the structure of the alkenyl group is such that a conjugated system can form.

$$C_6H_5OCH_2CH{=}C(CH_3)_2 \xrightarrow[\text{atm.}]{\text{distil}} C_6H_5OH + CH_2{=}C(CH_3)CH{=}CH_2$$

Of course, the Claisen rearrangement often interferes when the temperature reaches 200°.

Reductive cleavage

Phenol-alkyl ethers may be cleaved by catalytic reduction, but the aromatic ring is also reduced in the process.

$$C_6H_5OCH_3 \rightarrow C_6H_{11}OH + CH_4$$

Benzyl ethers of phenols, however, are most readily split by this method, giving the phenol and toluene. This transformation can be brought about in the presence of Raney nickel at 100° or by platinum at room temperature. A similar reaction is which sugar hydroxyls modified by benzyl ether groups are regenerated by reduction is well known and widely used.

Allyl-phenyl ethers are so readily cleaved by reduction that the cleavage competes with reduction of the allyl group. In the presence of palladium at room temperature, hydrogenation of phenyl allyl ether gives 25% of the cleavage product and 75% of propyl phenyl ether.

The combination sodium and alcohol effects an unusual cleavage, in which the aromatic carbon-oxygen bond is broken. For example, gallic acid trimethyl ether is reported to give m-methoxybenzoic acid. The presence of a carboxyl group is not essential to this type of reaction, for pyrogallol trimethyl ether is converted to resorcinol dimethyl ether under the same conditions. Although it has been postulated that this unorthodox cleavage is brought about by addition of sodium to the aromatic nucleus, followed by loss of the elements of sodium methoxide, it seems more likely that the process is one of reductive alcoholysis in a vinylogous system particularly susceptible to such attack.

Cleavage accompanied by dehydrogenation

Systems which are capable of ready transformation to quinoid form afford potential ether cleavage on dehydrogenation. Hydroquinone monomethyl ether is readily converted to quinone by silver nitrate or by nitric acid. Robinson and his coworkers (4) have shown that pyrogallol tribenzyl ether is oxidized by nitric acid to 2,6-dibenzyloxybenzoquinone in good yield.

Cleavage by means of mineral acids

The use of hydriodic acid to effect cleavage of aralkyl ethers is too well known to require elaboration. The ease with which this acid brings about the splitting of ethers is attributed to a combination of three factors:

1. Its relatively high dissociation constant
2. The high concentration of acid obtainable at the boiling point
3. Its rather good solvent properties

When complex molecules are to be dealkylated, it is often necessary to increase the solubility of the compounds in the medium. The solvents most frequently used for this purpose are acetic acid and phenol.

Within the past few years, 48% hydrobromic acid in acetic acid has become widely used a cleavage agent where hydriodic acid was found to bring about undesirable side reactions. In general, about a three-fold excess of hydrobromic acid is employed, together with sufficient acetic acid to bring the material into solution at the boiling point. If further dilution is desired, acetic anhydride is frequently used. A striking example of the successful use of this reagent was provided by John and Gunther (2), who opened the door to the synthesis of a large number of homologues and analogues of α-tocopherol (Vitamin E) by showing that highly substituted hydroquinone dimethyl ethers could be demethylated smoothly without bringing about other drastic changes in the molecule. Lack of a reagent to accomplish this purpose had previously held up Vitamin E research for some time.

Anhydrous mineral acid combinations have usually had little success. Dry hydrogen chloride and hydrogen bromide in acetic acid, for example, have proved poor reagents for the purpose. The reason for the failure of these reagents probably is that without water, the dissociation constant of the acid is too low to be effective.

Compounds showing resistance to cleavage under normal conditions may often be successfully split by the use of hydrochloric or hydrobromic acid in a sealed tube at elevated temperatures. A variety of concentrations, solvents and diluents have been employed in these reactions.

The sealed tube with its limitations and difficulties may often be avoided by using a reagent with which the ether is fusible and which liberates hydrogen halide at elevated temperatures. Aniline hydrochloride and also the hydrobromide have been used in this fashion. The molten mixtures prove good dissociation media for the liberated acid. On some occasions, hydrogen halide has been passed through the mixture during the course of the reaction. Prey (3) has recently shown that pyridine hydrochloride is a more effective reagent than the aniline salt. Whereas the latter will accomplish cleavage in molecules having more than one alkoxyl group, pyridine hydrochloride readily cleaves anisole itself, which is not attacked by aniline hydrochloride. Selective cleavage in molecules containing more than one ether linkage is uncertain and difficult, however.

Rate studies have shown that in solutions containing a tenfold excess of halogen acid in acetic acid, the relative rates of cleavage are HCl:HBr:HI equal 1:6: ∞ . Constitutive factors in the aromatic ethers are also highly influential. Substituted anisoles are split by halogen acids with an ease indicated by the following series: $CH_3O > CH_3 > OH > H > NH_2.HCl > Cl > Br$. The ease of cleavage also varies with the position of the substituent in the order p > o > m. In the cases of *meta*-orienting substituents, such as NO_2 and CH_3CO, however, the order of substitution effect is reversed.

In the case of mineral acid dealkylations as in others, it is found that benzyl ethers are much more readily removed than saturated aliphatic groups. Robinson and his coworkers (4) have made excellent use of this property in the synthesis of 2,6-dihydroxy-1,4-benzohydroquinone dimethyl ether from pyrogallol tribenzyl ether. As previously described, the latter compound was oxidized to 2,6-dibenzyloxyquinone by means of nitric acid. The quinone was reduced to the hydroquinone, converted to the dimethyl ether, and the two benzyl groups removed by means of hydrochloric acid in acetic acid at 65°.

Cases have been observed in which concentrated sulfuric acid and chlorosulfonic acid have split aromatic ether linkages. In most cases the dealkylation occurs simultaneously with sulfonation, although the dibutyl and diamyl ethers of catechol are cleaved and not sulfonated by both sulfuric acid and chlorosulfonic acid.

Throughout this discussion of ether cleavage by means of mineral acids, the degree of dissociation of these acids in the chosen medium has been singled out as an important requisite for the success of the cleavage. There is no question that these reactions involve an intermediate oxonium compound which decomposes with the formation of alkyl halide and phenol. Of course, the concentration of protons liberated by the mineral acids is a primary factor in oxonium compound formation; and it is, therefore, important to provide a medium which permits a high degree of dissociation of the mineral acid.

Cleavage by means of metallic salts, acid halides and acid anhydrides

Salts such as $AlCl_3$, $AlBr_3$, $ZnCl_2$, $FeCl_3$, $SnCl_4$, $SbCl_5$, BF_3, BBr_3, and others have been shown to bring about smooth cleavage of aralkyl ethers in many cases. Ethers of polyhydric phenols are sometimes selectively cleaved in good yield without the disturbance occasioned by numerous side reactions. It has been amply demonstrated that oxonium salt formation is intermediate in this reaction.

Aluminum bromide has proved a particularly useful reagent, since it is soluble in such indifferent solvents as benzene, petroleum ether, and carbon bisulfide. A number of the complex salts of this compound and phenolic ethers have been isolated from these solvents and well characterized.

The use of boron tribromide as a cleavage agent for aralkyl ethers has formed the subject of a recent paper (5). Although only preliminary results are available, the reagent appears promising.

<u>Cleavage by alkalis</u>

The use of sodium and potassium hydroxides in aqueous or alcoholic solution as well as sodium alkoxides as a reagent for the cleavage of aralkyl ethers has proven particularly effective in cases where acidic reagents attack the major portion of the molecule. By making a careful choice of concentration of alkali and temperature at which the reaction is carried out, one may often carry out a relatively clean-cut selective cleavage of polyhydric phenol ethers. For example, veratrole is converted to guaiacol by heating for three hours in alcoholic potassium hydroxide at 170°.

The most celebrated example of the application of this method is the synthesis of stilboestrol by Dodds and his co-workers (6). The demethylation of the dimethyl ether in the final step could not be carried out by means of any of the acidic reagents, but 30% potassium hydroxide in ethanol at 225° gave a practically quantitative yield of stilboestrol.

<u>Cleavage by alkali metals</u>

Shorigen showed that phenol ethers are cleaved by alkali metals at elevated temperatures, but there was such a variety of side reactions that the method had little practical value. Sodium-potassium liquid alloy has been occasionally used at room temperature, although the ordinary aralkyl ethers are not attacked by this reagent. Benzyl and trityl ethers are, however, quite readily split in this medium, although the reaction is more smoothly effected by sodium in liquid ammonia. Here again, however, there are side reactions, and it is usually impossible to stop the reaction at the desired stage.

$$C_6H_5OCH_2C_6H_5 \xrightarrow[2Na]{} C_6H_5ONa + NaCH_2C_6H_5$$

$$C_6H_5OCH_2C_6H_5 + NaCH_2C_6H_5 \rightarrow C_6H_5ONa \quad C_6H_5CH_2CH_2C_6H_5$$

Cleavage by organo-metallic compounds

During the course of his work Grignard observed that the stability of ethers toward alkylmagnesium halides rapidly diminishes at temperatures above 120°. At temperatures between 160° and 200°, cleavage occurs rapidly. Again, complex fomation is intermediate.

$$\begin{matrix}C_6H_5\diagdown \\ \quad O \\ R'\diagup\end{matrix} + R'MgX \rightarrow \left[\begin{matrix}C_6H_5\diagdown & & \diagup R' \\ & O \rightarrow Mg & \\ R'\diagup & & \diagdown X\end{matrix}\right] \rightarrow C_6H_5OMgX + RR'$$

Alkylmagnesium iodides give by far the best yields in this reaction. For example, phenetole gives but a 16% yield of phenol when methylmagnesium bromide is used, whereas an 80% yield is obtained when the iodide is employed.

Like the alkaline cleavage, the Grignard method is advantageous when acidic reagents lead to complex side reactions. However, the Grignard method is itself drastic and good yields are not often obtainable. An exception is found in the allyl ethers of phenols, which are cleaved at temperatures below 100°. Allyl phenyl ether itself is split at 55°.

The organo-alkali compounds effect ether cleavage at temperatures considerably below that required for successful Grignard splitting; however, these generally bring about extensive side reactions and are not practical for simple cleavage. Phenyl lithium does not cleave ordinary alkyl phenyl ethers at 50°; but benzyl phenyl ether is cleaved at temperatures even below 50°. Allyl phenyl ether reacts vigorously, with evolution of heat. Sodium and potassium alkyls and aryls cleave even alkyl phenyl ethers at 20°; but the side reactions are even more numerous than those occurring when lithium compounds are used. The products are, therefore, complex and quite unpredictable. The most usual types of side reactions exhibited by organo-alkalis in cleavage reactions are:

1. Metal exchanges

$$(C_6H_5)_3CH + KC(CH_3)_2C_6H_5 \rightarrow (C_6H_5)_3CK + C_6H_5CH(CH_3)_2$$

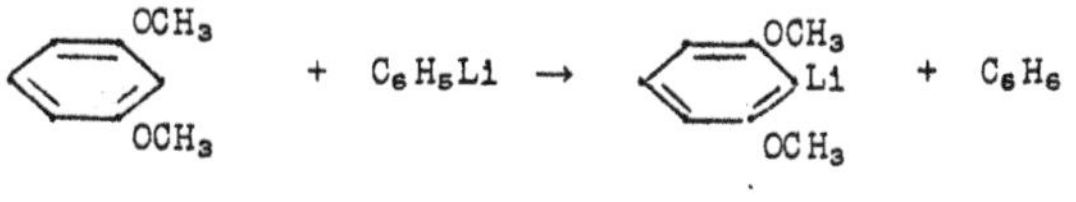

OCH3
OCH3
+ C6H5Li →
OCH3
Li
OCH3
+ C6H6

2. Secondary metallations

$$C_6H_5OCH_2C_6H_5 + C_6H_5Li \rightarrow C_6H_5OLi + (C_6H_5)_2CH_2$$

$$(C_6H_5)_2CH_2 + C_6H_5Li \rightarrow (C_6H_5)_2 + C_6H_6$$

The organo-alkali compounds ...
peratures considerably below th...
splitting; however, the ...
reactions and are not ...
lithium does not cleave ordinary ...
benzyl phenyl ether is cleaved ...
Allyl phenyl ether reacts ...
Sodium and potassium alkyls and aryls ...
ethers at 20°; but the side ...
than those occurring when lithium ...
products are, therefore, complex and ...
most usual types of side reactions ...
in cleavage reactions are:

1. Metal exchange

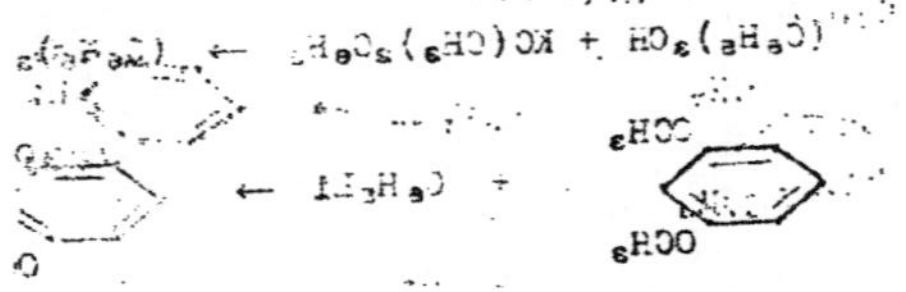

Bibliography:

(1) All material presented in this paper came from a review paper by Luttringhaus and von Saaf, Angew. Chem., 51, 915 (1938).
(2) John and Gunther, Ber., 72, 1649 (1939).
(3) Prey, ibid., 74, 1219 (1941).
(4) Baker, Nodzu, and Robinson, J. Chem. Soc., 1929, 79; 1938, 56.
(5) Benton and Dillon, J. Am. Chem. Soc., 64, 1128 (1942).
(6) Dodds, Goldberg, Lawson, and Robinson, Nature, 141, 247 (1938).

Reported by R. B. Carlin
September 23, 1942

THE ORGANIC CHEMISTRY OF

TELLURIUM

Although tellurium is usually classed among the rarer elements, it is commercially available in relatively large quantities. Recently new industrial developments have greatly increased the use of this little known element. However, most of these applications have been in the field of metallurgy and inorganic chemistry.

The organic compounds of tellurium which are known are relatively few and their chemistry is incompletely developed. The lack of research in this field is probably due to the fact that tellurium compounds, in general, are believed to be highly toxic. In addition, many of the reactions involving tellurium lead to mixtures and products which are difficult to handle.

The close proximity of tellurium to selenium and arsenic in the periodic table would lead one to believe that it might show reactions similar to these elements. Although tellurium exhibits physiological properties approximating those of selenium and arsenic, its chemical properties are more closely related to those of sulfur. At present, there are few practical uses of organic tellurium compounds. As yet, there seems to have been no systematic study made in this field. A number of organic derivatives and reactions of tellurium will be discussed in this report.

Aliphatic tellurium compounds

1. Type R_2Te

Although selenium yields derivatives of the types RSeH, R_2Se, and RSeR, only the type R_2Te is at present known with certainty in the case of tellurium. The dialkyl tellurides have been prepared by the following methods:

1) $K_2Te + Ba(RSO_4)_2 \rightarrow R_2Te + BaSO_4 + K_2SO_4$
2) $R_2TeI_2 + H_2O + Na_2SO_3 \rightarrow R_2Te + Na_2SO_4 + 2HI$
3) $Al_2Te_3 + 6ROH \rightarrow 3R_2Te + 2Al(OH)_3$

The dialkyl tellurides are heavy oils which tend to decompose on standing.

2. Types R_2TeX_2, R_3TeX and $RTeX_3$

Methods of preparation:

1) $Te + 2RX \rightarrow R_2TeX_2$

The dialkyltellurium dihalide yields the corresponding dihydroxy compound when treated with silver oxide.

$RETeX_2 + 2AgOH \rightarrow R_2Te(OH)_2 + 2AgX$

$$R_2Te(OH)_2 \xrightarrow{\Delta} R_2TeO$$

2) $RTeO_2H + R_3TeI \xleftarrow{HI} R_2TeO$

3) $CH_3TeOOH + HI \rightarrow CH_3TeI_3$
Considerable discussion has arisen concerning the structure of the dialkyl tellurium dihalides. Vernon held the opinion that dimethyl tellurium diiodide exists in two forms.

α-Diiodide β-Diiodide

Drew has since shown that the β-form is a complex involving two tellurium atoms.

3. Condensation products of tellurium tetrachloride and acid anhydrides.

Morgan and Drew have prepared a series of interesting tellurium derivatives by the interaction of various acid anhydrides with tellurium tetrachloride

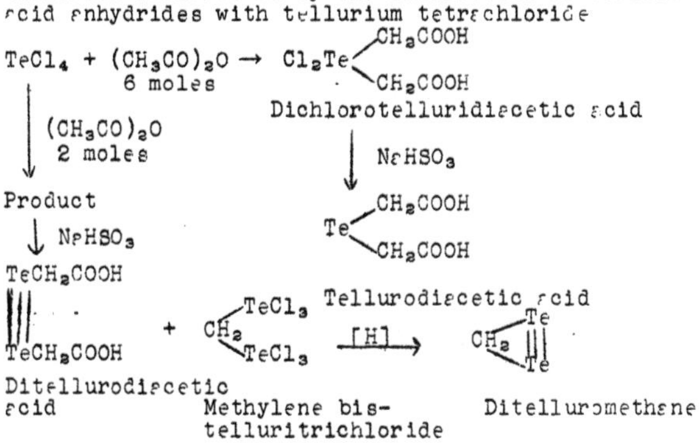
TeCl4 + (CH3CO)2O →
6 moles
Cl2Te
CH2COOH
CH2COOH
Dichlorotelluridiacetic acid
(CH3CO)2O
2 moles
Product
NaHSO3
TeCH2COOH
TeCH2COOH
NaHSO3
Te
CH2COOH
CH2COOH
Tellurodiacetic acid
TeCl3
CH2
TeCl3
[H]
CH2
Te
Te
Ditellurodiacetic
acid
Methylene bis-
telluritrichloride
Ditelluromethane

4. Cyclotelluro compounds

Several cycloorganic derivatives of tellurium have been investigated by Morgan and Burstall. They are prepared by condensing tellurium with the proper di-alkyl halide.

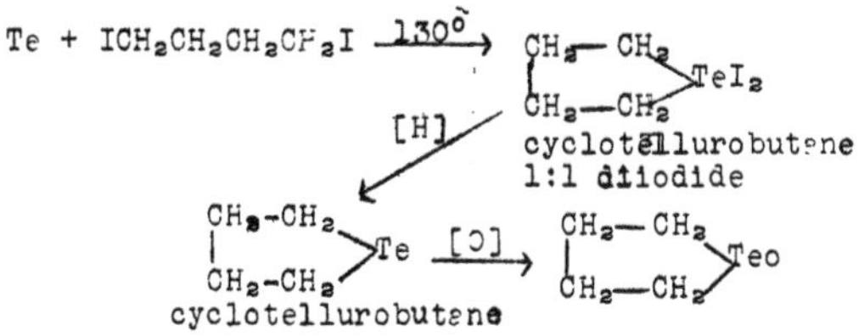
Te + ICH2CH2CH2CH2I
130°
CH2—CH2
CH2—CH2
TeI2
cyclotellurobutane
1:1 diiodide
[H]
CH2-CH2
CH2-CH2
Te
[O]
CH2—CH2
CH2—CH2
Teo
cyclotellurobutane

Aromatic tellurium compounds

It has been found that it is possible to introduce tellurium into an aromatic nucleus and a number of different types of derivatives have been prepared. Diaryl mercury compounds combine directly with tellurium to give diaryl tellurides.

$$Ar_2Hg + 2Te \rightarrow Ar_2Te + HgTe$$

An alternate procedure involves the use of the Grignard reagent.

$$2ArMgX + 2TeBr_2 \rightarrow Ar_2Te + 2MgBr_2$$

When diaryl tellurides are allowed to react with alkyl halides, they produce a new series of compounds.

$$Ar_2Te + RX \rightarrow Ar_2RTeX$$

The interaction of diaryl tellurides and halogens leads to compounds of the types R_2TeX_2, $R_2Te(OH)X$, and R_2TeO.

$$Ar_2Te + X_2 \rightarrow Ar_2TeX_2 \xrightarrow{HOH} Ar_2Te(OH)X \xrightarrow{NaOH} Ar_2TeO$$

Tellurium tetrachloride reacts with a number of aromatic compounds to form direct carbon to tellurium linkages. With dimethylaniline it yields an addition complex and 4,4' tetramethyldiaminodiphenyl telluridichloride.

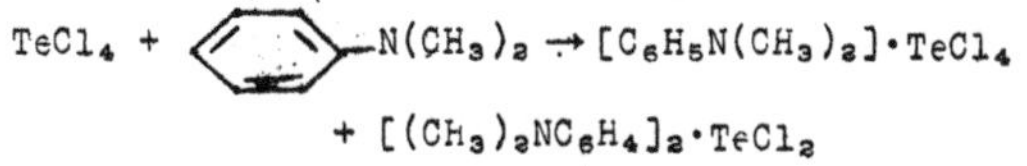
TeCl₄ +
N(CH₃)₂ → [C₆H₅N(CH₃)₂]·TeCl₄
+ [(CH₃)₂NC₆H₄]₂·TeCl₂

The cresols are likewise attacked by tellurium tetrachloride to give mono-di- and trichlorotellurium derivatives.

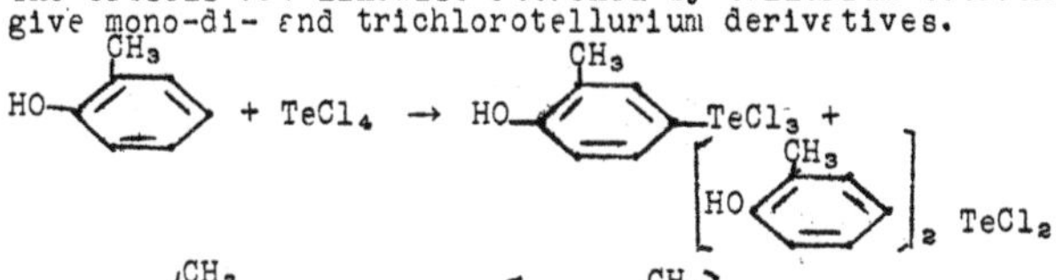
CH₃
HO
+ TeCl₄ →
TeCl₃ +
TeCl₂

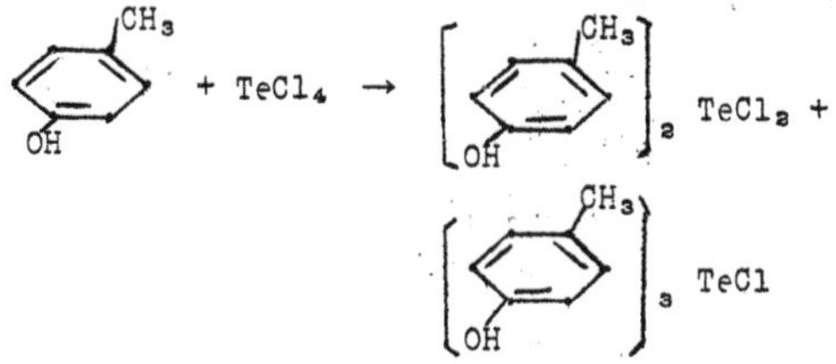
CH₃
OH
+ TeCl₄ →
TeCl₂ +
TeCl

The types of compounds available by the condensation of tellurium tetrachloride and mixed ethers may be exemplified by the case of phenetole.

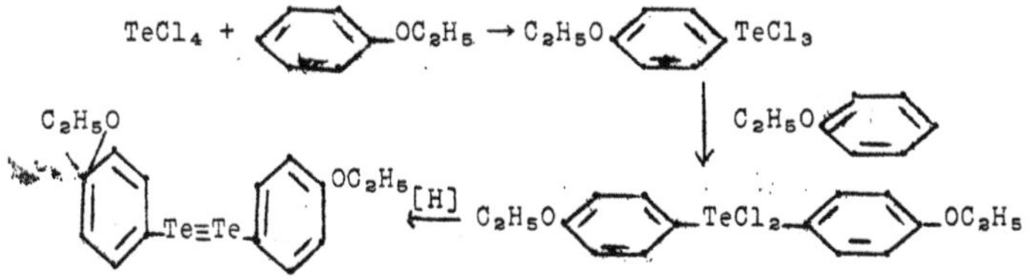
TeCl₄ +
OC₂H₅
C₂H₅O
TeCl₃
C₂H₅O
C₂H₅O
OC₂H₅
Te≡Te
[H]
C₂H₅O
TeCl₂
OC₂H₅

The aromatic ditellurides ArTe≡TeAr are intensely colored substances, often bearing a striking resemblance to the azo-compounds, and the prime cause of the color is the group -Te≡Te-.

An unusual series of compounds results from the reaction of tellurium tetrachloride with diphenyl ether; the most interesting product is phenoxtellurine.

$TeCl_4 + C_6H_5OC_6H_5 \rightarrow C_6H_5OC_6H_4TeCl_3 + HCl$

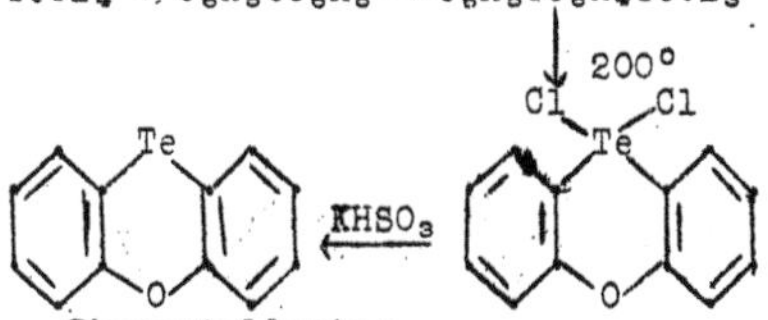
200°
Cl
Cl
Te
Te
KHSO₃
O
O

Phenoxtellurine

<u>Tellurium compounds derived from β-diketones</u>

The products of the interaction of the tetrachlorides of selenium and tellurium with acetylacetone differ from all previously known acetylacetone derivatives in containing a bivalent radical $C_5H_6O_2''$.

Cyclic Tellurium Derivatives of β-Diketones

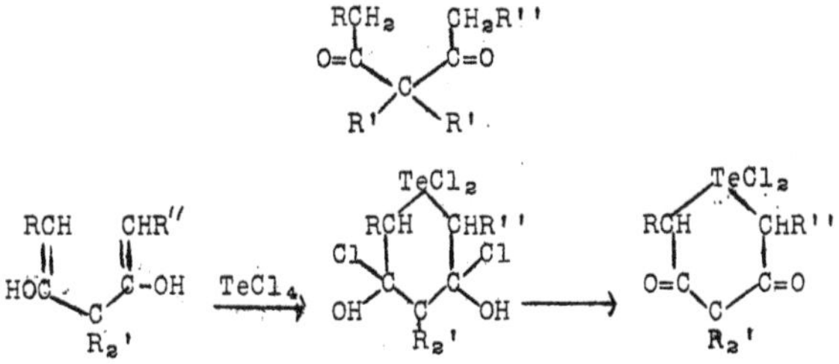
RCH₂
CH₂R''
O=C
C=O
C
R'
R'
RCH
CHR''
HOC
C-OH
C
R₂'
TeCl₄
TeCl₂
RCH
CHR''
Cl
Cl
OH
C
C
C
OH
R₂'
TeCl₂
RCH
CHR''
O=C
C=O
C
R₂'

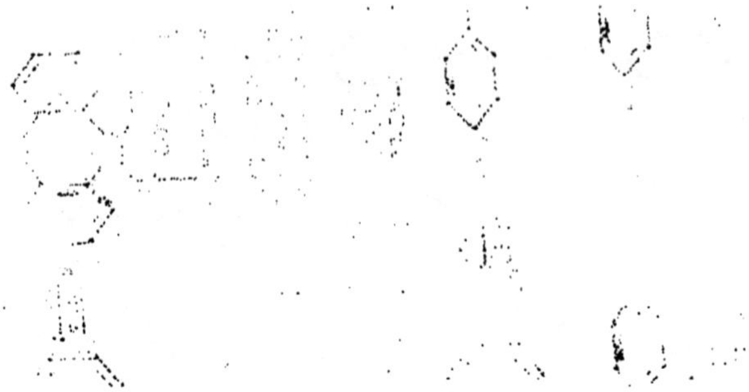

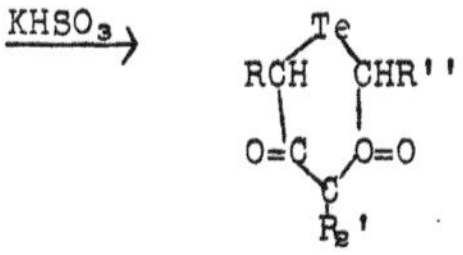
KHSO$_3$
Te
RCH
CHR''
O=C
C=O
C
R$_2$'

The cyclotelluropentane-3,5-diones are yellow substances sparingly soluble in water. In aqueous solution they are powerful germicides and the most potent member of the series, 2,6-dimethylcyclotelluropentane-3,5-dione is active against coliform organism in concentrations of one in 40,000,000. Unfortunately, these tellurium compounds are somewhat poisonous and induce haematuria.

Recently Fisher and Eisner attempted to use tellurium tetrachloride as the catalyst in Friedel-Crafts type reactions. The results in practically all cases were discouraging. Similarly they tried to substitute tellurium dioxide in place of selenium dioxide for specific types of organic oxidations. In all cases the tellurium dioxide was shown to be an inferior reagent.

Bibliography

Friend, A. Text-Book of Inorganic Chemistry, Charles Griffin Company, London, 1937, Vol. XI, Pt. IV.
Morgan, J. Chem. Soc., 1935, 554.
Fisher and Eisner, J. Org. Chem., 6, 169 (1941).
Cullinane, Rees, and Plummer, J. Chem. Soc., 1939, 151.
Reichel and Kirschbaum, Ann., 523, 211 (1936).
Waitkins, Bearse, and Schutt, Ind. Eng. Chem., 34, 899 (1942).

Reported by Norman Rabjohn
September 23, 1942

SYNTHETIC HORMONE-LIKE COMPOUNDS

Last year the synthetic substances, diethylstilbestrol and hexestrol, two compounds with potent estrogenic activity, were the subject of a seminar report.

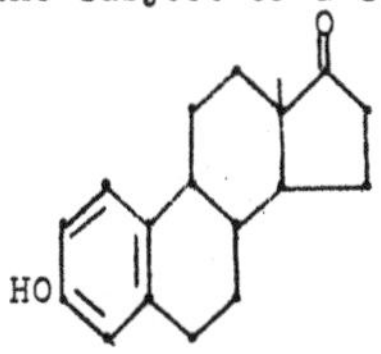

Estrone

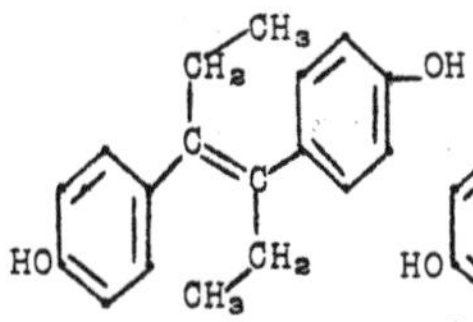

<u>trans</u>-Diethylstilbestrol

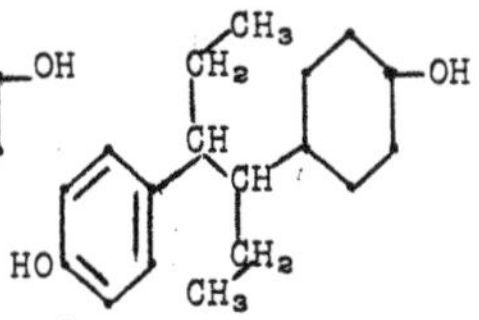

<u>meso</u>-Hexestrol

The successful utilization of these compounds in replacing the natural hormone has stimulated investigations into the possibility of the synthesis of substances possessing the physiological activity of other steroid hormones.

Of particular significance are the efforts to synthesize substance with the physiological activity of various hormones of the adrenal cortex. Investigation of extracts of these ductless glands has led to the isolation of over twenty closely-related crystalline compounds, (1) as well as an amorphous residue with marked activity. The compounds are closely related to progesterone and they possess in varying degree the ability to regulate at least four important physiological functions: Carbohydrate metabolism, muscle efficiency, distribution of electrolytes and function of the kidney.

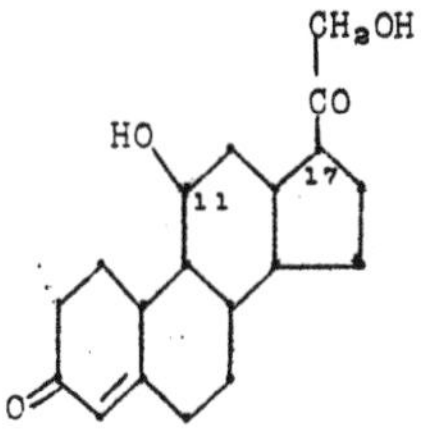

Corticosterone has been reported to be effective in treatment of operative shock (2) and cortical hormones have been utilized in the treatment (but not cure) of Addison's disease, which results from damage to the adrenal glands. The most characteristic functional group of these hormones is the hydroxyacetyl group on carbon 17 and attempts at preparation of synthetic hormones have all included this group. ω-Hydroxyacetophenone has been reported by Linnell and Roushdi (3) to have 1/2500 the activity of desoxycorticosterone in maintaining the life of adrenalectomized dogs. The most successful synthetic prepared by these investigators was 4-hydroxy-3'-hydroxacetyl-α,α'-diethylstilbene, closely related to diethylstilbestrol; this compound had 1/200 the activity of desoxycorticosterone.

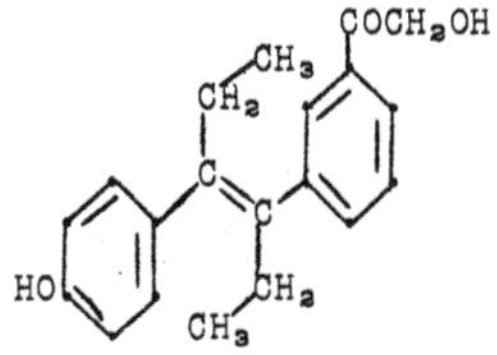
COCH$_2$OH
CH$_3$
CH$_2$
C
C
HO
CH$_2$
CH$_3$

Walker (4) has reported similar derivatives based on the diphenylether nucleus rather than stilbene.

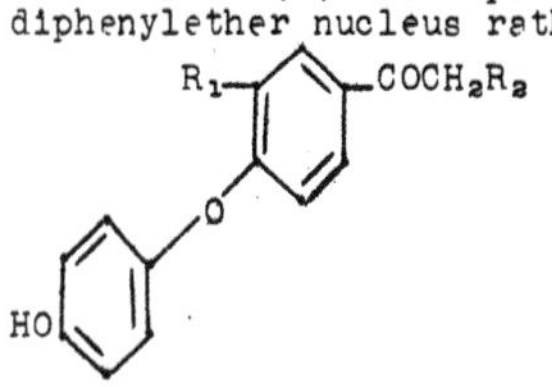
R$_1$
COCH$_2$R$_2$
O
HO

R_1 and R_2 = H or OH

The compounds with R_2 = H were inactive when tested for progestational activity. Incomplete tests for the activity of compounds with R_2 = OH have indicated no cortical hormone activity.

In connection with the activity of stilbene derivatives as estrogenic and cortical hormones, the important physiological action of 4,4'-diamidinostilbene is of some note.

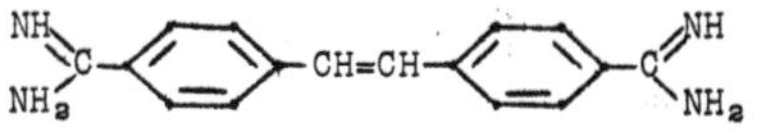
NH
C
NH$_2$
CH=CH
NH
C
NH$_2$

In an extensive investigation of the action of a large number of aliphatic and aromatic diamidines and related derivatives, Yorke (5) has found this substance to be the most effective in the treatment of such protozoan diseases as sleeping sickness, malaria, etc.

<u>Bibliography</u>

(1) See Reichstein, Helv. Chim. Acta, <u>24</u>, 247 (1941).
(2) Selye and Dosne, Lancet, <u>1940</u>, <u>II</u>, 70.
(3) Linnell and Roushdi, Quart. J. Pharm. Pharmacol, <u>14</u>, 270 (1941).
(4) Walker, J. Chem. Soc., <u>1942</u>, 347.
(5) Yorke, Trans. Roy. Soc. Trop. Med. Hyg., <u>33</u>, 463 (1940).
(6) May and Baker, Ltd., Brit. Pat. #510,097, July 27, 1939.
(7) Sah, J. Am. Chem. Soc., <u>64</u>, 1487 (1942).

Reported by C. C. Price
September 30, 1942

STEREOCHEMISTRY OF CATALYTIC HYDROGENATION

THE PERHYDRODIPHENIC ACIDS

The stereochemistry of the perhydrophenanthrenes is important because this group of compounds bridges the gap between the decalins, of which the stereochemistry is well-known, and the naturally-occurring steroids.

It has been found possible, in studying the perhydrophenanthrenes, to convert certain of their derivatives into the corresponding perhydrodiphenic acids. It is therefore valuable to know the absolute configuration of the latter compounds.

Linstead and his co-workers in a recent series of sever papers have reported the results of a stereochemical study of the catalytic hydrogenation of a number of aromatic compounds related to the phenanthrenes. The first paper (I) is a summary of the other six; Papers II-V are an account of the stereochemistry of the perhydrodiphenic acids; and the last two (VI and VII) relate the hydrogenation products of 9-phenanthrol and 9,10-phenanthraquinone to the perhydrodiphenic acids. This seminar report deals only with the work on the perhydrodiphenic acids (Papers II-V).

The Six Theoretical Perhydrodiphenic Acids.-

I

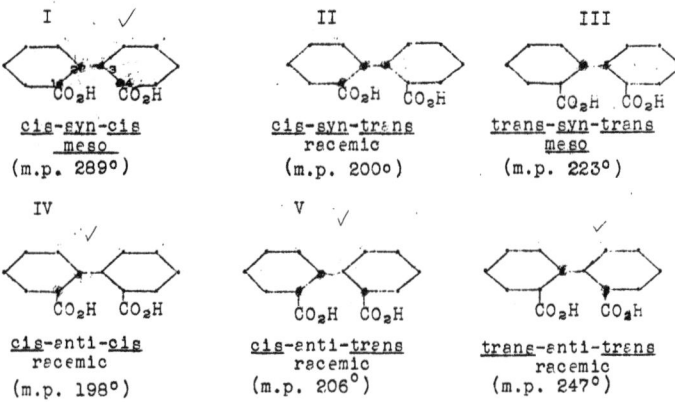

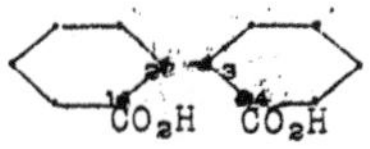
3
4
CO_2H CO_2H

cis-syn-cis
meso
(m.p. 289°)

II

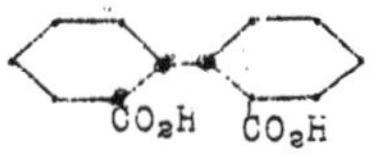
CO_2H CO_2H

cis-syn-trans
racemic
(m.p. 200°)

III

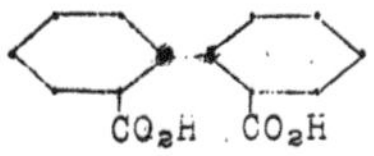
CO_2H CO_2H

trans-syn-trans
meso
(m.p. 223°)

IV

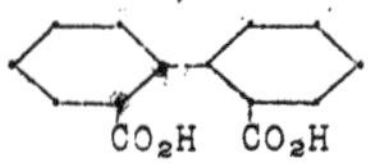
CO_2H CO_2H

cis-anti-cis
racemic
(m.p. 198°)

V

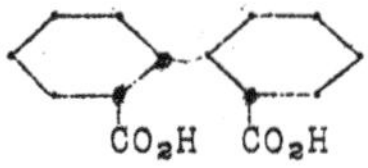
CO_2H CO_2H

cis-anti-trans
racemic
(m.p. 206°)

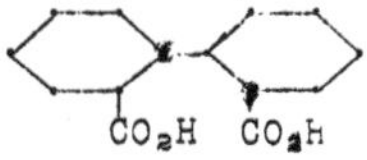
CO_2H CO_2H

trans-anti-trans
racemic
(m.p. 247°)

Nomenclature and conventions. When diphenic acid is completely hydrogenated, four asymmetric carbon atoms appear, numbered 1, 2, 3, 4 in Formula I. The configuration of the groups around C_1-C_2 and around C_3-C_4 are designated cis and trans, according to whether the pair of hydrogen atoms on C_1 and C_2 (or C_3 and C_4) are on the same side of the molecule. The configuration of the C_2-C_3 bond (the backbone) is designated syn or anti, depending on the

positions of the C_2-C_3 hydrogen atoms. The positions of the hydrogen atoms are represented by black dots, a dot indicating that a hydrogen atom is above the molecule. By convention a dot is always placed at C_2.

Preparation of the Six Perhydrodiphenic Acids.-

Linstead and his co-workers have succeeded in preparing all the perhydrodiphenic acids and in assigning to them their absolute configurations:

A. Diphenic acid $\xrightarrow[\text{Adams' cat.}]{H_2}$ 3 Perhydrodiphenic acids m.p. 289° (main product)
200°
198°

B. Dimethyl diphenate $\xrightarrow[\text{Pd}]{H_2}$ Perhydrodiphenic acid dimethyl ester, m.p. 73°

C. 73° dimethyl ester $\xrightarrow{NaOCH_3}$ 57° isomer $\xrightarrow[\text{hydrol.}]{\text{alk.}}$ perhydrodiphenic acid, m.p. 223°

73° dimethyl ester $\xrightarrow{\text{Acid hydr.}}$ 289° acid; 289° acid $\xrightarrow{CH_2N_2}$ 73° dimethyl ester

D. Monomethyl ester of 198° acid $\xrightarrow[\text{KOH}]{\text{alc.}}$ Perhydrodiphenic acid, m.p. 206°

E. Dimethyl ester of 198° acid $\xrightarrow[\text{KOH}]{\text{alc.}}$ Perhydrodiphenic acid, m.p. 247°.

The mixtures of acids were separated by precipitation through successive additions of dilute hydrochloric acid to their sodium salt solutions.

The 247° acid had previously been obtained by hydrogenation by Vocke, and also by Linstead and Walpole by oxidation of a 9-keto-perhydrophenanthrene prepared by Rapson and Robinson.

Three other compounds which appeared to be perhydrodiphenic acids had been obtained by Vocke (m.p. 213°), Linstead and Walpole (m.p. 203°), and Marvel and White (m.p. 174°). In the light of present evidence, these are probably, respectively, impure 223° acid, a dibasic acid derived from a spirane rather than a perhydrophenanthrene, and a dimorphic modification of the 200° acid.

Grouping of Acids into Two Series (Syn and Anti).-

Reaction C above indicates that an inversion takes place when the 73° ester is converted to the 57° ester. Since the "backbone configuration" (C_2C_3) is fixed by the hydrogenation, this inversion

must take place around the C_1 and/or C_2 positions, according to three possibilities:

1. Both C_1 and C_4 have the same initial configuration, and both centers are inverted.

2. Both have the same initial configuration and only one center is inverted.

3. They have different initial configurations and only one center is inverted.

It has been shown that the first possibility is the correct one, by the following evidence:

The important fact was utilized that the 289° acid (corresponding to the 73° ester) could not be converted to the 223° acid by alkali, but only through intermediate formation of the dimethyl ester. Thus inversion cannot be brought about at a carbon atom containing a free carboxyl groups (due to lack of enolization).

The monomethyl ester of the 289° acid was prepared and treated with alkali. In the light of the above fact it could only undergo inversion at one center, and if possibility (1) were correct, the inversion product could not correspond to the 223° acid, since (1) postulates a double inversion. Actually the inversion product yielded the 200° acid on hydrolysis and not the 223° acid.

The above evidence enables us to declare that the 289° and 223° acids must either be I and III or IV and VI, depending on the configuration of the backbone, and that the 200° acid must belong to the same backbone (syn or anti) series, being either II or V.

The other backbone series must then by elimination consist of the 198°, 206°, and 247° acids. It seemed probable, from reactions D and E above, that the same type of inversion by alkali must be taking place as in the other series. Thus the 206° acid, the product of inversion and hydrolysis of the monomethyl ester of the 198° acid, must be intermediate in the series (either II or V) and the 198° and 247° acids must be I and II or IV and VI.

Differentiation between the Syn and Anti Series.-

As apparent from Formulas I-VI, one series (syn) contains two meso isomers and one racemic mixture, while the other (anti) contains three racemic mixtures. Therefore:

1. That series which can be shown to have more than one member capable of resolution must be the anti-series.

2. That series which can be proven to have a meso form must have the syn-configuration.

Both these tests have been successfully carried out.

The 198°-206°-247° series has been shown to be <u>anti</u> by resolution of the 198° and 247° acids.

In the 289°-200°-223° series attempts to resolve the 289° and 223° acids were fruitless, which indicated they were meso. The 289° acid was proven to be meso by an ingenious procedure (first employed by Stoermer and Steinbeck on cis-hexahydrophthalic acid): Asymmetry was produced in the molecule by formation of the monomethyl ester. This was then resolved by means of cinchonidine. Hydrolysis of the active monoester and also formation of the dimethyl ester from the active monoester by means of mild diazomethane treatment both gave inactive products, showing that the asymmetry was again destroyed.

Further proof, that the 200° acid was intermediate between the 289° and 223° acids, was also provided. The l-monomethyl ester of the resolved 289° acid was inverted and hydrolyzed by alkali (mono-inversion, thus yielding an enantiomorph of the 200° acid). The d-monomethyl ester was treated likewise and the other enantiomorph was obtained. A 50-50 mixture of these gave the 200° racemic acid.

Thus the 289°-200°-223° series is syn. and the 200° acid is cis-syn-trans.

Differentiation between cis-syn-cis and trans-syn-trans and between cis-anti-cis and trans-anti-trans Acids.-

It was now only necessary to prove whether the 289° and 223° acids in the syn-series were I and III or vice versa, and whether the 198° and 247° acids in the anti-series were IV and VI or vice versa.

This has been accomplished in the syn-series as follows:
An intermediate product, which was isolated in most of the hydrogenation reactions, is hexahydrodiphenic acid (VII).

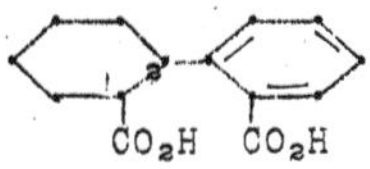

VII

This acid was shown to yield the 289° acid exclusively (77%. with no other acids isolated) on hydrogenation. Therefore the configuration around C_1-C_2 must be the same as that in the 289° acid.

VII was then degraded by ozonolysis to a hexahydrophthalic acid which is definitely known to be cis. Thus the 289° acid must be cis-syn-cis, and by elimination the 223° acid must be trans-syn-trans.

In the *anti*-series the *trans-anti-trans* configuration has been assigned to the 247° acid, partly on the basis of a degradation analogous to that applied to the *syn*-series, and also because this is the stable modification (a large amount of evidence has been gathered in this series of papers and elsewhere that hydrogenation yields the *cis*-form which can then be inverted to *trans*).

Generalization concerning Catalytic Hydrogenation.-

A general conclusion drawn from this work is that hydrogenations over platinum catalysts yield largely *cis* and *syn* products (at least in compounds of this type).

Bibliography

Linstead, Chem. and Ind., 56, 510 (1937).
Linstead and Walpole, J. Chem. Soc., 1939, 842, 850.
I. Linstead, Doering, Davis, Levine and Whetstone, J. Am. Chem. Soc., 64, 1985 (1942).
II. Linstead and Doering, *ibid.*, 64, 1991 (1942).
III. Linstead and Doering, *ibid.*, 64, 2003 (1942).
IV. Linstead, Davis and Whetstone, *ibid.*, 64, 2006 (1942).
V. Linstead, Davis and Whetstone, *ibid.*, 64, 2009 (1942).
VI. Linstead, Whetstone and Levine, *ibid.*, 64, 2014 (1942).
VII. Linstead and Levine, *ibid.*, 64, 2022 (1942).

~~Reported by R. L. Frank~~
~~September 30, 1942~~

MUSTARD GAS

A. Introduction

β',β'-Dichlorodiethyl sulfide, known as Mustard Gas, Yperite, "H S", or "yellow cross", still holds its position as the most dangerous and feared among war gases since it was first used with devastating effect in the battle of Ypres.

In the pure state it is a nearly odorless, oily liquid boiling at 217.5°C (at 760 mm.) and melting at 14.4°C. Its specific gravity compared to water at 15° is 1.279. In regard to solubility, it acts like a typical organic substance being only slightly soluble in water but dissolving completely in organic substances such as kerosene, ethyl alcohol, and carbon tetrachloride.

The highly toxic nature of "HS" to humans is indicated by the pronounced blistering and edemia of external skin areas and cellular destruction in lung tissue, leading to pneumonia. Taken internally, it causes pronounced loss in weight and eventual death.

The first recorded synthesis of the compound was made by Riche in 1854. Later, Guthrie, in 1860, prepared mustard oil by passing ethylene into sulfur chloride:

$$2CH_2{=}CH_2 + SCl_2 \rightarrow S\begin{cases} CH_2CH_2Cl \\ CH_2CH_2Cl \end{cases}$$

Nieman also prepared the substance by a similar process at the same date. Other pre-War I investigators were Meyer, who made a special study of the compound, and Clarke. During World War I, the investigations dealt largely with improved methods of manufacture and modes of protection. Then, following the war, considerable research was carried on to determine the reactivity of the compound toward organic and inorganic reagents. In recent years the trend has been toward determining the mechanism of the physiological effect, new methods of detection, and physico-chemical determinations leading to an understanding of its reaction mechanisms.

B. Preparation

The old industrial preparation of β',β'-dichlorodiethyl sulfide is carried out according to either of the following processes:--

I. The Meyer Process:

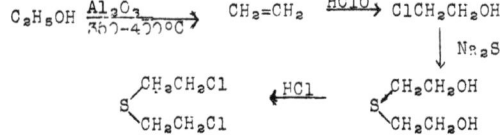

$$C_2H_5OH \xrightarrow[350-400^\circ C]{Al_2O_3} CH_2{=}CH_2 \xrightarrow{HClO} ClCH_2CH_2OH \xrightarrow{Na_2S} S\begin{cases} CH_2CH_2OH \\ CH_2CH_2OH \end{cases} \xrightarrow{HCl} S\begin{cases} CH_2CH_2Cl \\ CH_2CH_2Cl \end{cases}$$

II. The Guthrie Process: (Gibson-Pope modification)

$$CH_2{=}CH_2 + S_2Cl_2 \rightarrow S\left<\begin{matrix}CH_2CH_2Cl\\ CH_2CH_2Cl\end{matrix}\right. + S$$

Newer processes or improvements on the foregoing preparations were soon forthcoming due to the fact that the Meyer process, though yielding a pure product, was too expensive, and the product obtained by the Guthrie method contained sulfur in excessive amount. These new methods are summarized as follows:

III. The Nenitzescu-Scarlatescu Method (1935)

$$\left.\begin{matrix}CH_2\\ |\\ CH_2\end{matrix}\right>O + H_2S \xrightarrow{40-60^{\circ}C} S\left<\begin{matrix}CH_2CH_2OH\\ CH_2CH_2OH\end{matrix}\right. \xrightarrow{HCl} S\left<\begin{matrix}CH_2CH_2Cl\\ CH_2CH_2Cl\end{matrix}\right.$$

[yield = 90%]

IV. The Pompei Method (1940)

$$CaS + ClCH_2CH_2OH \xrightarrow[\text{C as cat}]{100.5^{\circ}\ (5\text{-}6\ hr)} (CH_2CH_2OH)_2S + (ClCH_2CH_2)_2S$$

[yield = 23-30%]

V. The Pie-Ying Chao Method (1940)

$$S\left<\begin{matrix}CH_2CH_2OH\\ CH_2CH_2OH\end{matrix}\right. + COCl_2 \rightarrow S\left<\begin{matrix}CH_2CH_2Cl\\ CH_2CH_2Cl\end{matrix}\right. + CO_2 + H_2O$$

[Yield = 83%]

C. General Reactivity

At normal temperatures, mustard gas is quite stable. However, at elevated temperatures the following reactions will occur.

$$2\ S\left<\begin{matrix}CH_2CH_2Cl\\ CH_2CH_2Cl\end{matrix}\right. \underset{}{\overset{180^{\circ}}{\rightleftarrows}} S\left<\begin{matrix}CH_2CH_2\\ CH_2CH_2\end{matrix}\right>S + 2C_2H_4Cl_2$$

$$\xrightarrow{<500^{\circ}} HCl + \text{unknown toxic gases}$$

The reactions of "HS" with many reagents have been extensively studied. A few of the important samples of its unusual behaviour are worth reviewing:

20% alcoholic KOH

$$S\left<\begin{matrix}CH_2CH_2Cl\\ CH_2CH_2Cl\end{matrix}\right. \xrightarrow{KOH} S\left<\begin{matrix}CH_2{=}CH\\ CH{=}CH_2\end{matrix}\right. \xrightarrow[hrs.]{48} \text{polymerized jelly}$$

$$S\left<\begin{matrix}CH_2{=}CH\\ CH{=}CH_2\end{matrix}\right. \xrightarrow{HCl} S\left<\begin{matrix}CH_2CH_2Cl\\ CH_2CH_2Cl\end{matrix}\right.$$

Halogens

$$S \overset{CH_2CH_2Cl}{\underset{CH_2CH_2Cl}{<}} \xrightarrow{Cl_2} Cl_2 \cdot S \overset{CH_2CH_2Cl}{\underset{CH_2CH_2Cl}{<}} \xrightarrow{H_2O} OS \overset{CH_2CH_2Cl}{\underset{CH_2CH_2Cl}{<}} + 2HCl$$

$$\xrightarrow{Br_2} (ClCH_2-CH_2) \cdot S \cdot 2Br_2$$

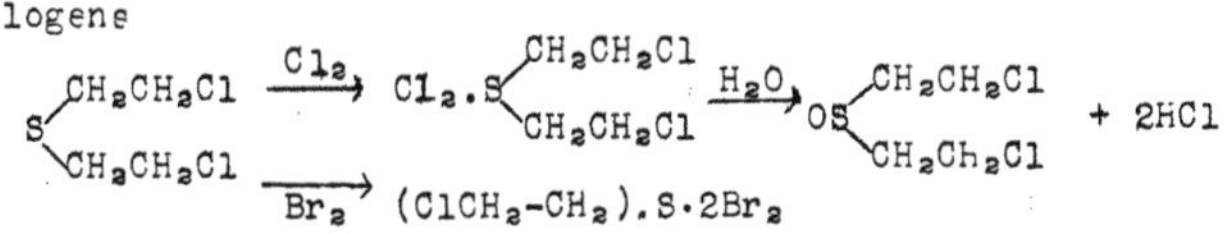

Ammonia and aliphatic amines

$$S \overset{CH_2CH_2Cl}{\underset{CH_2CH_2Cl}{<}} + RNH_2 \rightarrow S \overset{CH_2CH_2}{\underset{CH_2CH_2}{<}} NR + 2HCl$$

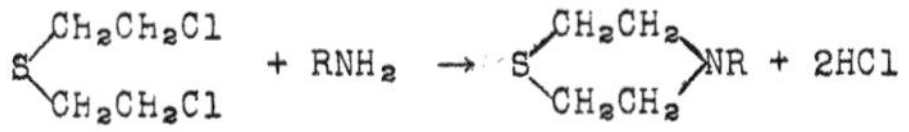

Oxidizing agents

$$S \overset{CH_2-CH_2-Cl}{\underset{CH_2CH_2-Cl}{<}} \overset{\substack{dil. \\ HNO_3 \\ (O)}}{\xrightarrow{ H_2O_2 }} OS \overset{CH_2CH_2Cl}{\underset{CH_2CH_2Cl}{<}}$$

$$\xrightarrow[dil'HNO_3]{KMnO_4} O_2S \overset{CH_2CH_2Cl}{\underset{CH_2CH_2Cl}{<}}$$

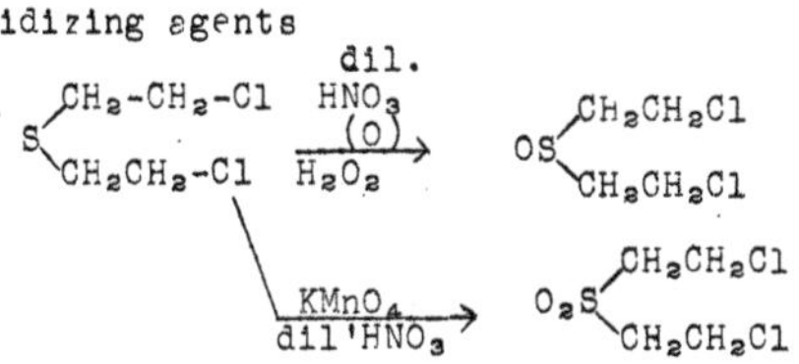

-Hydrolysis-

The effect of water or dilute alkali on aqueous mustard gas
has received considerable attention due to its importance in de-
contamination and physiological studies. According to Peters and
Welker (1923) the reaction takes place in two distinct phases:

$$S \overset{CH_2CH_2Cl}{\underset{CH_2CH_2Cl}{<}} + H_2O \rightleftharpoons \left[S \overset{CH_2CH_2Cl}{\underset{CH_2CH_2OH}{<}} \right] + HCl$$
$$[I]$$

$$S \overset{CH_2CH_2Cl}{\underset{CH_2CH_2OH}{<}} + H_2O \rightarrow S \overset{CH_2CH_2OH}{\underset{CH_2CH_2OH}{<}} + HCl$$
$$[II]$$

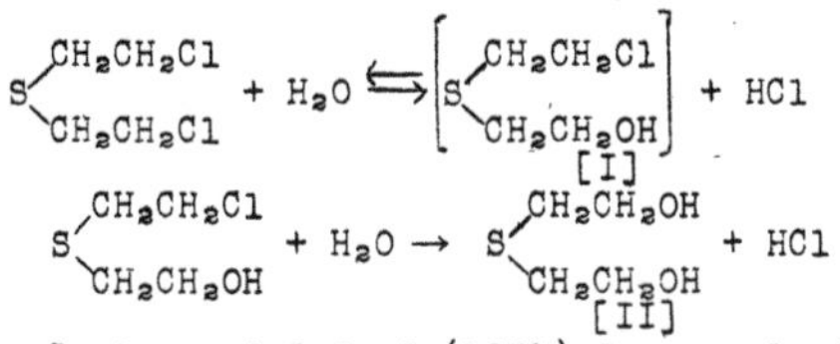

Later, Davies and Oxford (1931) from analysis of the decomposition
products proposed a similar, yet more complete, mechanism to form
complex hydrolysis products from the intermediate [I] and final
product [II] (above).

Recent evidence by Mohler and Hartnagel, based on kinetic
studies of the hydrolysis rate, indicates the following process.
Solvation, coupled with the presence of negative hydroxyl groups,
ionizes the polar C-Cl bond:

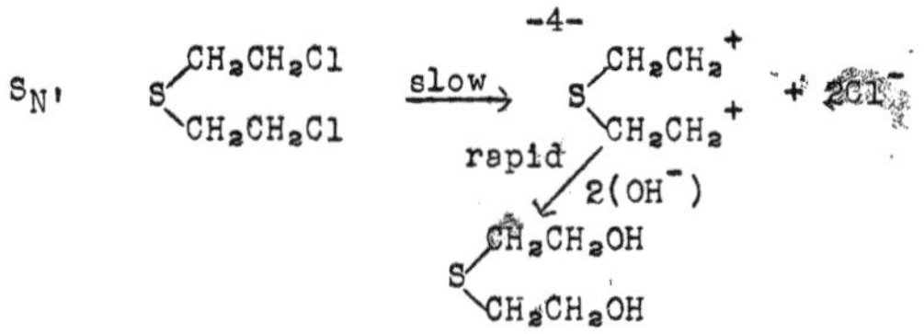

S_N'

$$S\begin{cases}CH_2CH_2Cl\\CH_2CH_2Cl\end{cases} \xrightarrow{slow} S\begin{cases}CH_2CH_2^+\\CH_2CH_2^+\end{cases} + 2Cl^-$$

$$\xrightarrow[rapid]{2(OH^-)} S\begin{cases}CH_2CH_2OH\\CH_2CH_2OH\end{cases}$$

Reactions with Detecting Agents

Mustard gas reacts with many detecting agents to form colored or easily identified products:

Selenious Acid

$$H_2SeO_3 + S\begin{cases}CH_2CH_2Cl\\CH_2CH_2Cl\end{cases} \xrightarrow[H_2SO_4]{dil.} O_2S\begin{cases}CH_2CH_2Cl\\CH_2CH_2Cl\end{cases} + H_2O + Se$$

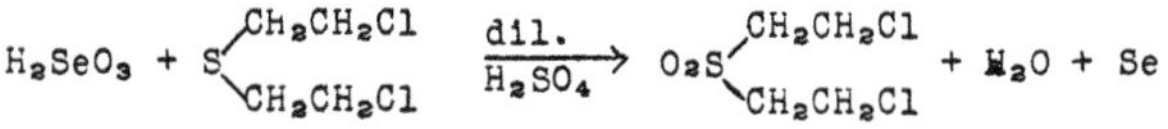

Sodium monosulfide

$$Na_2S + S\begin{cases}CH_2CH_2Cl\\CH_2CH_2Cl\end{cases} \to S\begin{cases}CH_2-CH_2\\CH_2-CH_2\end{cases}S \xrightarrow{48\ hr.\ polym.\ jelly}{HCl\ C_4H_8Cl_2S}$$

dithiane

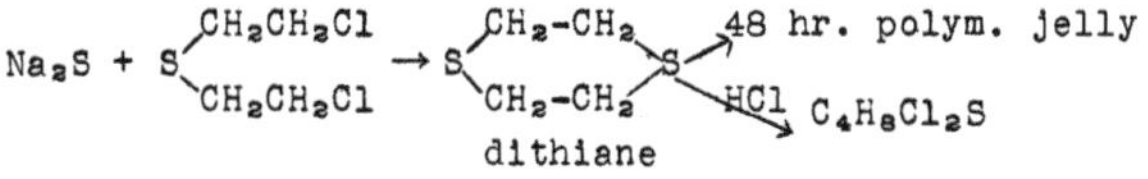

Gold Chloride

$$AuCl_3 + \ddot{S}\begin{cases}CH_2CH_2Cl\\CH_2CH_2Cl\end{cases} \to AuCl_3 \cdot \ddot{S}\begin{cases}CH_2CH_2Cl\\CH_2CH_2Cl\end{cases}$$

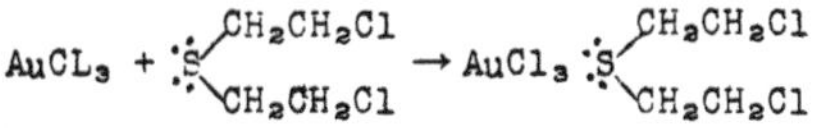

Physiological Effects:--

The chief physiological effect of "mustard" is the vesicant action. The production of blisters was at one time believed to be due to the absorption of the compounds with the production of thiodiglycol and hydrochloric acid. In refutation to this, it has been shown that hydrolysis products of mustard gas have no effect on simple cells. Kling believes that the toxic effect is probably due to the specific action of mustard on the cell surface. Other theories are that the highly toxic sulfone may be formed or that the action is due to the special configuration of the Cl atom.

Neutralization:--

Neutralization of skin areas affected by mustard may be accomplished by using hexamethylene tetramine which reacts with the mustard and is diffusible in the tissues:

$$2S(CH_2CH_2Cl)_2 + 10\ H_2O + C_6H_{12}N_4 \to$$

$$2S(CH_2CH_2OH)_2 + 4NH_4Cl + 6CH_2O$$

Chloramine T, acid derivatives of amines and hydrazines have also proven useful. The use of potassium permanganate of chloride of lime on skin areas intensifies the action to an extraordinary degree.

Bibliography

Sartori, "The War Gases".
Mohler and Hartnagel, Helv. Chim. Acta, 24, 564-570 (1941).
Pie Ying Chao, J. Chin. Chem. Soc., 7, 102-104 (1940).
Scarlatescu, Chem. Warfare Bull., 24, 94-100 (1938).
Kling, Compte Rendu, 208, 1679-81 (1939).
Riche, Ann. chim. phys., (3) 42, 283 (1854).
Bruere and Bouchereau, C. A., 14, 2165
Bales and Nickelson, J. Chem. Soc., 123, 2486.
Davies and Oxford, Ibid., 1931, 224
Helfrich and Reid, J. Am. Chem. Soc., 42, 1208 (1920).

Reported by C. S. Benton
October 7, 1942

CHROMATOGRAPHIC ADSOPPTION

In 1906 the Russian botanist, M. Twsett, first used the principle of chromatographic adsorption to separate a mixture of compounds (1). Tswett was able to resolve the pigments extracted from leaves into more components than had previously been possible. However, for several reasons this important discovery of a new method of analysis was almost entirely neglected for twenty-five years. In the first place, many chemists thought that the reactive leaf xanthophyll which Tswett resolved was actually one compound that had been isomerized by the action of the adsorbent. Secondly, at that time, only a few milligrams of material were isolable using this method, and, without the micro-technique developed later, it was impossible to analyze such small quantities.

A quarter of a century later, in 1931, Kuhn and Lederer (2) were able to separate carotene--for a hundred years thought to be a single compound--into α- and β-carotene by chromatographic adsorption and to identify these substances. In the same year, Petter (3), using a more tedious method, independently arrived at the same result. The importance of these and similar results obtained by Kuhn and co-workers at this time firmly established the value of this method as a powerful chemical tool. Since then materials and technique have been enormously improved and today chromatographic adsorption finds use in almost every field of chemistry. It has been used to separate isotopes, elements, inorganic ions, hydrocarbons, alcohols, ketones, aldehydes, acids, esters, sulfonic acids, heterocyclic compounds, dyes, drugs, and natural products (1).

Although there have been many modifications of Tswett's original apparatus, the basic principles of design and operation have been the same. A tube, from 0.5 to 10 cm. in diameter and 2 to 100 cm. long, is constricted at the lower end. It is attached to a filter flask below the constriction and placed in a vertical position. A plug of glass wool or cotton is placed above the constriction and the column carefully filled with the adsorbent. Another plug is usually placed above the adsorbent. The original solution to be chromatographed is then poured into the top of the column and sufficient suction applied to give the desired rate of percolation or flow. After this solution has run through, there will be a series of bands quite close together at the top of the column. The solution is immediately followed by the developing solvent, which spreads out the bands and separates them. After the chromatogram has been sufficiently developed, the adsorbate--the adsorbent and the substances adsorbed--is pushed out of the column, the various bands separated, and each pure substance eluted with a suitable solvent.

A few words of explanation are necessary concerning adsorbent and solvents.

The choice of a proper adsorbent is the most difficult part of any separation by chromatographic adsorption. Since there is not time for a complete discussion of this subject, only a few of the more important criteria will be presented. The adsorbent selected should hold back large amounts of material, yet not hold it so strongly as to prevent development and elution by the proper solvents; it should be inert to all other chemicals involved in the operation; it should have particles of such size as to permit rapid filtration without channelling, and the particles should be non-porous and colorless; it should exert a highly selective adsorptive action, since two substances which are adsorbed equally strongly would not be separated. Some of the chemicals which have proved most satisfactory as adsorbents are the alkaline earth oxides, hydroxides and salts, starch, inulin, sucrose, and various charcoals. Even when the correct material has been chosen, results are widely different depending on the amount and character of activation.

The selection of a suitable solvent depends both on the solubility of the compounds and on the adsorbent. The best and most widely used have been the petroleum ethers, but if the material is too strongly adsorbed from these, a more polar solvent is employed. Usually, the same solvent is tried as a developer, but if it is not satisfactory, the polarity is again increased. For elution, a very polar substance such as alcohol, water, etc., which tends both to deactivate the adsorbent and dissolve the compound, is used. To aid in choosing the proper solvent, a so-called elutropic series has been developed: low petroleum ether, high petroleum ether, carbon tetrachloride, cyclohexene, carbon disulfide, ether, acetone, benzene, toluene, esters, alcohols, water, acids. By picking the proper solvent or mixture of solvents from this series, almost any required polarity may be obtained.

Some examples of how chromatographic adsorption has recently been used will best illustrate its many possibilities.

trans-Stilbene contains strongly fluorescent contaminants, which cannot be removed by ordinary methods, but these impurities have been removed by adsorption from benzene-petroleum ether solutions on an activated alumina column, and development with benzene (4). In this case development was continued until all the pure compound had been washed into the filtrate. A sample of trans-stilbene in benzene was irradiated with ultra-violet light, evaporated to dryness, the residue dissolved in petroleum ether and the solution chromatographed on alumina. Colored impurities remained at the top of the column, while further down there was a zone of the trans-isomer 42 mm. long, separated by 6 mm,, from a 48 mm. band of cis-stilbene. In a typical experiment, 160 mg. of material in 25 ml. of petroleum ether was developed with 90 ml. of the same solvent. The yield was 99 mg. of trans- and 52 mg. of cis-isomer. Since the stilbenes are colorless, they were

located by brushing the column with a narrow streak of a 1% permanganate solution, which turned brown in a few seconds in the stilbene zones and not for several minutes elsewhere. Similar experiments have been performed with p-methyl and p-methoxystilbene.

β-Carotene contains five double bonds capable of a trans-cis shift:

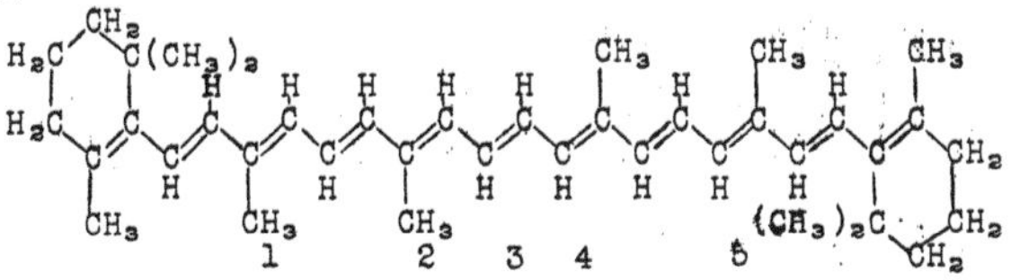
1
2
3
4
5

β-carotene with double bonds available for shift numbered

Treatment of β-carotene with small amounts of iodine in petroleum ether solution causes trans-cis shifts to take place with the formation of various geometrical isomers (5). The solution was chromatographed on a calcium hydroxide column and thirteen different isomers isolated.

Samples of cis and trans-benzoin oxime have been prepared and purified by adsorption (6). The trans-isomer forms a green complex with ammoniacal copper sulfate, the cis-isomer a brown complex. This color difference was made the basis of locating the two zones containing the isomers, using the brush method as discussed above.

A substance "erysocine" isolated from four different species of the plant Erythrina was thought to be a single compound, since its melting point, specific rotation, and analysis all indicated this to be the case (7). However chromatography on an activated alumina column showed that erysocine was actually a mixture of equal parts of two previously identified alkaloids, erysodine and erysovine.

Using a column of optically active quartz powder, it has been possible to separate several pairs of enantiomorphs (8).

Bibliography

1. Strain, "Chromatographic Adsorption Analysis.
2. Kuhn and Lederer, Ber., 64, 1349 (1931).
3. Petter, Amsterd. Akad. Wiss. 34, 10 (1931).
4. Zechmeister and McNeely, J. Am. Chem. Soc., 64, 1919 (1942).
5. Polgar and Zechmeister, J. Am. Chem. Soc., 64, 1856 (1942).
6. Zechmeister, McNeely, and Solyom, J. Am. Chem. Soc., 64, 1922 (1942
7. Folkers and Shavel, J. Am. Chem. Soc., 64, 1892 (1942)
8. Karagunis and Coumolos, Atti X° Congr. intern. chim 2, 278 (1938) C. A., 33, 7165 (1939)

Reported by O. H. Bullitt, Jr.
October 7, 1942

ORGANOBISMUTH COMPOUNDS

Introduction

The stability of organometallic compounds depends in part upon the strength of the carbon--metal bond; the less electropositive the metal, the stronger the covalent bonds it forms with carbon. The elements of the B family of group V in the periodic table are weakly electropositive metals, and form a number of stable types of organometallic compounds. Since in any group the electropositive character of the elements increases with increasing atomic number, bismuth does not form all the types of organometallic compounds known for antimony and arsenic, and those which it does form are generally less stable. Organobismuth compounds resemble organomercury, organolead, and organotin compounds in reactivity; a number of attempts have been made to compare their relative reactivities by reactions common to all of them, but these have generally failed because of the difficulty of choosing conditions which would permit the reactions to proceed to true equilibrium. Bismuth exhibits valences of three and five in its organic compounds; since the electropositive character of an element decreases with increasing valence, the pentavalent bismuth compounds are found to be much more stable than the trivalent ones.

Symmetrical Trialkylbismuth Compounds.--Symmetrical trialkylbismuth compounds may be prepared by the reaction of a bismuth-alkali metal alloy with an alkyl halide

$$3RX + BiK_3 \rightarrow R_3Bi + 3KX,$$

or by the reaction of a Grignard reagent with a bismuth halide,

$$3RMgX + BiX_3 \rightarrow R_3Bi + 3MgX_2;$$

a number of less general methods, such as the reaction of aluminum carbide with bismuth chloride in hydrochloric acid, electrical discharge through a gaseous hydrocarbon between bismuth electrodes, reaction between free radicals and bismuth mirrors, and spontaneous radioactive disintegration of tetramethylradium D, have also been employed in the preparation of trimethylbismuth and triethylbismuth. The trialkylbismuth compounds are heavy, highly refractive, nearly colorless oily liquids; they decompose before boiling when heated at atmospheric pressure, but can be distilled under diminished pressure. They ignite spontaneously in air, burning to bismuth oxide, and explode in pure oxygen. They are cleaved by chlorine and bromine, even at 0°C, forming R_2BiX compounds. They are cleaved quantitatively by inorganic acids, explosively by concentrated oxidizing acids. They are also cleaved by hydrogen sulfide, silver nitrate, and mercuric chloride. Triethylbismuth is readily cleaved by compounds containing the -SH group, but is not affected by -NH, -C≡CH, -OH, or most of the weaker carboxyl hydrogens; it has therefore been used as a test for thioenolization by the

ordinary Zerewitinoff technique,

$$(C_2H_5)_3Bi + RSH \rightarrow C_2H_6 + (C_2H_5)_2BiSR.$$

Thus the trialkylbismuth compounds can form no stable pentavalent derivatives, but are cleaved by reagents ordinarily used in their preparation.

Dialkylbismuth Halides.--Dialkylbismuth halides, of the type R_2BiX (where X is Cl or Br), may be prepared by treating the trialkylbismuth with the calculated quantity of the halogen, in a cooled ether solution. They are crystalline solids which burn spontaneously in air.

Alkylbismuth Dihalides.--Alkylbismuth dihalides, of the type $RBiX_2$, may be prepared by reaction between the calculated quantities of bismuth halide and trialkylbismuth; while alkylbismuth dichlorides and dibromides are readily prepared in this way, the diiodides are prepared by treating the dichlorides with potassium iodide. Alkylbismuth dihalides are high-melting solids, stable in air. They react with aqueous ammonium hydroxide or sodium hydroxide to form alkylbismuth oxides, RBiO, which are easily oxidized further by air.

Symmetrical Triarylbismuth Compounds.--Symmetrical triarylbismuth compounds may be prepared by the reaction of a bismuth-alkali metal alloy with an aryl halide,

$$3ArX + BiK_3 \rightarrow Ar_3Bi + 3KX;$$

by the reaction of a Grignard reagent with a bismuth halide,

$$3ArMgX + BiX_3 \rightarrow Ar_3Bi + 3MgX_2;$$

or by the decomposition of an aryldiazonium chloride-bismuth chloride complex, followed by treatment with ammonia or hydrazine.

$$(ArN_2Cl)_2 \cdot BiCl_3 + 4Cu \rightarrow Ar_2BiCl + 4CuCl + 2N_2$$

$$Ar_2BiCl \xrightarrow{H_2NNH_2} Ar_3Bi$$

Triarylbismuth compounds are colorless crystalline solids, soluble in organic solvents, insoluble in water, and stable toward atmospheric oxygen. They react readily with chlorine or bromine to form triarylbismuth dihalides; triphenylbismuth diiodide may also be prepared at -78°C, but decomposes at higher temperatures. The aromatic groups in triarylbismuth compounds are held rather loosely by bismuth, and are cleaved readily by inorganic and organic acids,

$$Ar_3Bi + HCl \rightarrow ArH + Ar_2BiCl,$$

$$Ar_3Bi + CH_3CO_2H \rightarrow ArH + Ar_2BiO_2CCH_3;$$

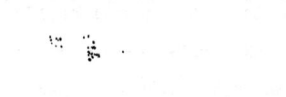

by thiophenol,

$$Ar_3Bi + 2C_6H_5SH \rightarrow ArBi(SC_6H_5)_2 + 2ArH;$$

by inorganic halides,

$$Ar_3Bi + PCl_3 \rightarrow Ar_2BiCl + ArPCl_2,$$

$$Ar_3Bi + HgCl_2 \rightarrow Ar_2BiCl + ArHgCl;$$

and by halogen halides and halogen pseudohalides,

$$Ar_3Bi + ICl \rightarrow Ar_2BiCl + ArI,$$

$$Ar_3Bi + CNBr \rightarrow Ar_2BiBr + ArCN,$$

the more negative group or atom going to the bismuth, the more positive one to the aromatic nucleus eliminated. Migration of phenyl groups from triphenylbismuth to either more or less electropositive elements can be effected to some extent, but since equilibrium is rarely reached, the course of the reaction usually depends primarily upon such specific factors as solubility, solvent, stability, catalyst, and other conditions of the reaction.

Unsymmetrical Triarylbismuth Compounds.--Unsymmetrical triarylbismuth compounds were first prepared by the reaction of a diarylbismuth halide, or an arylbismuth dihalide, with a Grignard reagent; a low-temperature modification of this method is still used. Recently, however, it was found that diarylbismuth halides can be treated with sodium, in liquid ammonia solution, to form first the diarylbismuth free radical,

$$Ar_2BiX + Na \rightarrow Ar_2Bi + NaX,$$

and then the diarylbismuth sodium compound,

$$Ar_2Bi + Na \rightarrow Ar_2BiNa;$$

the diarylbismuth sodium reacts readily with an aryl halide,

$$Ar_2BiNa + Ar'X \rightarrow Ar_2BiAr' + NaX,$$

forming the unsymmetrical triarylbismuth. These unsymmetrical triarylbismuth compounds are quite stable, and can be distilled without undergoing disproportionation, provided catalysts are not present. Studies of preferential cleavage of radicals in unsymmetrical triarylbismuth compounds have been used to determine the relative stability of the bond between bismuth and the different aryl radicals; for example, the reaction of *n*-butyllithium with triarylbismuth compounds,

$$Ar_2BiAr' + n\text{-}C_4H_9Li \rightarrow Ar_2BiC_4H_9\text{-}n + LiAr',$$

has placed the radicals in the following order of decreasing ease of cleavage from bismuth: (*p*-chlorophenyl, *p*-bromophenyl,

p-fluorophenyl, α-naphthyl, phenyl, p-tolyl, p-ethoxyphenyl, o-tolyl, mesityl, o-chlorophenyl).

Diarylbismuth Halides and Arylbismuth Dihalides.--Diarylbismuth halides and arylbismuth dihalides may be prepared conveniently by reaction between the calculated quantities of bismuth halide and triarylbismuth. They are high-melting crystalline solids, and most of them are decomposed by moisture, alcohol, and ammonia.

Trivalent Organobismuth Salts of Organic Acids.--Studies of the reactions of triarylbismuth compounds with thiophenols and certain carboxylic acids has indicated that preferential cleavage of two aryl groups, rather than one, generally occurs:

$$(C_6H_5)_3Bi + 2C_6H_5SH \rightarrow C_6H_5Bi(SC_6H_5)_2 + 2C_6H_6,$$

and

$$(C_6H_5)_3Bi + 2C_6H_5CO_2H \rightarrow C_6H_5Bi(O_2CC_6H_5)_2 + 2C_6H_6.$$

These are the only types of trivalent organobismuth salts of organic acids which have been studied.

Pentavalent Bismuth Compounds.--The pentavalent aryl bismuth derivatives are the most stable organobismuth compounds. They are fairly high-melting, crystalline solids. Triarylbismuth dichlorides and dibromides are prepared by treating a cold solution of the triarylbismuth with the halogen. They are so resistant to cleavage by acid that nuclear sulfonation and nitration are possible at fairly low temperatures. The halogens in triarylbismuth dihalides are quite reactive, and undergo double decompositions readily. For example, when triphenylbismuth dichloride is treated with ammonium hydroxide, first triphenylbismuth hydroxychloride and then triphenylbismuth dihydroxide are formed:

$$(C_6H_5)_3BiCl + NH_4OH \rightarrow (C_6H_5)_3Bi(OH)Cl + NH_4Cl$$

$$(C_6H_5)_3Bi(OH)Cl + NH_4OH \rightarrow (C_6H_5)_3Bi(OH)_2 + NH_4Cl.$$

These triarylbismuth dihydroxides react with acids, forming salts. They also act as oxidizing agents, converting the common alcohols to aldehydes and ketones.

Pentavalent Organobismuth Salts of Organic Acids.--Salts of pentavalent organobismuth compounds with organic acids may be prepared by reactions between triarylbismuth dihydroxides and organic acids,

$$Ar_3Bi(OH)_2 + 2RCO_2H \rightarrow Ar_3Bi(O_2CR)_2 + 2H_2O$$

triarylbismuth carbonate and organic acids,

$$Ar_3BiCO_3 + 2RCO_2H \rightarrow Ar_3Bi(O_2CR)_2 + H_2O + CO_2$$

triarylbismuth dihalides and alkali metal salts of organic acids,

$$Ar_3BiX_2 + 2RCO_2Na \rightarrow Ar_3Bi(O_2CR)_2 + 2NaX$$

or benzoyl peroxide and triarylbismuth compounds,

$$Ar_3Bi + (C_6H_5CO)_2O_2 \rightarrow R_3Bi(O_2CC_6H_5)_2 + O_2$$

Summary.--Certain organobismuth compounds are used in analytical work. The use of organobismuth compounds in therapeutic work has thus far been limited by the need for a water-soluble organobismuth compound having low toxic effect upon the human body.

Bibliography

Gilman, H. and Yale, H. L., Chem. Rev., 30, 281 (1942).

Reported by Rudolph Deanin
October 14, 1942

Symmetrical trinitrotoluene was prepared first by Wilbrandt in 1863 at Göttingen University by treating toluene with fuming nitric and sulfuric acids. 1,3,5-Trinitrobenzene was prepared in 1876 by Hepp who treated m-dinitrobenzene with a large excess of nitric acid and fuming sulfuric acid for several days.

Commercially TNT is usually prepared by a two or three stage nitration of toluene with nitric and sulfuric acids. Jackson and Phinney prepared TNT by heating trinitrophenylacetic acid with water or alcohol to eliminate carbon dioxide. The substituted acetic acid was synthesized from picryl chloride and malonic ester.

Trinitrobenzene is prepared commercially by the oxidation of TNT with sodium dichromate and sulfuric acid. It can also be prepared by reducing picryl chloride with copper powder in hot aqueous alcohol, and by treating 2,4,6-trinitrobenzaldehyde or 2,4,6-trinitrobenzoic acid with alcoholic ammonia. Hill and Torrey prepared TNB in 20% yields by the decomposition of nitromalonic aldehyde.

$$3\ O_2N\text{-}CH(CHO)_2 \rightarrow TNB + 3\ HCOOH$$

Action of Alkali.

The Organic Syntheses preparation of 3,5-dinitroanisole utilizes the action of sodium methoxide on trinitrobenzene.

$$C_6H_3(NO_2)_3 + CH_3ONa \xrightarrow{CH_3OH} 1,3,5\text{-}C_6H_3(NO_2)_2(OCH_3)$$

By boiling TNB with sodium carbonate and alcohol Lobry de Bruyn and van Leent obtained 3,3',5,5'-tetranitroazoxybenzene and some 3,5-dinitrophenol.

Meisenheimer prepared the red crystalline addition product of sodium methoxide and TNB by the action of sodium hydroxide in methyl alcohol.

TNB + CH_3OH + NaOH → 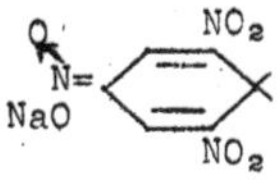 $OCH_3 \cdot 1/2\ H_2O$

Di- and tri-alcoholates and TNB·NaOH have been obtained also as colored crystalline compounds. Even after acidification of aqueous solutions of these salts, red water soluble compounds are left which will precipitate with heavy metals giving primary explosives. For this reason polynitro aromatic hydrocarbons should not be purified with alkali.

Trinitrotoluene will also form colored addition products with KOH, $KOCH_3$, etc. which will explode upon heating or sometimes spontaneously.

Condensations.

According to Vender, trinitrotoluene will condense with formaldehyde in the presence of sodium hydroxide or sodium carbonate to yield β-trinitrophenylethyl alcohol.

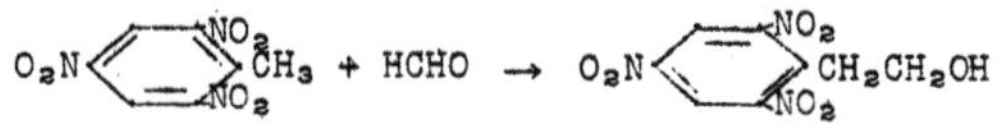
O_2N
NO_2
CH_3
+ HCHO →
O_2N
NO_2
CH_2CH_2OH
NO_2

Under the influence of a few drops of piperidine TNT will condense with benzaldehyde to form a trinitrostilbene. A solvent such as alcohol or benzene is used so that the heat of reaction will not cause the mixture to take fire. This is another reason why alkali must be excluded in the preparation of TNT. Aldehydic substances are usually present from the action of the nitrating acid on the wooden nitrator.

TNT may also condense with aldehydes to form substituted propanes in a reaction usually catalyzed by piperidine. Piperonal will unite with two moles of TNT in pyridine solution to yield 2,4,6,2",4",6"-hexanitro-3',4'-methylenedioxytriphenylpropane.

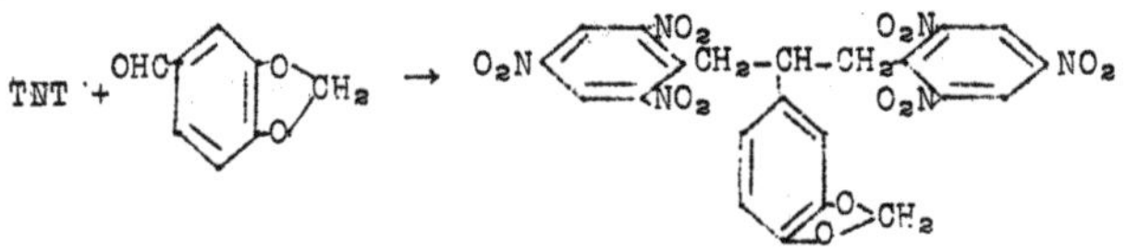
TNT +
OHC
CH_2
→
O_2N
NO_2
CH_2-CH-CH_2
NO_2
O_2N
NO_2
O_2N
CH_2

Pastak also condensed TNT with benzaldehyde, p-tolualdehyde, and cuminaldehyde to obtain similar substituted propanes. He found that the stilbene is the main product when the reaction is carried out without a solvent or with alcohol as the solvent, while the propane is formed when pyridine is the solvent and the reaction is run in the cold. The influence of the aldehyde is shown by the fact that anisaldehyde gives the stilbene exclusively while heliotropin (piperonal) gives mainly the substituted propane.

TNT will condense with itself under the influence of alkali to give azoxy and complex azo compounds. Hexanitrodibenzyl and hexanitrostilbene may be isolated on careful alkaline oxidation.

$$TNT + (O) \xrightarrow{NaOH} C_6H_2(NO_2)_3CH_2CH_2C_6H_2(NO_2)_3 + C_6H_2(NO_2)_3CH{=}CHC_6H_2(NO_2)_3$$

In their investigation of mordant dyes, Sachs and coworkers condensed TNT with 1,2-naphthoquinone-4-sulfonic acid.

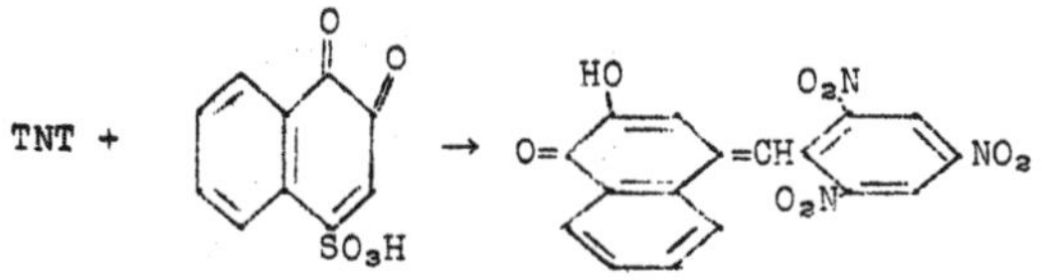
TNT +
O
O
SO_3H
→
HO
O=
=CH
O_2N
NO_2
O_2N

Trinitrotoluene will react with aryl nitroso compounds to give products which Sachs and other later workers believed to be azomethines. However, these condensation products have been shown to be nitrones by Tanasescu and Nanu.

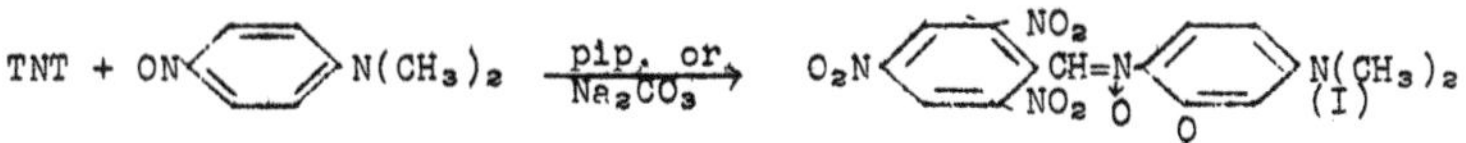
TNT + ON
N(CH₃)₂
pip. or
Na₂CO₃
O₂N
NO₂
CH=N
NO₂
O
N(CH₃)₂
(I)

The nitrones which are similar to N-methyl ethers, $RCH=\overset{O}{N}CH_3$, of aldoximes will likewise undergo a Beckman rearrangement by heating with acetic anhydride or acetyl chloride.

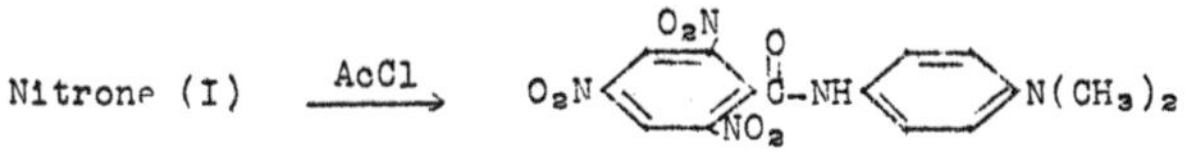
Nitrone (I)
AcCl
O₂N
O₂N
O
C-NH
NO₂
N(CH₃)₂

Picramide was prepared by Meisenheimer and Patzig using hydroxylamine and trinitrobenzene.

$$TNB + NH_2OH \xrightarrow{\text{cold alc.}} C_6H_2(NO_2)_3NH_2 + H_2O$$

Decomposition.

Although TNT may be distilled under reduced pressure, b.p. 210-12°/10-20 mm., without decomposition, it will decompose slowly with the evolution of gas at 150° or when exposed to sunlight in an open container. However, at 281° C, it decomposes violently.

Smith gives some possible reactions for the explosion of TNT.

$$TNT \rightarrow CO + H_2 + N_2 + C$$
$$\text{or} \rightarrow CO + H_2 + N_2 + CH_4$$

Diazomethane.

Heinke obtained the addition products of both trinitrotoluene and trinitrobenzene with diazomethane, but he was unable to assign a structure to either. Both of the compounds give the "Liebermann's nitroso" reaction.

$$TNT + 3\ CH_2N_2 \rightarrow C_{10}H_{11}O_6N_5\ (\text{m.p. } 177^\circ) + 2\ N_2$$

$$TNB + 4\ CH_2N_2 \rightarrow C_{10}H_{11}O_6N_5\ (\text{m.p. } 194^\circ) + 3\ N_2$$

Halogenation.

All attempts to halogenate the side chain of TNT have failed, and it can even be recrystallized from bromine according to Davis. MacKerrow obtained symmetrical dibromonitrobenzene by heating TNB with bromine at 230° and hexabromonitrobenzene by using bromine and ferric bromide.

Molecular Compounds.

Both TNT and TNB will form colored molecular compounds with organic bases, phenols, phenolic ethers, cyclic oxygen compounds, and aromatic hydrocarbons. Sudborough and his co-workers have prepared a great many of these compounds. The following table gives the ratio, color, and melting point of some of these molecular compounds.

TNB	Adding Substance	m.p.	Color
1:1	α-naphthol	178°	orange-yellow
1:1	piperonal	79°	golden-yellow
1:1	coumarone	103°	pale yellow
TNT			
3:2	carbazole	160°	yellow
1:1	aniline	83°	red

The structures of these molecular compounds are not known, and some are so unstable that they are detected only by m.p.-composition curves although a great many of them are stable crystalline compounds.

Oxidation.

Hepp oxidized trinitrobenzene to picric acid by using potassium ferricyanide.

$$TNB + K_3Fe(CN)_6 + Na_2CO_3 \rightarrow C_6H_2(NO_2)_3OH$$

Reduction.

Tin and hydrochloric acid will reduce trinitrotoluene or trinitrobenzene to the triamino compound. Weidel found that alcoholic hydrogen sulfide will reduce one nitro group of either TNB or TNT to a hydroxylamino group.

By using hydrogen sulfide in pyridine Brady and co-workers reduced TNT to 2,4-diamino-6-nitrotoluene.

Blanksma obtained 3,3',5,5'-tetranitroazoxybenzene and some 3,5-dinitroaniline by the reduction of trinitrobenzene.

$2\ TNB + Na_2S_2 \xrightarrow{alc.}$ 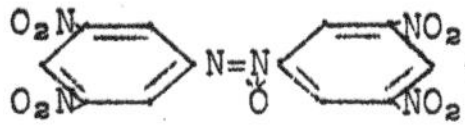 $+ Na_2S_2O_3$

Bibliography

Blanksma, Rec. trav. chim. 28, 112 (1909).
Brady, Day, and Reynolds, J. Chem. Soc. 2266 (1929).
Copisarow, Chem. News 112, 283 (1915).
Davis, "The Chemistry of Powder and Explosives," John Wiley and Sons, Inc., New York City (1941), Vol. I.
Gilman and Blatt, Organic Syntheses, John Wiley and Sons, Inc., New York City (1941), Col. Vol. I, p. 219, 542.
Heinke, Ber. 31, 1399 (1898).
Hepp, Ann. 215, 344 (1882).
Hepp, Ber. 9, 403 (1876).
Hill and Torrey, Am. Chem. J. 22, 97 (1899).
Jackson and Phinney, Am. Chem. J. 21, 430 (1899).
Lobry de Bruyn and van Leent, Rec. trav. chim. 13, 148 (1894).
MacKerrow, Ber. 24, 2944 (1891).
Meisenheimer, Ann., 323, 214, 241 (1902).
Meisenheimer and Patzig, Ber. 39, 2534 (1906).
Pastak, Bull. soc. chim. 39, 77 (1926).
Sachs, Berthold and Zaar, Chem. Zentr. 78, 1131 (1907).
Smith, "Trinitrotoluenes and Mono- and Dinitrotoluenes," D. Van Nostrand Co., New York City (1918).
Sudborough and Beard, J. Chem. Soc. 99, 209 (1911).
Tanasescu and Nanu, Ber. 72, 1083 (1939).
Vender, Gazz. chim. ital. 45 [II], 97 (1915), (C.A. 10, 1513[2]).
Weidel, Monatsh. 19, 224 (1898).
Willbrandt, Ann. 128, 178 (1863).

Reported by M. E. Chiddix
October 14, 1942.

THE INTRODUCTION OF FLUORINE INTO ORGANIC COMPOUNDS

In the course of the last few years organic fluorine compounds have attained a steadily increasing importance. The chemistry of these compounds is nevertheless--in contrast to the chemistry of the other halogens--still quite deficient. An important reason for this has to do with the necessity of special apparatus and elaborate precautions in working with elementary fluorine. Mixtures of fluorine with organic gases and vapors often explode with great violence, and the element itself is extremely poisonous.

Addition of fluorine to the double bond.--

The entrance of elementary fluorine into the double bond is theoretically possible. By this reaction the heat of formation of a C-C bond is appreciably exceeded, and the molecule is ruptured. If, however, the reaction is run in solution or takes place as a wall reaction the heat of reaction is dissipated. For example, at -80° in difluorodichloromethane solution, Bockmuller was able to obtain sym-difluorotetrachloroethane from tetrachloroethylene and fluorine.

$$CCl_2{=}CCl_2 + F_2 \rightarrow CCl_2F{-}CCl_2F$$

Similarly, fluorine adds to oleic acid, hexadecene, crotonic acid and others to give the expected products. The yields are never high, however, since substitution and dimerization occur as side reactions.

Methods of adding fluorine to the double bond by an indirect route have also been utilized. For example, Bockmuller used lead tetrafluoride at 0° in carbon tetrachloride solution and succeeding in fluorinating a,a-diphenylethylene in 40% yield.

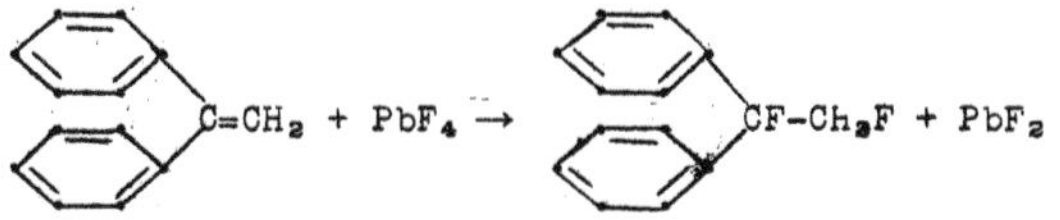
C=CH₂ + PbF₄ →
CF-Ch₂F + PbF₂

Phenyliododifluoride or para-tolyliododifluoride may also be used.

Addition of hydrogen fluoride to the double bond.--

The addition of hydrogen fluoride to the double bond goes with astonishing ease. The reaction, which requires no catalyst, goes best with anhydrous hydrogen fluoride at 90° for ethylene; lower temperatures are used with the higher olefins to avoid

polymerization. Propylene and hydrogen fluoride combine in the usual way to give isopropyl fluoride.

$$CH_3CH{=}CH_2 + HF \rightarrow (CH_3)_2CHF$$

The cyclopropane ring is split by hydrogen fluoride at 25°, forming n-propyl fluoride. Unsaturated acids take up hydrogen fluoride in chloroform or carbon tetrachloride solution to give the anticipated fluorine derivatives, and with unsaturated alcohols the hydroxyl group remains intact.

Acetylene and its derivatives add hydrogen fluoride as well. In the presence of mercuric ion as a catalyst vinyl fluoride is formed, which then goes to 1,1-difluoroethane.

$$HF + CH{\equiv}CH \rightarrow H_2C{=}CHF \xrightarrow{HF} H_3C{-}CHF_2$$

Similarly, when hydrogen fluoride is passed into a solution of stearolic acid in methylene chloride, 9,10 difluorostearic acid is formed.

$$CH_3(CH_2)_7C{\equiv}C(CH_2)_7COOH + 2HF \rightarrow CH_3(CH_2)_7CHF{-}CHF(CH_2)_7COOH$$

Hydrogen fluoride adds to ketene to give acetyl fluoride in quantitative yield. Cyanic acid, dissolved in ether at -78°, yields carbamic acid fluoride

$$NH{=}C{=}O + HF \rightarrow NH_2{-}\overset{O}{\overset{\|}{C}}{-}F$$

Substitution with hydrogen fluoride.--

The substitution of fluorine for another halogen atom proceeds by several different methods. Elementary fluorine displaces the other halogens attached to a carbon, but this procedure has no preparative significance. Hydrogen fluoride converts the $-CCl_3$ group to $-CF_3$ without a catalyst, and the analogous reaction of benzotrichloride is quite important.

$$C_6H_5CCl_3 + 3HF \rightarrow C_6H_5CF_3 + 3HCl$$

According to the patent, the reaction takes place at 110° under pressure.

Meta and para $-CCl_3$ groups are transformed simultaneously to compounds of the type $C_6H_4(CF_3)_2$. The chlorination of two ortho-situated methyl groups, as in o-xylene, stops with the formation of ortho-trichloromethyl benzal chloride. Substitution with hydrogen fluoride replaces the chlorine atoms.

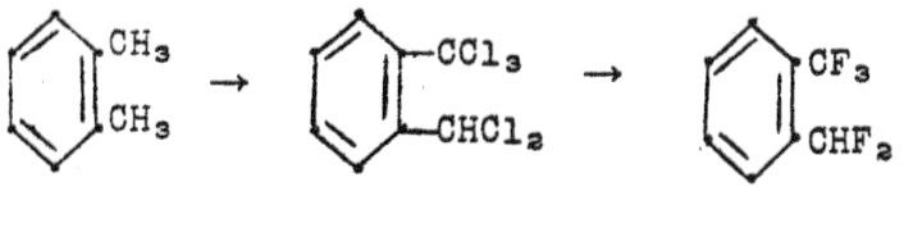
CH3
CH3
CCl3
CHCl2
CF3
CHF2

In this isolated case the $-CHCl_2$ group may be transformed to $-CHF_2$, a reaction which is not feasible with benzal chloride, since resinous products are formed. Substitution of hydrogen fluoride in the $-CCl_3$ group also occurs when *meta*- or *para*-situated -COCl groups are present and leads to the corresponding trifluoromethyl benzoyl fluorides. The presence of phthalimide groups in the benzotrichloride molecule does not hinder the substitution, and *o*-toluidine can be transformed into *o*-trifluoromethyl aniline in the following manner:

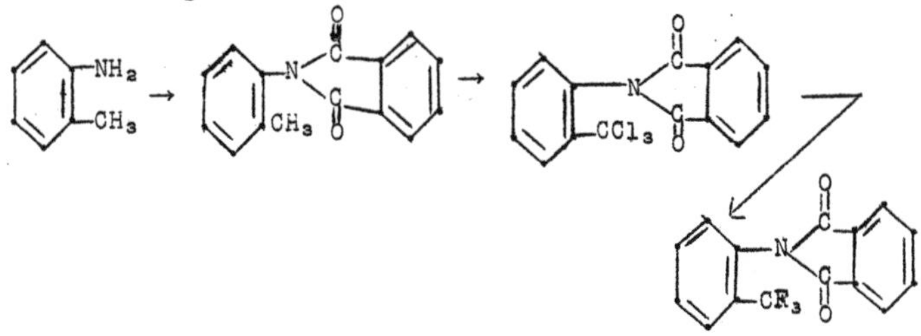
NH₂
CH₃
N
C
O
CH₃
CCl₃
CF₃

If the $-CCl_3$ group is attached to sulfur, as in trichloromethyl phenyl sulfide, the chlorine may be replaced by hydrogen fluoride.

$$C_6H_5\text{-}S\text{-}CCl_3 + 3HF \rightarrow C_6H_5\text{-}S\text{-}CF_3 + 3HCl$$

Substitution with metallic fluorides.--

One of the most important contributions to the organic chemistry of fluorine was made by Swarts, who discovered that compounds which have at least two chlorine or bromine atoms on a carbon may exchange these halogen atoms when treated with antimony trifluoride in the presence of bromine or antimony pentachloride as a catalyst. In the hands of Henne this procedure has been applied to numerous polyhalogen derivatives of the lower paraffins, and several of these compounds have been used as refrigerants.

In general, it has been ascertained that the $=CCl_2$ group reacts most easily with antimony trifluoride. The $-CHCl_2$ group, when attached to a $-CCl_2-$ group, is quite slow to react. Groups carrying only one chlorine atom cannot be substituted with this reagent. An exception to this rule is found, of course, in the acid chlorides. here the chlorine atom is sufficiently active to undergo substitution by the above method.

$$3R\text{-}COCl + SbF_3 \rightarrow 3R\text{-}COF + SbCl_3$$

For the preparation of the monofluoroparaffins certain other metallic fluorides have been found useful. Silver fluoride,

mercurous fluoride or mercuric fluoride may be used in the substitution of fluorine for bromine or iodine. Mercuric fluoride is the best reagent, and the substitution, which proceeds according to the scheme

$$2R\text{-}X + HgF_2 \rightarrow 2R\text{-}F + HgX_2$$

is almost quantitative for the lower alkyl bromides and iodides. Methylene chloride and chloroform will not react at ordinary temperatures and may be used as solvents. Alcohol, ethyl acetate or ether stop the reaction completely.

In many cases where a sufficiently reactive halogen atom is present, fluorine may be substituted with the help of thallous fluoride, and beryllium fluoride has been used to substitute fluoride for the hydroxyl group in aliphatic alcohols.

Introduction of fluorine in the aromatic nucleus.--

The introduction of fluorine into the aromatic nucleus takes place almost always through the diazotized amine. The reaction takes place by heating the diazonium fluoride that is formed from aniline by diazotization in aqueous hydrofluoric acid.

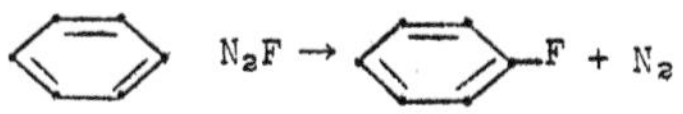
N₂F →
F + N₂

This reaction takes place without copper as a catalyst, and in many instances the presence of copper or cuprous salts is detrimental. Better yields are obtained through the use of concentrated hydrofluoric acid, and an industrial process makes use of almost anhydrous hydrogen fluoride.

Another method which has proved to be excellent is the diazonium borofluoride method of Balz and Schiemann. It involves the thermal decomposition of dry diazonium borofluoride.

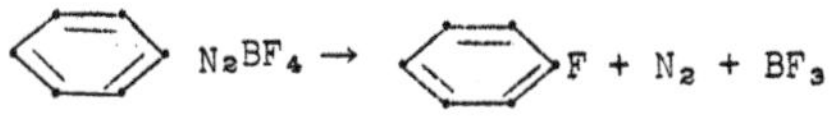
N₂BF₄ →
F + N₂ + BF₃

In contrast to most diazonium salts, the borofluoride is nearly insoluble, and quickly separates out when the hydroborofluoric acid or sodium borofluoride is added to the diazonium chloride solution. For this reason, the nucleus must not contain hydroxyl or carboxyl groups which would increase the solubility of the salt. In addition, the diazonium borofluorides are insensitive to shock and decompose quietly upon heating. The yields are excellent, and the procedure requires only paraffin-coated glass vessels.

Bibliography

Bochmuller, Angew. Chem., 53, 419 (1940).
Swarts, Bull. Soc. Chim., 35, 1533 (1924).
Swarts, ibid., 39, 444 (1930).
Bancroft and Wherty, Nat. Acad. Sci., 17, 183 (1931).
Finger and Reed, Trans. Ill. State Acad. Sci., 29, 89 (1936).

Reported by R. J. Dearborn
October 21, 1942

MYDRIATICS

The Influence of Structure on Physiological Activity

A mydriatic is a substance causing dilation of the pupil of the eye. The solanaceous alkaloids, especially atropine, have long been used for this purpose. In order to see how these drugs induce physical reactions in the eye, let us note a few salient features in the structure of the latter. The iris is regulated through two sets of nerves belonging to the involuntary nervous system. Two opposing sets of muscles, the dilator and the sphincter actually operate the aperture. The sympathetic nerve controls the former; the parasympathetic nerve the latter. Excitation of one of these nerves causes contraction of the muscle it controls. The two muscles are continually in balance and opposing each other.

Adrenaline excites the sympathetic nerve, causing dilation of the pupil. Atropine paralyzes the endings of the parasympathetic nerve and effects dilation by relaxing the sphincter. Mydriatics with adrenaline-like action are "active" while those with paralytic action are "passive". Atropine, and many other passive mydriatics also paralyze the ciliary muscles which control accomodation; a good active mydriatic is therefore to be preferred.

The potency of these drugs is quite easy to test. Usually the compounds are made up to 1% aqueous solutions as the amine salt or quaternary ammonium salt. Such a solution is then placed in the eye of a cat or similar test animal; at exactly the same time the dropper also delivers a drop of equal size of a comparison solution into the other eye of the animal. By such technique atropine is found active at dilutions of 1 part in 130,000. Test animals and human beings often react differently; the negro is more resistant than the Chinese, who is more resistant than the white man to mydriasis.

The exact mode of action of a substance displaying such physiological activity may be more difficult to determine. We have seen that these drugs may effect the muscles in an exciting or paralyzing manner. Duration of mydriasis is determined by the velocity of absorption of the drug (1); this in turn could be a property of the eye or of the drug. Furthermore, Bernard has found that lowering of surface tension is the chief factor contributing to the mydriatic activity of strophanthin, digitalis, sodium taurocholate, and others (2). An illustration of the confusion caused by these different effects may be furnished by the fact that the mydriasis produced by adrenaline is augmented by cocaine, while that produced by ephedrine is altered little. All three are active mydriatics.

Active mydriatics are essentially the same compounds to which we attribute pressor activity; a few of them are adrenaline, ephedrine, β-phenylethylamine, 2-amino-1,2,3,4-tetrahydronaphthalene

ω-amino-acetophenone, 2-amino-1-phenylpropanol, 2-amino-1-furylpropanol, and cocaine. The presence of cocaine in this group is surprising in view of its relation to atropine and its well-known paralytic activity.

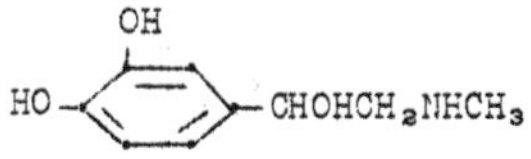

Adrenaline

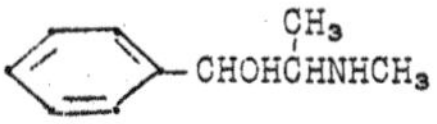

Ephedrine

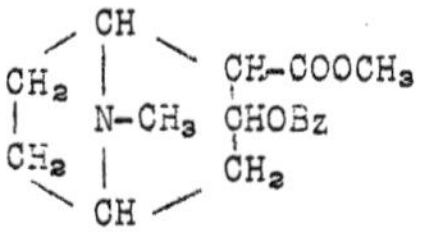

Cocaine

The β-phenylethylamine structure is common to all of these compounds except cocaine. In the more active compounds an α-hydroxyl is present in the side chain. Tiffeneau (3) concludes that these structural features are essential in active mydriatics.

Passive mydriatics are esters of which atropine is characteristic. It is the most active of the tropeines. Concentrated solutions of tropine and tropic acid have but feeble activity; esterification is necessary.

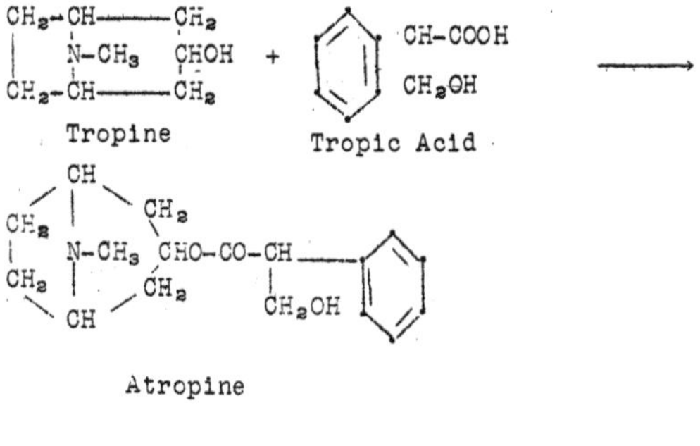

Atropine

Variations of both the alcohol and acid groups of this ester have been investigated in order to find the structural features necessary for mydriatic activity.

Various alcohols have been prepared which contain either the piperidine or the pyrrolidine ring, and some of these are nearly isomeric with tropine by the inclusion of methyl or other alkyl groups.

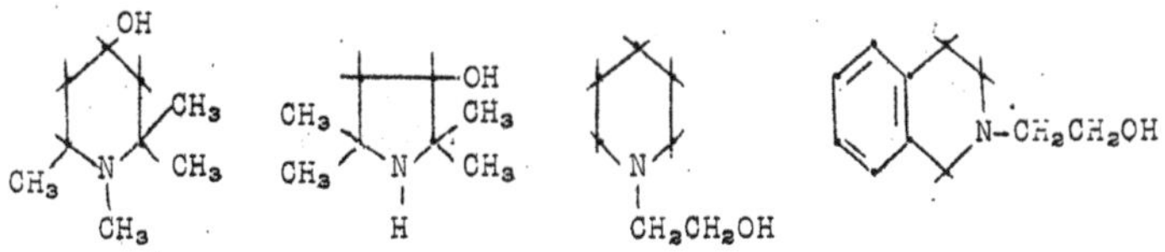

I II III IV

The mandelate of I is the useful drug known as "euphtalamine"; its potency is vastly greater than the similar ester of II. Von Braun and his coworkers found III-benzoate much more effective than the similar pyrrolidine compound; likewise with IV.

The activity of β-[2-propylpiperidino]-ethyl benzoate is slightly greater than III-benzoate. However, IV-benzoate is less active than the latter, and the same difference is present between β-pyrrolidinoethyl- and β-[3,4-benzopyrrolidino]-ethyl benzoate.

The piperidine ring is, therefore, the essential ring in tropine, hydrocarbon substituents not being of great significance in either raising or lowering the activity.

The amine group is unquestionably responsible for much of the activity of the tropine complex. Fourneau and Ehrlich felt that this group (haptophoric) linked the molecule to the nervous tissue. Von Braun found the benzoates of γ-amino alcohols and tropates of β-amino alcohols to be most active; however, atropine is the tropate of a γ-amino alcohol. Nor-atropine is one-eighth as active as the tertiary amine, atropine. This is also true for l-hyoscyamine. Methylatropine nitrate, "eumydrine", is of activity intermediate between that of atropine and homatropine (mandelyltropeine), atropine being the more active. However, the quaternary salts are more toxic.

We have noted Von Braun's conclusions concerning β- and γ-aminoalcohols. The particular molecules he used were of formula V, where R = $-(CH_2)_n-OH$. The β-γ unsaturated compounds behaved no differently. Blicke and Maxwell (5) found the benzilate of III to be the most active of twenty similar amino alcohols.

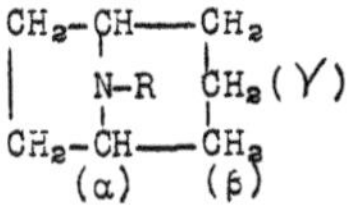

V

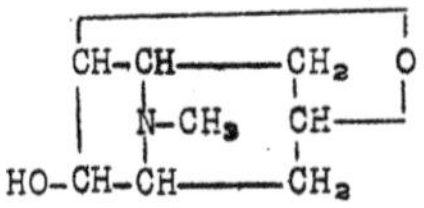

VI

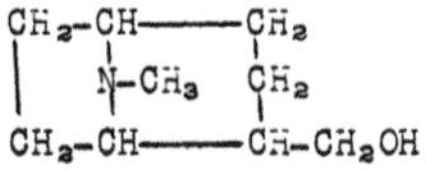

VII

In the ring the β and γ positions are equivalent as was shown (3) by comparing the tropate of scopaline, VI, scopolamine, with atropine. Homotropine, VII, forms a tropate, "mydriasine", which is a good mydriatic and does not impair accommodation.

Thus, whether the hydroxyl is on the ring or on a side chain, the only apparent requirement for activity is that it be in the β or γ position from the nitrogen.

As is frequently true in the chemistry of physiologically active compounds, stereoisomerism exercises a profound influence in certain of these compounds. Pseudotropine, in which the N-methyl and hydroxyl groups are trans, is the more stable diastereoisomer of tropine. Its tropate is inactive. Hyoscyamine, the natural alkaloid, is levorotatory and one hundred times more active in mydriasis than d-hyoscyamine. Atropine is made by racemizing the former. It is the tropyl complex, whether found in hyoscyamine or scopolamine, that renders the molecule as a whole levorotatory. This tremendous variance in the activity of enantiomorphs is not observed as between d- and l-homatropine.

Though Pyman (6) concluded after studying eight members that tropeins of aliphatic acids were inactive, terebyltropeine, VIII, was later found to produce mydriasis (7). Also Blicke and Maxwell, using their "most active" β-piperidinoethyl alcohol to esterify twenty acids, concluded that besides tropic, dicyclohexylgylcollic acid gave the most active ester.

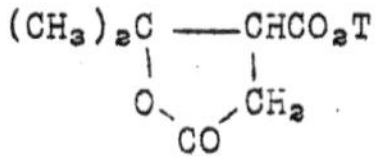

VIII

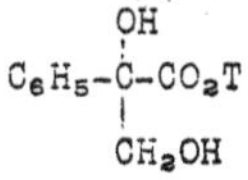

IX

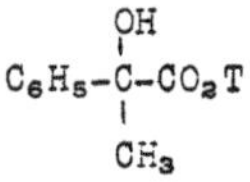

X

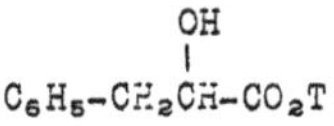

XI

T = tropyl

Pyman classes β-phenyl-α-hydroxypropionyltropein, XI, with atropine for potency; he found atroglyceryltropeine, IX, nearly as active and atrolactyltropeine, X, slightly less active than homatropine. Replacement of phenyl by 2-pyridyl in XI also gave an active compound. In tropic acid the replacement of the hydroxyl with chlorine gave an active tropeine; also similar substitutions in homotropine by -Cl, -Br, or $-NH_2$ yielded mydriatics.

When a substituent, e.g., hydroxyl or methyl, is present on the benzene ring in benzoyl- or mandelyltropeine, the isomer containing the group in the *para* position has the least activity and the *ortho* isomer the most.

Bibliography

(1) von Oettingen, The Therapeutic Agents of the Pyrrole and Pyridine Group, Edwards Bros. Inc., Ann Arbor, Mich., 1936, pp. 130-168.
(2) Barnard, Am. J. Physiol., 84, 407 (1928).
(3) Tiffeneau, Rev. gen. sci., 33, 544, 583 (1922).
(4) von Braun, Braunsdorf, Rath, Ber., 55, 1666 (1922).
(5) Blicke and Maxwell, J. Am. Chem. Soc., 64, 429 (1942).
(6) Pyman, J. Chem. Soc., 111, 1103 (1917).
(7) Jowett and Hahn, J. Chem. Soc., 89, 357 (1906).

Reported by George Mueller
October 21, 1942

FREE RADICAL BEHAVIOR OF DIAZONIUM COMPOUNDS

Much experimental evidence has been gathered in recent years to substantiate the view that there is a transient existence of aryl and halogen free-radicals when a diazonium salt decomposes with the liberation of nitrogen. The present report is concerned with recent work in this field, particularly with the series of papers by W. A. Waters.

I. Nitrosoacetylaryamines.--

Haworth and Hey have shown that a nitrosoacetylarylamine may react with a hydrocarbon RH in either (or both) of two ways.

(I) $ArN(NO)COCH_3 + RH \rightarrow ArR + N_2 + CH_3COOH$

(II) $ArN(NO)COCH_3 + RH \rightarrow ArH + N_2 + CH_3COOR$

When R is aryl Reaction I (biaryl formation) is predominant. Reaction II predominates when a non-aromatic hydrogen-containing solvent is used.

Reaction I: This is the well-known reaction of Gomberg for the synthesis of diaryls, in which the neutral radical from the aryl diazoacetate enters the nucleus in the ortho and para positions irrespective of the nature of the so-called directive group.

Hey has widely extended the reactions of various nitrosoacetanilides with aromatic solvents for preparative purposes. Examples of this reaction from his work are:

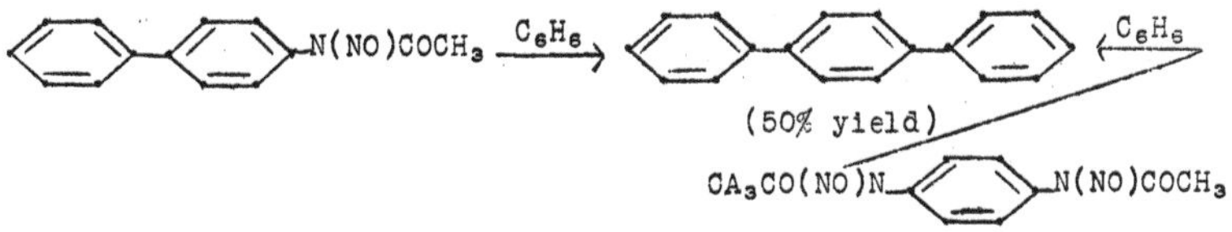

N(NO)COCH₃
C₆H₆
C₆H₆
(50% yield)
CA₃CO(NO)N
N(NO)COCH₃

Reaction II: Waters decomposed dry benzenediazoacetate (prepared as its tautomer, nitrosoacetylaniline) at room temperature under several dry, organic solvents. In hydrogen containing solvents in all cases a 5 to 40% yield of a simple organic material, and a complex tar, were produced. In hexane or cyclohexane only benzene was isolated, but acetaldehyde and acetic acid were additional products when diethyl ether, dioxane, or acetone were solvents. With alkyl halides the halogenbenzenes were formed, and with acetic anhydride carbon dioxide was evolved.

Grieve and Hey had previously found that the rate of evolution of nitrogen during such decompositions was unimolecular and independent of the solvent. The rate determining process apparently was:

$C_6H_5N(NO)COCH_3 \rightarrow C_6H_5\cdot + N_2 + CH_3COO\cdot$

The free phenyl radical then combined with the solvent present according to Reaction II:

$$C_6H_5\cdot \quad \xrightarrow{EtX} C_6H_5X \qquad \xrightarrow{C_6H_{12}} C_6H_6 \qquad \xrightarrow{EtOEt} C_6H_6 \ (5\text{-}40\%) \qquad + \ \text{a complex tar consisting of polyphenyls}$$

Since the acetate anion never loses carbon dioxide except at an anode during electrolysis the formation of this gas was attributed to a disproportionation of the neutral acetate radical:

$$CH_3COO\cdot \rightarrow CO_2 + CH_3\cdot$$

The formation of complex tars as major products is understandable since free radicals are extremely reactive and capable of attacking any molecule with which they come in contact.

II. Diazonium Chlorides.--

A. Reaction with non-ionizing solvents.

When benzene diazonium chloride was decomposed by warming it under acetone, hexane, carbon tetrachloride, ethyl iodide, or ethyl acetate, the products were hydrogen chloride and chlorobenzene. If the reaction under acetone were kept neutral with chalk, chloroacetone was also a product. The similar decomposition reaction with alcohols is more familiar, yielding usually benzene and either an aldehyde or a phenyl ether.

Hantzsch had proposed an initial addition complex, ArClN=NH(OR), for the reaction between diazonium halides and alcohols. This complex supposedly broke up into the observed products. However, Prey showed that the decomposition rate of benzene diazonium chloride was almost identical in different alcoholic solutions, although the overall reactions differed considerably. The rate determining process cannot, therefore, be decomposition of an addition complex for this would vary with the alcohol used.

Waters has proposed as a mechanism the molecular rearrangement of the benzene diazonium chloride to unstable benzene diazochloride ($PhN_2^+Cl^- \rightarrow PhN{=}NCl$) which then decomposes into free radicals, Ph· and Cl·, and nitrogen. Subsequent attack of the solvent by the free radicals would explain all of the observed products:

1) Formation of benzene with alcohol and with acetone

$$Ph\cdot + RH \rightarrow PhH + R\cdot$$

2) Formation of hydrogen chloride in all reactions

$$Cl\cdot + RH \rightarrow HCl + R\cdot$$

3) Formation of chlorobenzene

$$Ph\cdot + Cl\cdot \rightarrow PhCl$$

$$Ph\cdot + HCl \rightarrow PhCl + H\cdot$$

$$Ph\cdot + CCl_4 \rightarrow PhCl + \cdot CCl_3$$

4) Chlorination of acetone

$$Cl\cdot + CH_3COCH_3 \rightarrow ClCH_2COCH_3 + H\cdot$$

B. Reaction with metals.

Benzene diazonium chloride, under acetone and in the presence of chalk, was decomposed in the presence of 38 different elements. Both gold and palladium were reactive though metals like chromium and aluminum, which have oxide films, were practically inert. Elements of metallic character yielded chlorides; aromatic compounds were formed by Hg, Sn, As, Sb, S, Se and Te, but not by Au, Tl, Ge, Pb, Bi, Mg, or other metals.

In the case of antimony, compounds of different structural types were produced simultaneously: triarylstibine dichlorides, Ar_3SbCl_2; triarylstibines, Ar_3Sb; and diarylstibnous chlorides, Ar_2SbCl. This was in accord with Water's theory, because neutral aryl radicals and chlorine radicals would be expected to combine, independently of each other, with any available substance containing an uncompleted electronic shell. It cannot be the diazonium cations which react with the antimony since in water or alcohol no reaction takes place. From aniline, p-chloroaniline, and p-bromoaniline the chief products were triarylstibine dichlorides. With amines containing ortho substitutents e.g., 5-chloro-o-toluidine, the chief products were triarylstibines. Products with sulfur, selenium, and tellurium were PhSSPh, PhSePh, and PhTePh, respectively.

With mercury, phenylmercuric chloride, PhHgCl, was formed. This was also the product when Dichloroamine-T and mercury were refluxed in acetone. The fact that benzenediazonium chloride reacted to give the same product as a compound which has a known positive chlorine, >N-Cl, was evidence that the change $(PhN_2^+Cl^- \rightarrow PhN{=}NCl)$ does occur.

Electrons in atoms of other than zero valency tend not to pair with each other. Consequently stable bonding pairs of electrons can be formed by union of atoms and the odd electron of free radicals. The reaction of a free radical would be easier with an atom than with a molecule which has all of its electrons already shared in covalent bonds. It was found that benzenediazonium chloride did not react with red phosphorous, amorphous boron, carbon, or silicon. These non-metals possess crystal structures built up by covalent linkages, and consequently have no electrons available for union with free radicals.

C. Reaction with acetonitrile

Diazonium chlorides were decomposed under acetonitrile at 40° in the presence of chalk. The main reaction product was always a black tar, but small quantities of simpler products were usually separated and showed that there were four types of reactions occurring.

$$ArN_2^+Cl^- + CH_3CN \begin{cases} \rightarrow ArH & (A) \\ \rightarrow ArCl & (B) \\ \rightarrow ArNHCOCH_3 & (C) \\ \rightarrow ArCOCH_3 & (D) \end{cases}$$

Ar was phenyl, o-tolyl, p-tolyl, p-chlorophenyl, 4-chloro-o-tolyl, and 5-chloro-o-tolyl.

Reactions (C) and (D) are peculiar to nitriles and can be explained as follows:

$$ArN_2^+Cl^- \rightarrow ArN{=}NCl \rightarrow Ar\cdot + N_2 + Cl\cdot$$

$$(C)\ Ar\cdot + Cl\cdot + CH_3C{\equiv}N \rightarrow CH_3\underset{\displaystyle Cl}{C}{=}NAr \xrightarrow{H_2O} CH_3CONHAr$$

$$(D)\ Ar\cdot + Cl\cdot + CH_3C{\equiv}N \rightarrow CH_3\underset{\displaystyle Ar}{C}{=}NCl \xrightarrow{H_2O} CH_3COAr$$

Propionitrile was much less reactive than acetonitrile and only reactions (A) and (B) were proven to occur.

III. Diazohydroxides.--

The benzenediazohydroxides (PhN=NOH) should split up, according to Water's theory, into free phenyl and hydroxyl radicals; these should then react as did those from benzenediazonium chloride and benzenediazoacetate. This analagous behavior was observed. Cyclohexane and carbon tetrachloride yielded benzene and chlorobenzene, respectively, when an alkaline solution of sodium benzene diazotate was allowed to decompose in the presence of these solvents.

Substances which form two neutral radicals should be good oxidizing agents, and this behavior was, in fact, observed. One example is the Bart reaction for the preparation of aryl arsonic acids in which antimony is oxidized from a valence of three to five:

$$ArN{=}NOH + Na_2HAsO_3 \rightarrow N_2 + ArAsO(ONa)_2 + H_2O$$

Additional evidence that diazonium compounds break up into free radicals in alkaline solution has been given by Neunhoeffer and Weise. They described the different behavior of 2-hydroxy-1,4-naphthoquinone when reacted with diazonium chlorides in alkaline and in acid solutions.

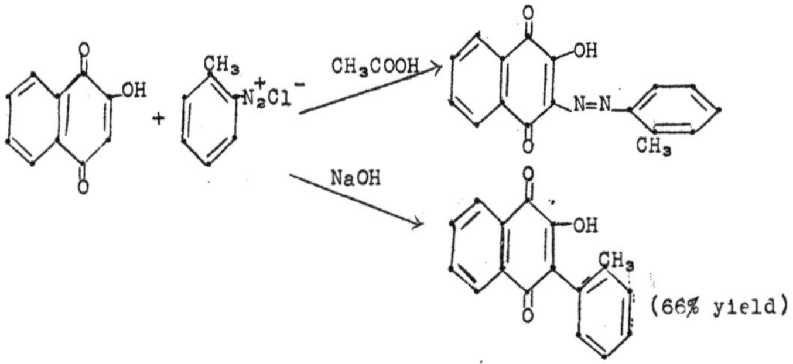
CH3COOH
NaOH
(66% yield)

Further evidence that the reaction goes by a non-ionic fission of the covalent structure, ArN=NOH, was given by Haworth, Heilbron, and Hey. Aqueous p-nitrobenzenediazonium chloride was added to excess pyridine at 40°. The three isomers, α-, β-, and γ-4.-nitrophenylpyridine, were formed. If the reaction were ionic the expected product would be 4-nitrophenylpyridinium chloride, or a mixture of the α- and γ-isomers by a subsequent migration of the aryl group on this salt. The production of all three isomers points to the presence of a reagent which can react kationoid or anionoid as the occasion demands. Such a reagent is the free aryl radical with its one unshared electron:

$$C_6H_5N + p\text{-}NO_2C_6H_4\overset{+}{N_2}Cl^- + H_2O \rightarrow C_6H_5N\cdot HCl + NO_2C_6H_4N{=}NOH$$

$$NO_2C_6H_4N{=}NOH \xrightarrow{-N_2} NO_2C_6H_4\cdot + \cdot OH$$

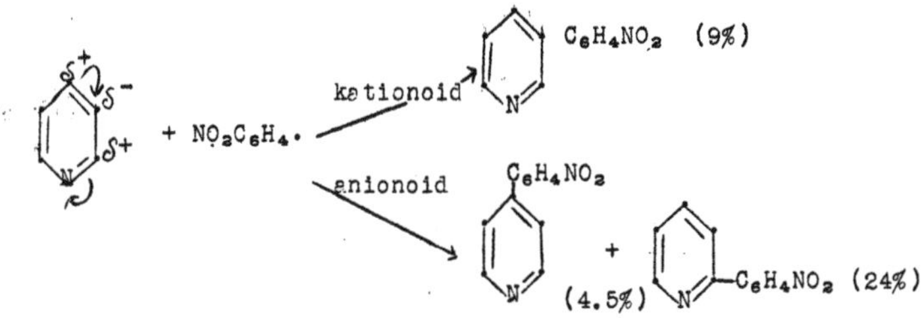
+ NO2C6H4·
kationoid
C6H4NO2 (9%)
anionoid
(4.5%)
C6H4NO2 (24%)

IV. <u>Syn-diazo cyanides</u>.--

Solutions of syn-aryldiazocyanides do not decompose in dry, non-ionizing solvents, but are changed into the more stable anti-aryldiazo cyanides. By adding copper powder to the syn-compound decomposition can be made to occur, and the products are HCN, tars, and various aromatic compounds depending on the solvent used. For the case of syn-o-chlorobenzenediazo cyanide, the isolated yields

in the various solvents were as follows:

$$o\text{-}ClC_6H_4N{=}N\text{-}CN \xrightarrow[Cu]{-N_2} o\text{-}ClC_6H_4\cdot$$

$$o\text{-}ClC_6H_4\cdot \xrightarrow{CCl_4} o\text{-}C_6H_4Cl_2 (14\%) + o\text{-}ClC_6H_4CN\ (4\%)$$

$$o\text{-}ClC_6H_4\cdot \xrightarrow{C_6H_6} o\text{-}ClC_6H_4CN\ (7\%)$$

$$o\text{-}ClC_6H_4\cdot \xrightarrow{EtOH} C_6H_5Cl\ (20\%)$$

$$o\text{-}ClC_6H_4\cdot \xrightarrow{EtOEt} C_6H_5Cl\ (6\%) + o\text{-}ClC_6H_4CN\ (3\%)$$

The formation of nitriles by the Gatterman reaction was in a large measure repressed in these dry solvents. This is evidence that the Gatterman reaction occurs only when ionization is possible, and possibly that it is a reaction of the diazonium salt, rather than of the free radical.

Since syn-aryldiazo cyanides obviously have the covalent structure, and since they give exactly the same products as previously observed for the nitroso-acetanilides and the aryldiazonium salts, it is likely that the latter two types shift over to the covalent structure before reacting. From the variety of products obtained it is also evident that the two radicals, which result from the removal of the azo group as nitrogen gas, do not unite at once but react with the solvent molecules after a transient independent existence.

Bibliography

Waters, J. Chem. Soc., 1937, 113, 2007, 2014; 1938, 843, 1077; 1939, 864, 1792, 1796.
Grieve and Hey, J. Chem. Soc., 1934, 1797.
Pray, J. Phys. Chem., 30, 1477 (1926).
Haworth and Hey, J. Chem. Soc., 1940, 361.
Neunhoeffer and Weise, Ber., 71, 2703 (1938).
Haworth, Heilbron, and Hey., J. Chem. Soc., 1940, 349.

Reported by Robert Gander
October 28, 1942

THE OZONIZATION REACTION

The earliest reported uses of ozone as a synthetic method were made by Otto and Trillat in 1898. They described the commercial preparation of vanillin from isoeugenol and of piperonal from isosafrole. Improved methods have led to the preparation of many substances hitherto unknown. Noller and Adams in 1926 reported the preparation of aldehyde esters by ozonization of esters of unsaturated acids.

$$CH_3(CH_2)_7CH{=}CH(CH_2)_7COOCH_3 \rightarrow CH_3(CH_2)_7CHO + CHO(CH_2)_7COOCH_3$$

For example, methyl ω-aldehydooctanoate was made by ozonization of methyl oleate. New methods of reduction of ozonides have led to the preparation of even such sensitive dialdehydes as glutaraldehyde, adipaldehyde, and pimelaldehyde.

The most important use of ozonization has been in the proof of structure of unsaturated compounds by cleavage of the chain. Probably the most obvious disadvantage is the explosive nature of certain ozonides. This may be avoided in almost every case by exercising special precautions in technique.

Various structural configurations have a marked effect on the course of the reaction. Ozone reacts more rapidly with isolated open chain double bonds than with two or more conjugated double bonds or those present in an aromatic ring. It reacts more readily with a carbon to carbon double bond than with a carbon to nitrogen double bond. In general the presence of a carbon to nitrogen double bond leads to complications.

Often there is interaction, oxidation, or further decomposition of the primary products so that interpretation of the reaction may be difficult, as in the ozonization of pulegone. It was not possible to identify 1-methyl-3,4-cyclohexanedione since β-methyl adipic acid was formed immediately.

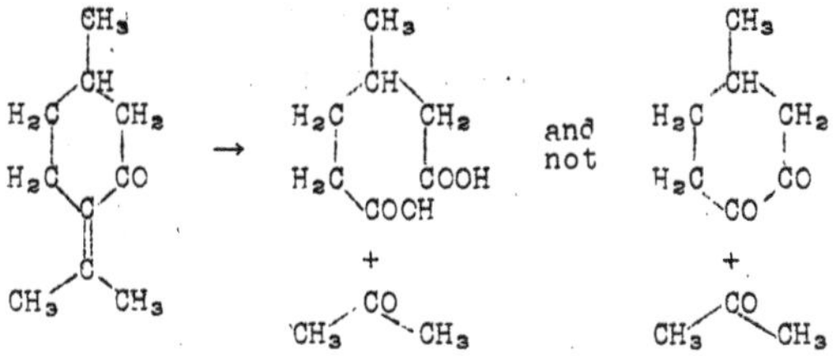

Many papers have been published by Harries, Staudinger, Rieche, Pummerer, Briner and others on the structures of the intermediate products and the mechanism of the reaction. The most favored theory assumes the formation of primary products called molozonides which may either polymerize or undergo rearrangement to ozonides.

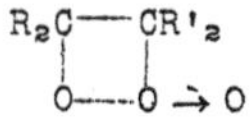

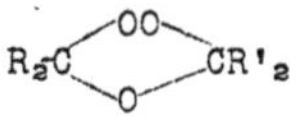

ozonide

molozonide

The formation of polymeric ozonides usually occurs where rearrangement of the molozonide to ozonide is difficult e.g., where the double bond is in a ring. The solvent may also affect the rearrangement. In acetic acid, where association of molecules does not readily take place because of its polar character, monomeric ozonides are obtained. In non-polar solvents, polymerization often occurs.

Ozone is prepared by passing dried oxygen through silent electric discharge tubes of the Berthelot type. Ozone in rather high concentration may be obtained. The rate of flow may be calculated from a flowmeter and the concentration may be found by titrating a potassium iodide solution through which the ozone has been passed.

A number of methods of passing the ozone into the solution to be ozonized have been used. The usual method is to use a gas bubbling tube. Pummerer and Richtzenheim used a counter current flow of ozone and the solution through a tower packed with small glass rings. Spencer in a recent paper reported the ozonization in the vapor phase of substances with an appreciable vapor pressure.

The concentration of the ozone may be varied with the nature of the compound to be ozonized. A high concentration (14-15%) facilitates addition of the reagent to aromatic compounds and substances with conjugated double bonds, whereas a low concentration (1-5%) is essential for the isolation of certain aldehydes which are sensitive to oxidation. Excessive ozonization must be avoided because of the oxidative effect of the ozonized oxygen on the reaction products.

A number of solvents have been used and no general rules can be given. Some compounds may be ozonized in the pure state but this is more likely to lead to an explosion. Dilute solution and low temperatures give more favorable results. Fischer stated that pure dry ethyl acetate was the best solvent for a number of alicyclic and straight-chained unsaturated compounds. Whitmore in the ozonization of olefines used a low boiling paraffin hydrocarbon (b.p. 0-30°) kept at about -10° in an ice salt mixture. Glacial acetic acid, ethyl bromide, ethanol, hexane, chloroform, carbon tetrachloride and water have been used successfully.

Decomposition of the ozonides may lead to a variety of compounds necessitating rather exact methods. In general the ozonides of the higher aliphatic unsaturated compounds like those of the hydroaromatic compounds are very stable. Those of the

doubly unsaturated aliphatic hydrocarbons decompose readily. Aliphatic ozonides containing oxygen in other parts of the molecule react readily with ice water. Similarly, ozonides of benzal compounds and their oxygen derivatives decompose readily. Ozonides of six- and seven-membered ring compounds are stable compared to those of five-membered ring compounds.

Oxidative cleavage of ozonides leading to acids has found few applications. Decomposition with alkaline permanganate or with hydrogen peroxide in alkaline solution give best results.

Methods of reductive cleavage have been investigated extensively. Treatment of the ozonide with the reducing agent without delay after the ozonization is essential for avoiding the occurence of acidic decomposition products.

Whitmore and his coworkers have made a thorough study of various methods of decomposing ozonides including the use of zinc dust and water, concentrated and dilute sodium bisulfite solutions, glacial acetic acid and zinc dust, liquid ammonia and hydrazine hydrate. The best method they found was treatment with water and zinc in the presence of traces of silver and hydroquinone. However, their method involves isolation of a pure ozonide and a complicated apparatus which is hard to replace in case of explosions. At present Whitmore is investigating the use of Raney nickel and has had some promising results. Ozonides react with Raney nickel to give aldehydes and ketones and NiO. The reaction is vigorous at 35° when a pentene solution of the ozonide is added to a suspension of nickel in pentane. Yields are as good as previously obtained by other methods.

The method of catalytic hydrogenation discovered by Fischer appears to be the best method of reductive decomposition. The hydrogenation usually proceeds very quickly and with much evolution of heat. If warming of the solution is allowed during the hydrogenation low yields may result due to rearrangement of the ozonide. Yields of 50-90% of the theoretical have been obtained. Five per cent palladium on calcium carbonate was used as the catalyst.

The hydrogenation of thick polymeric ozonides such as solid cyclohexene-ozonide did not proceed at room temperature but was accomplished by Fischer by warming in an autoclave with hydrogen under pressure. Decomposition of the resulting aldehyde was retarded by the use of methanol and ethanol as solvents, whereby unreactive acetals were formed. It was found preferable to ozonize in a solvent in which polymerized insoluble ozonides did not form. Ethyl acetate was found particularly useful.

BIBLIOGRAPHY

Cook and Whitmore, J. Am. Chem. Soc., 63, 3540 (1941).
Church, Whitmore, and McGrew, J. Am. Chem. Soc., 56, 176 (1934).
Fischer, Ann., 464, 82 (1928).
Long, Chem. Rev., 27, 437 (1940).
Noller and Adams, J. Am. Chem. Soc., 48, 1074 (1926).
Spencer, et al, J. Org. Chem., 5, 610-617 (1940).
Smith, J. Am. Chem. Soc., 47, 1844 (1925); ibid. 55, 4327 (1933).

Reported by W. R. Hatchard
October 28, 1942

OLEFINIC PEROXIDES

The mechanism of atmospheric oxidation of olefins is of great importance in the chemistry of drying oils and rubber.

The purpose of this seminar report is to review briefly the subject of peroxides in general and then to report on articles by Criegee and by Farmer concerning olefin peroxides. By these articles ideas of autoxidation of olefins have been considerably advanced.

Review of Peroxides

Classification:

I. Ordinary peroxides of the type C-O-O-C
   A. Acyl peroxides of the type of benzoyl peroxide.
   B. Free radical peroxides such as $\phi_3C-O-O-C\phi_3$
   C. Diethyl peroxide $C_2H_5-O-O-C_2H_5$. This compound is ether-like and very hard to reduce.
   D. Transannular (ring) peroxides
      1. Transannular peroxides of aromatic rings such as

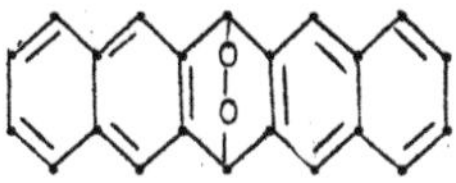

O
O

and

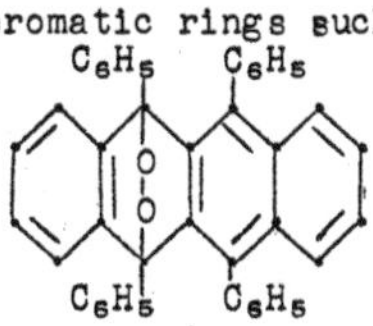

C_6H_5
C_6H_5
O
O
C_6H_5
C_6H_5

Hydrocarbons of the anthracene-naphtnacene-pentacene series will peroxidize on irradiation in the presence of air to give similar products. On heating oxygen is usually lost and the parent hydrocarbon regenerated.

2. Transannular peroxides of non-aromatic ring systems such as

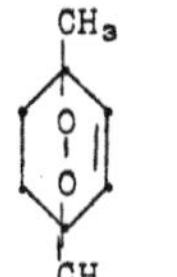

CH_3
O
O
CH
CH_3
CH_3

ascaridole

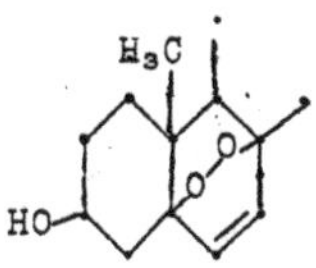

H_3C
O
O
HO

ergosterol peroxide

Compounds of this type whose structures have been elucidated are not numerous.

OLEFINIC PRECURSORS

[illegible]

[illegible]

[illegible]

Class II. Hydroperoxides containing C-O-O-H groups

Examples:

$C_6H_5C(=O)-OOH$

perbenzoic acid

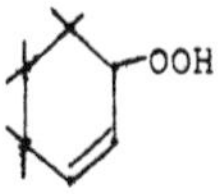
OOH

cyclohexene peroxide

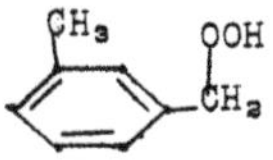
CH3
OOH
CH2

m-xylene peroxide

$C_2H_5OC(H)(O-O-H)-CH_3$ (?)

ethyl ether peroxide

Properties shared by peroxides:-

1. With the exception of acyl peroxides, per-acids and ethyl peroxide, which are prepared directly or indirectly from sodium or hydrogen peroxide, the peroxides are prepared by atmospheric oxidation of some substance in the presence of light. Irradiation is known not to be absolutely necessary for certain of these reactions, but it greatly accelerates the reaction especially during the early stages. Temperature and catalysts also play a large role. At 70° in the presence of ferrous phthalocyanine, methylcyclohexene on irradiation in the presence of air gives a mixture which is chiefly isomeric methylcyclohexenones. At 35° under similar conditions but without the catalyst methyl cyclohexene peroxides are produced in good yield. Rate of peroxide formation varies from the very rapid rate of the free radicals to the formation of only several % of ether peroxide over a period of several months.

2. The carbon-hydrogen bonds which were peroxidized are reactive bonds.

3. Heating most peroxides results in more or less violent decompositions which often completely disrupt the molecule.

4. The C-OOH and C-OO-C groups are capable of reduction to C-OH and -COH HOC- groups, respectively, by such reducing agents as aldehydes, olefins and sodium sulfite.

A discussion of the recent work of Criegee and of Farmer is now entered upon.

The basis of this work is the proof that the structure of cyclohexene peroxide is

OOH

instead of

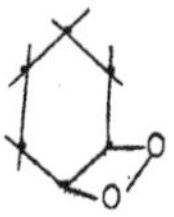
O
O

as was formerly supposed.

Criegee proved this by the following facts:

1. The sodium sulfite reduction product of the peroxide is cyclo-hexen-3-ol.
2. The peroxide takes up one mol of bromine.
3. It contains one active hydrogen atom.

The best theory previously proposed for the atmospheric oxidation of olefins:

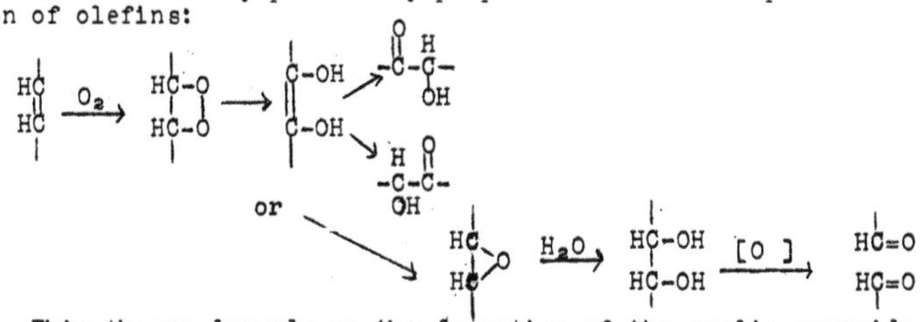

O_2
or
H_2O
[O]

This theory depends on the formation of the cyclic peroxide intermediate. The support for the existence of such an intermediate is limited to the fact that unsaturation decreases progressively (though not quantitatively) as oxidation occurs, and to the ease of thereby accounting for scission products which are formed from such reactions.

Using the new-found structure of the peroxide as a tool, it was desirable to inquire into the mechanism of its reactions, and those of 1-methylcyclohexene and 1,2-dimethylcyclohexene peroxides.

The peroxide was so reactive that it could not be prepared without traces of impurities. It could be distilled.

It was found possible to predict most likely points of greatest autoxidative attack which are indicated by the asterisks below.

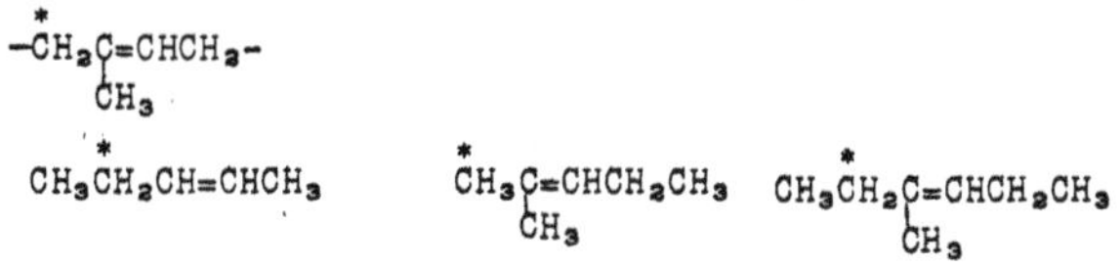

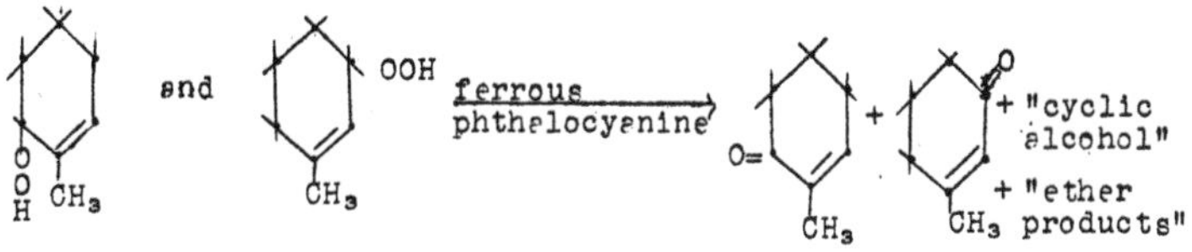

and
OOH
ferrous
phthalocyanine
+ "cyclic alcohol"
+ "ether products"

substantial yield

Ketone formation as in the above equation could be obtained through thermal or catalytic decomposition but not by isomerization.

Reaction of cyclohexene peroxide with dilute sulfuric acid leads to large quantities of 1,2,3 cyclohexene triol and small amounts of cyclopentenealdehyde, the latter being considered due to presence of impurities. Formation of the trial was explained by the following mechanism:

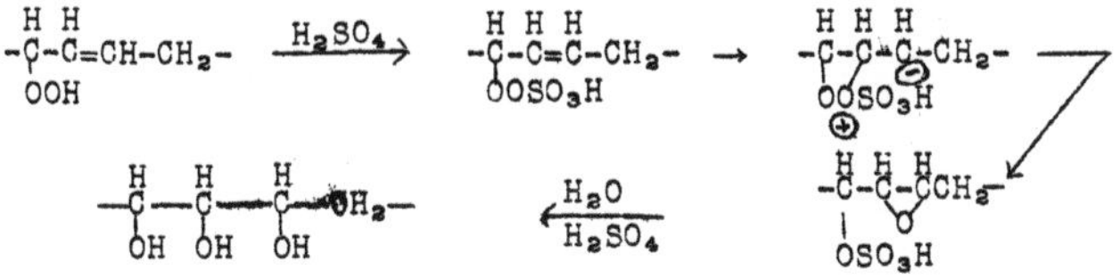

up to 80% yield + lower molecular weight acids

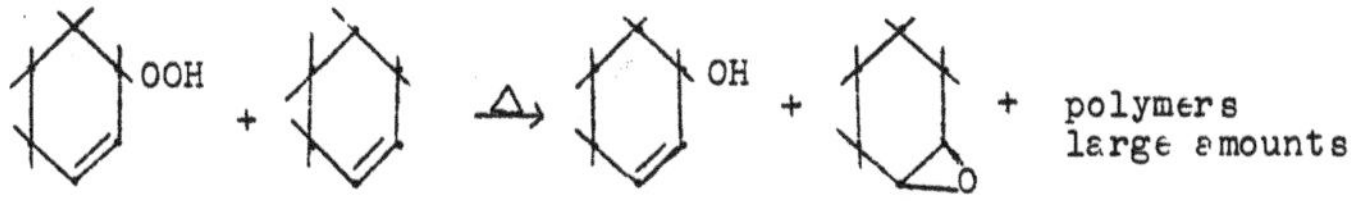

small amounts

The polymers obtained were mostly dimeric, but were not homogeneous. The most likely source appears to be reaction between hydroperoxide or hydroxyl groups and oxide groups, probably preceded or followed by some oxidation to carboxylic acids.

Bibliography

Bergmann and McLean, Chem. Rev., 28, 367-94 (1941).
Farmer and Sundralingham, J. Chem. Soc., 1942, 121.
Farmer and Sutton, ibid., 1942, 139.
Criegee, Pilz and Flygare, Ber., 72, 1799 (1939).
Clover, J. Am. Chem. Soc., 44, 1107 (1922).

Reported by H. F. Kauffman, Jr.
November 4, 1942.

CHARACTERIZATION OF CARBOXYLIC ACIDS AS THE UREIDES WITH THE AID OF CARBODIIMIDES

In 1901, Schall discovered that carboxylic acids react with carbodiphenylimide to form the corresponding monoacyl diphenylurea (or ureide). He formed such ureides with acetic, benzoic and thiobenzoic acids. In spite of the apparently smooth course of the reaction, the only use made of this reaction before 1938 was by Busch who, in 1909, made ureide derivatives of the three aminobenzoic acids and of phthalic acid. In 1938, Zetzsche and his co-workers started a study of this reaction in connection with their studies on the fatty acids of cork and the structure of sporollenine. They pointed out the advantages of its use, especially with sensitive acids. Unlike most reagents, a derivative can be formed at fairly low temperatures without the use of acids, alkalis, or any other strong reagents.

$$C(=NR)_2 + R'COOH \rightarrow RNHC(=O)-N(R)-C(=O)-R'$$

The first investigations were made with carbodiphenylimide. It was found that there is only a slight increase in the melting point of the ureide over the acid, and this fact, together with the instability of the carbodiimide, made it unsatisfactory as a reagent. They then studied the use of carbodi-p-tolylimide, carbodi-p-iodophenylimide, carbodi-p-bromophenylimide, and carbodi-β-naphthylimide. Of these, the tolyl compound was found to be the most useful since it is a crystalline solid at room temperature, reacts readily with a very wide variety of acids, and forms derivatives with fairly high melting points. The halogen compounds form derivatives with higher melting points than those of the tolyl derivatives, but the reaction does not occur for all of the acids. Later studies showed that carbobis-(4-dimethylaminophényl)-imide has superior properties to those of the carboditolylimide.

It was found that the reaction proceded readily even in dilute solution with the time of reaction varying from a few seconds to a few hours. The yields were practically quantitative in the absence of side reactions. There were three principle side reactions observed.

(1) Conversion of the ureide into the isocyanate and the acid amide.

$$RNHC(=O)-NR-C(=O)-R' \rightarrow RNCO + R'C(=O)-NHR$$

The isocyanate can react with the unreacted acid to form the anhydride, the diarylurea and carbon dioxide.

(2) Direct formation of the acid anhydride and urea.

$$C(=NR)_2 + 2HOC(=O)-R' \rightarrow O=C(NHR)_2 + O(C(=O)R')_2$$

(3) Polymerization of the carbodiimide.

In reactions with dibasic acids, it was found that, in most cases, the product of dehydration was formed instead of, or in addition to, the ureide. For instance, oxalic acid gave only carbon monoxide and carbon dioxide. Glutaric and adipic acids formed diureides.

It was found that the α,β-unsaturated acids gave colored ureides with carbobis-(4-dimethylaminophenyl)-imide. Zetzsche has proposed that this might be used as a test for α,β-unsaturation in acids. Substituents in the α-position decreased the intensity of the color, the so-called "α-effect".

Aromatic acids were also found to give colored ureides. If benzoic acid is considered as the standard, *o*-substituents decreased the intensity of the color, *p*-substituents increased the intensity of the color and *m*-substituents had no effect on the color of the ureide. If two groups were present, the group in the *o*-position determined the color effect. The directing orientation of the substituents had no influence; the cyano group was as effective as the methoxy group. The only exceptions observed in twelve series of compounds were the nitrobenzoic acids, in which no color difference was noticed between the *o*- and *p*-isomers.

Similarly, the α-haloaliphatic acids formed colored ureides with the basic carbodiimides. The bromo- and iodo- acids gave about the same color which was considerably darker than for the corresponding chloro-compounds. The intensity of the color decreased from α-halopropionic acid through α-halomelissic acid, but the α-haloacetic acids showed very little color. The same was true of the basic ureide of phenylchloroacetic acid, and Zetzsche and Röttger concluded that the presence of a hydrogen atom on the β-position was necessary for color. It was also found that β-haloaliphatic acids, in the absence of a hydrogen atom in the γ-position, formed colored ureides with the basic carbodiimides.

An interesting type of derivative was formed by the reaction of acids with optically active carbodiimides, principally those obtained from menthyl amine and bornyl amine. The carbodimenthyl imide was an unsatisfactory reagent, since, with most acids, the acid anhydride and urea were formed instead of the desired ureide. The carbodibornylimide was more satisfactory, giving the ureides with a larger number of the acids tested. More satisfactory reagents were the unsymmetical diimides, *N*-menthyl-*N*'-p-dimethylaminophenylimide and *N*-bornyl-*N*'-dimethylaminophenylimide. These reacted quite satisfactorily with all of the acids tested. All of the ureides formed with the optically active carbodiimides had characteristic rotations as well as characteristic melting points.

Bibliography

C. Schall, J. prak. chem. (2) 64, 261 (1901).
M. Busch, J. prak. chem. (2) 79, 539 (1909).
Zetzsche, Lüscher and Meyer, Ber. 71B, 1088-93 (1938).
Zetzsche, Meyer, Overbeck and Lindlar, Ber. 71B, 1516-21 (1938).
Zetzsche, and Röttger, Ber. 72B, 1599-612 (1939).
Zetzsche, and Fredrich, Ber. 72B, 1735-40 (1939).
Zetzsche, and Röttger, Ber. 72B, 2095-8 (1939).
Zetzsche, and Röttger, Ber. 73B, 50-6 (1940); C.A. 34, 3250.
Zetzsche, and Röttger, Ber. 73B, 465-7 (1940); C.A. 34, 6593.
Zetzsche, and Neiger, Ber. 73B, 467-77 (1940); C.A. 34, 6594.
Zetzsche, and Fredrich, Ber. 73B, 1114-23 (1940); C.A. 35, 2480.
Zetzsche, and Voigt, Ber. 74B, 183-8 (1941); C.A. 35, 3992 (1941).
Zetzsche, and Lindlar, Ber. 71B, 2095-2109 (1938).

Reported by R. S. Ludington
November 4, 1942

REPLACEMENT REACTIONS IN ALLYLIC SYSTEMS

It has long been known that reactions involving a three carbon system of the structure (A) or (B), where the R group may be alkyl, aryl, or hydrogen, are apt to yield rearranged products on replacement of X.

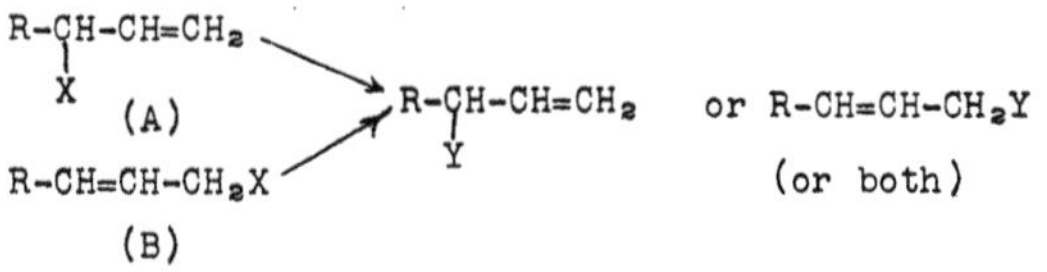

The reactions are complicated in many cases by a thermal rearrangement of the reagent or product under the conditions of the experiment or during purification. Extensive studies of replacement reactions

in such systems have been carried out to determine the nature of the reaction product RC_3H_4Y, i.e., to determine whether it is a pure compound or a mixture, and if a mixture, whether the same mixture is obtained starting with either the primary or the secondary isomer.

According to the views of Ingold, shared by Kenyon and Young, replacement reactions in allylic systems of the type R_1-CH=CH-CH(X)-R_2, may go by a combination of two mechanisms:

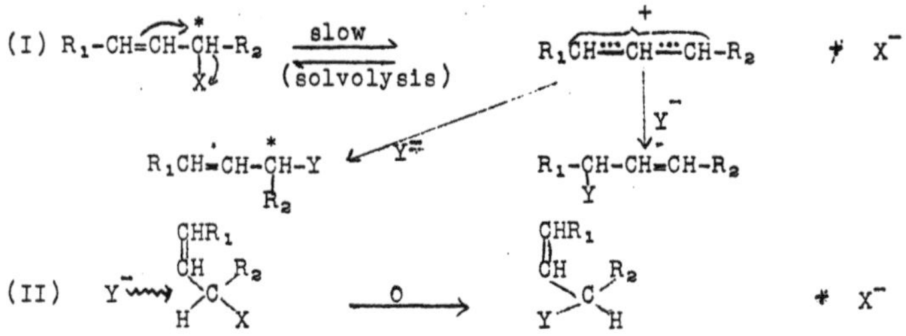

(I) involves the formation of a resonating carbonium ion by the solvolytic removal of X^-, leading to an equilibrium mixture of allylic isomers on reaction with Y^-, and considerable racemization at the asymmetric carbon atom. (Unimolecular S_N1 reaction kinetics)

(II) is a normal, second order reaction (S_N2), leading to one product, and resulting in a high degree of inversion at the asymmetric carbon atom.

The relative importance of these two mechanisms in a given reaction will be determined by the nature of the medium, the reagent, and the nature of R_1 and R_2. If only method (I) operated, one would expect the same mixture of products starting with either isomer; the greater the percentage of reaction by (II), the greater will be the difference in the mixtures obtained as products from the two different isomers. Roberts, Young and Winstein have shown that in the conversion of crotyl chloride and methylvinylcarbinyl chloride to their respective acetates and ethyl ethers, it was possible by appropriate choice of solvent, to nearly completely suppress the S_N1 type of reaction, obtaining practically 100% of the normal product:

$$CH_3\text{-}CH{=}CH\text{-}CH_2Cl \xrightarrow[EtOH]{NaOEt} CH_3\text{-}CH{=}CH\text{-}CH_2OC_2H_5 \quad (100\%)$$

$$CH_3\text{-}CH{=}CH\text{-}CH_2Cl \xrightarrow{KOAc\ in\ AcOAc,\ 100^\circ} CH_3\text{-}CH{=}CH\text{-}CH_2\text{-}OAc \quad (100\%)$$

$$CH_3\text{-}CH(Cl)\text{-}CH{=}CH_2 \xrightarrow[EtOH]{NaOEt} CH_3\text{-}CH(OC_2H_5)\text{-}CH{=}CH_2 \quad (96\%)$$

$$CH_3\text{-}CH(Cl)\text{-}CH{=}CH_2 \xrightarrow{Et_4NOAc\ in\ (CH_3)_2CO,\ 58^\circ} CH_3\text{-}CH(OAc)\text{-}CH{=}CH_2 \quad (100\%)$$

With silver ion, however, which promotes an ionic process similar to the S_N1, practically the same product was formed from either chloride:

$$\left.\begin{array}{l} CH_3\text{-}CH{=}CH\text{-}CH_2Cl \\ CH_3\text{-}CH(Cl)\text{-}CH{=}CH_2 \end{array}\right\} \xrightarrow[HOAc]{Ag^+} AgCl + [CH_3\text{-}CH \overset{\cdot\cdot}{-} CH \overset{\cdot\cdot}{-} CH_2]^+$$

$$\xrightarrow{OAc^-} \underset{(56\text{-}65\%)}{CH_3\text{-}CH{=}CH\text{-}CH_2\text{-}OAc} + \underset{(44\text{-}35\%)}{CH_3\text{-}CH(OAc)\text{-}CH{=}CH_2}$$

Young and Lane suggest that the most likely mechanism for the solvolytic process is an exchange reaction in which the reactant ion (Y^-) coordinates with the molecule before the complete rupture of the C-X linkage:

$$R\text{-}CH{=}CH\text{-}CH_2X + nROH \rightarrow R\text{-}CH{=}CH\text{-}\overset{+}{C}H_2 \rightarrow \overset{-}{X}\cdot(ROH)_n$$

If X^- is far enough removed that resonance may set in, then a mixture of products is obtained, while if Y^- coordinates before resonance can set in, then so-called normal replacement would ensue. Assuming this for the S_N1 reaction, and a normal S_N2 reaction between bromide ion and the oxonium salt of the alcohol, Young and Lane have developed a method of accounting quantitatively for the composition of the mixtures of bromides obtained in several <u>controlled</u> reactions of the butenols and hydrogen bromide under varying conditions:

$$CH_3\text{-}CH{=}CH\text{-}CH_2OH \xrightarrow{HBr} CH_3\text{-}CH{=}CH\text{-}CH_2Br \quad (79\text{-}86\%)$$

$$CH_3\text{-}\underset{OH}{CH}\text{-}CH{=}CH_2 \xrightarrow{HBr} CH_3\text{-}CH{=}CH\text{-}CH_2Br \quad (72\text{-}83\%)$$

With reagents other than hydrogen bromide, different results are obtained. With phosphorous tribromide in pyridine, the mixtures are more nearly half and half (Young and Lane); while with thionyl chloride, Meisenheimer and Link report an 80% yield of ethylvinylcarbinyl chloride ($Et\text{-}\underset{Cl}{CH}\text{-}CH{=}CH_2$) from ethallyl alcohol, and a 75% yield of the primary chloride from the secondary alcohol (54-57% in ϕNEt_2). This latter can be very easily explained by the assumption that the chlorosulfinic ester formed in the first step undergoes an intramolecular, second-order reaction (S_Ni):

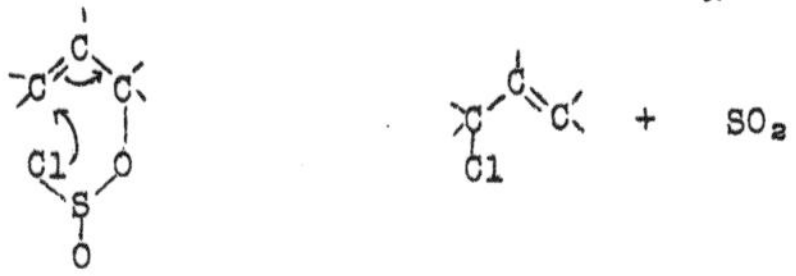

This suggests the possibility of an abnormal S_N2 reaction, in which replacement and inversion of structure occur simultaneously:

$$Y^- \rightsquigarrow C{=}C\text{-}C\text{-}X \rightarrow Y\text{-}C\text{-}C{=}C \quad + X^-$$

Reactions of Allyl Halides with Metals

It has long been known that benzyl magnesium chloride and closely related Grignard reagents react abnormally with formaldehyde and certain similar compounds. Benzylmagnesium chloride possesses a modified allylic system in which the double bond is in a benzene ring. It was not, however, until Gilman devised a method (1930), later modified by Young and his students, for obtaining the Grignard reagents instead of the hydrocarbon coupling product from normal allyl halides, that study was directed to the reactions of allyl magnesium halides.

Gilman and Harris concluded that the Grignard reagent of cinnamyl chloride existed in the form $\phi\text{-}\underset{MgCl}{CH}\text{-}CH{=}CH_2$, since

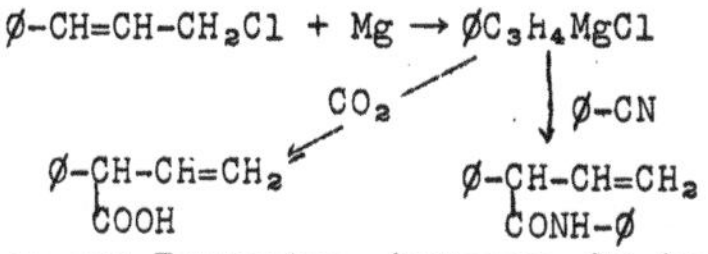
Ø-CH=CH-CH₂Cl + Mg → ØC₃H₄MgCl
CO₂
Ø-CN
Ø-CH-CH=CH₂
COOH
Ø-CH-CH=CH₂
CONH-Ø

Coleman and Forrester, however, by treating $\phi\text{-}C_3H_4MgCl$ with chloramine, a reagent known to react normally with benzylmagnesium bromide, obtained cinnamylamine. In the most recent work, Young, Ballou, and Nozaki have shown the Grignard reagent to be a mixture of 73% $\phi\text{-}\underset{MgCl}{CH}\text{-}CH{=}CH_2$ and 27% $\phi\text{-}CH{=}CH\text{-}CH_2MgCl$. Both forms occur in

sufficient quantities to account for the results of both Gilman and Harris and Coleman and Forrester.

In a recent series of papers Young and his co-workers have reported studies on the reactions of crotyl and methylvinylcarbinyl chlorides and bromides with various metals. The method used in these studies is interesting: The butenyl halides were treated with the metal in a boiling 80% ethanol solution. The mixture of butenes resulting from this treatment was analyzed by a method developed by Dillon, Young and Lucas; it consisted in passing the butenes into bromine, measuring the rate, K_2, of the reaction, and measuring the density of the resulting mixture of dibromides. From this data, the composition of the dibromide mixture could be obtained, and from that, the composition of the butene mixture. The same butene mixture was always obtained from either the primary or the secondary halide. With magnesium or zinc in absolute ethanol or propanol, the mixture of butenes was practically the same as when the reactions were run in 80% ethanol. There is apparently some relationship between the character of the butene mixture obtained and the molal electrode potential of the metal, the amount of 1-butene in the mixtures decreasing in the order chromium (95%), tin (85%), cadmium (79%), zinc (67%), magnesium (55%), and aluminum (40%).

The fact that in the reactions

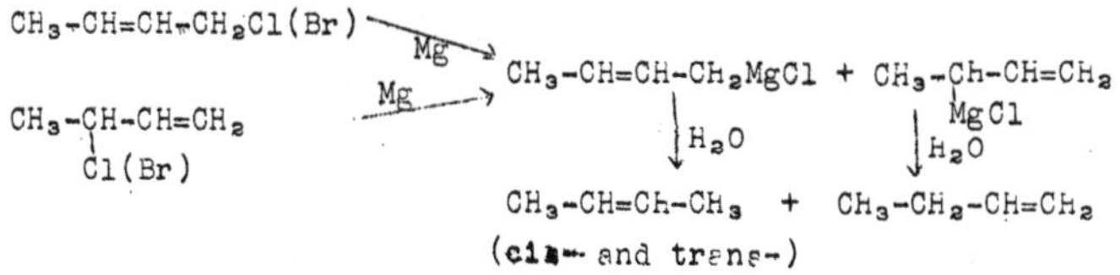
CH_3-CH=CH-CH_2Cl(Br)
Mg
CH_3-CH-CH=CH_2
Cl(Br)
Mg
CH_3-CH=CH-CH_2MgCl + CH_3-CH-CH=CH_2
MgCl
H_2O
H_2O
CH_3-CH=CH-CH_3 + CH_3-CH_2-CH=CH_2
(cis- and trans-)

the same butene mixture is obtained whether one starts with the primary or secondary halide, suggests two, perhaps three possible explanations:

(A) The formation of the same mixture of butenylmagnesium halides at the instant of reaction; the radicals, once united with magnesium, being held in a more or less stable manner,

(B) The establishment of an equilibrium between primary and secondary butenylmagnesium halide, after the Grignard has been formed,

(C) Rearrangement during the reaction with water.

Presumably (B) would result from the equilibrium

$$2RMgX \rightleftharpoons R_2Mg + MgX_2$$

In order for a rearrangement to occur in this step, an ionic intermediate, a resonating butenyl carbanion, must be postulated. Ionically this equilibrium may then be represented:

Cr-MgBr ⇌ Bu⁻ + MgBr ⇌ MVC-MgBr

$$Bu^- + Cr\text{-}MgBr \rightleftharpoons \begin{bmatrix} Cr\text{-}Mg\text{-}Cr \\ or \\ Cr\text{-}Mg\text{-}MVC \end{bmatrix} + Br^-$$

$$Bu^- + MVC\text{-}MgBr \rightleftharpoons \begin{bmatrix} MVC\text{-}Mg\text{-}Cr \\ or \\ MVC\text{-}Mg\text{-}MVC \end{bmatrix} + Br^-$$

where MVC- is methyvinylcarbinyl, Cr- is crotyl, and

Bu^- is 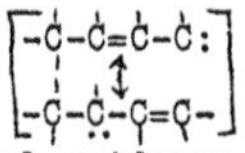

Experimental evidence indicates that (A) is the most likely possibility. Young and Pokras have shown that the operation of mechanism (B), observed in the rearrangement brought about by the forced conversion of RMgX to R_2Mg by dioxane, leads to an allylic rearrangement of the butenyl radicals, and produces a different mixture of butenes upon hydrolysis of the R_2Mg fraction than is obtained from the Grignard reagent itself. Furthermore, the evidence shows that it is not simply a case of redistribution of the radicals between the RMgX and R_2Mg. The data is as follows:

Source of Composition Butene Mixture	% 1-butene	% _cis_-2-butene	% _trans_-2-butene
Butenylmagnesium bromide	56.4	26.5	17.2
Butenylmagnesium bromide dioxane precipitate	54.9	28.0	17.1
Dibutenyl magnesium	44.5	32.2	23.2

<u>Bibliography</u>

Arcus and Kenyon, J. Chem. Soc., 1912 (1938)
Coleman and Forrester, J. Am. Chem. Soc., <u>58</u>, 27 (1936)
Dillon, Young and Lucas, <u>ibid</u>., 52, 1953 (1930)
Gilman and Harris, <u>ibid</u>., <u>53</u>, 3541 (1931)
Meisenheimer and Link, Ann., <u>479</u>, 211 (1930)
Ogg, J. Am. Chem. Soc., <u>61</u>, 1946 (1939)
Roberts, Young and Winstein, <u>ibid</u>., <u>64</u>, 2158 (1942)
Young, Ballou, and Nozaki, <u>ibid</u>., <u>61</u>, 12 (1939)
Young and Eisner, <u>ibid</u>., <u>63</u>, 2113 (1941)
Young, Kaufman, Loshokoff, and Pressman, <u>ibid</u>., <u>60</u>, 900 (1938)
Young and Lane, <u>ibid</u>., <u>60</u>, 847 (1938)
Young and Pokras, J. Org. Chem., <u>7</u>, 233 (1942)
Young and Winstein, J. Am. Chem. Soc., <u>58</u>, 441 (1936)
Young, Winstein, and Prater, <u>ibid</u>., <u>58</u>, 289 (1936)

Reported by John A. McBride
November 11, 1942

BORON TRIFLUORIDE IN ORGANIC CHEMISTRY

Boron trifluoride has recently found extensive use as a catalyst in organic reactions. It has shown properties similar to $AlCl_3$, $ZnCl_2$, $FeCl_3$, $SnCl_4$, etc. According to Kastner (1) its activity is decidedly inferior to that of $AlCl_3$, but BF_3 is often superior to the more commonly used catalysts in that its use does not result in the formation of undesirable by-products. Because of its extremely great tendency to form coordination-compounds, BF_3 can be used with considerable success in condensation reactions in which water, alcohols, or organic acids are formed as by-products.

Several coordination compounds of BF_3 due to their great stability have also found use as catalysts in organic reactions. The most suitable of these are $H_2O \cdot BF_3$, $CH_3OH \cdot BF_3$, $(CH_3COOH)_2 \cdot BF_3$ and $(C_2H_5)_2O \cdot BF_3$. As Meerwein (2) has pointed out, the acidity of weak electrolytes is raised so much through complex-formation that the coordination compounds are similar in their behavior to H_2SO_4 or perhaps still better, $HClO_4$. The BF_3 complexes with H_2O, CH_3OH and CH_3COOH can therefore, in many cases, be used with greater success than H_2SO_4 or $HClO_4$, which often bring about side reactions.

Preparation of Esters.

Aliphatic and aromatic acids may be esterified with alcohols (3) or olefins (e.g. propylene, butylene) (4) in the presence of BF_3. With aliphatic acids the yields vary greatly and increase with the acidity of the acid. Only a small amount of BF_3 is needed; a larger amount does not increase the yields of esters. With the substituted aromatic acids a larger amount of BF_3 is necessary, but the yields are in general higher than in the aliphatic series. Since a carboxylic acid adds to an olefin according to Markownikow's rule, secondary and tertiary alkyl esters are always formed. If cyclopropane is used, the n-propyl ester is formed.

Esters are also obtained by the reaction of $CH_3CONH_2 \cdot BF_3$ with alcohols or phenols (5).

$$CH_3CONH_2 \cdot BF_3 + ROH \longrightarrow CH_3COOR + NH_3 \cdot BF_3$$

The nitriles with equimolecular amounts of BF_3 react with alcohols to form esters (6). The yields are usually about 35%.

$$R'CN + \begin{matrix} HOR \\ HOR \\ HOR \end{matrix} + BF_3 \longrightarrow \left[R'C\begin{matrix} OR \\ OR \\ OR \end{matrix} \right] + NH_3 \cdot BF_3$$

$$\downarrow$$

$$R'C \overset{O}{<}OR \quad + \quad ROR$$

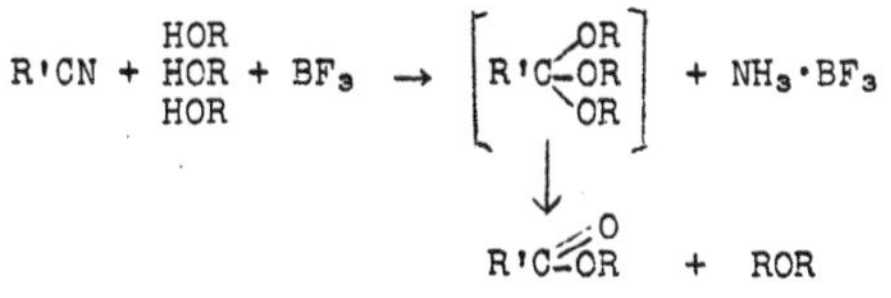

The reaction of BF_3 with ethyl orthoformic ester to form ethyl ether and ethyl formate indicates that the above reaction involves the formation of the orthoformic ester as an intermediate.

Esters of formic acid, propionic acid, benzoic acid, and salicylic acid are converted to the corresponding acetates by heating for one hour at 100° with $CH_3COOH \cdot BF_3$ (7). In acetolysis reactions, BF_3 has proven much more effective than $ZnCl_2$ and H_2SO_4.

Preparation of Alcohols and Ethers

Water may be added to olefins in the same manner as the carboxylic acids. Thus isopropyl alcohol and diisopropyl ether are produced from propylene and $BF_3 \cdot 2H_2O$ (8). The reaction is limited only to olefins of the types $RCH{=}CH_2$ and $RCH{=}CHR$ and usually does not proceed too smoothly. With olefins of the types $R_2C{=}CH_2$, $R_2C{=}CHR$ and $R_2C{=}CR_2$, only polymerization products are formed with $BF_3 \cdot 2H_2O$. Alcohols may be added to olefins at 100° in the presence of BF_3. In this manner methyl tert-amyl ether is obtained from methyl alcohol and trimethylethylene.

Preparation of Aliphatic Acids

The patent literature (9,10) describes an interesting synthesis of aliphatic acids from alcohols and carbon monoxide which may be catalysed by various complexes of boron trifluoride with water, phosphoric acid, sulfuric acid and hydrochloric acid.

$$ROH + CO \rightarrow RCOOH$$

The alcohol may be replaced by an olefin, ether, or ester. The reaction is preferably carried out at 150-400° C, under 350-700 atm.

Preparation of Nitriles and N-Substituted Amides

Since esters are formed by the reaction of $RCONH_2 \cdot BF_3$ and alcohols, it might be expected that acid anhydrides could be synthesized by an analogous reaction using carboxylic acids in place of alcohols. However, such is not the case; nitriles are obtained in excellent yields. While the yield of acetoni-

$$CH_3CONH_2 \cdot BF_3 + CH_3CONH_2 \xrightarrow{CH_3COOH} CH_3CN + CH_3COOH + NH_3 \cdot BF_3$$

trile in the above reaction is 98%, only a 15% yield is obtained in the absence of acids (11).

The boron trifluoride compounds of amides also react when boiled for 30 minutes with primary and secondary aliphatic and aromatic amines to form N-substituted amides (11).

$$R'CONH_2 \cdot BF_3 + R_2NH \rightarrow R'CONR_2 + NH_3 \cdot BF_3$$

Preparation of Acetals

A mixture of BF_3 and HgO is an excellent catalyst for making ethylidene compounds from acetylene and substituted acetylenes. It does not remove halogens or destroy ether or ester

groups. It has enabled the preparation with good yields of a great many acetals and ketals which hitherto could not be synthesized.

$$2ROH + CH{\equiv}CH \rightarrow CH_3CH\begin{matrix} OR \\ OR \end{matrix}$$

Acetylene reacts according to the above reaction with aliphatic and certain aromatic and cyclic alcohols to form acetals (12). Mono- and dialkyl acetylenes (13) react like acetylene with methyl alcohol, but with other monohydric alcohols will not give ketals. Ethyl alcohol reacts quite violently with amyl acetylene, but only polymerization products are obtained.

Acetylene and alkyl acetylenes react with polyhydric alcohols and alpha-hydroxy acids to form cyclic acetals and ketals (12).

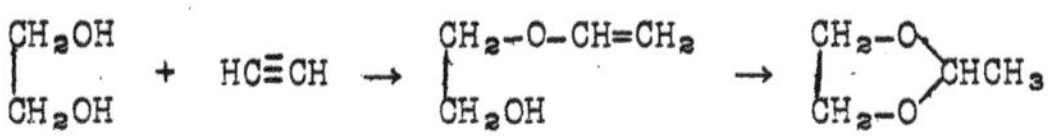

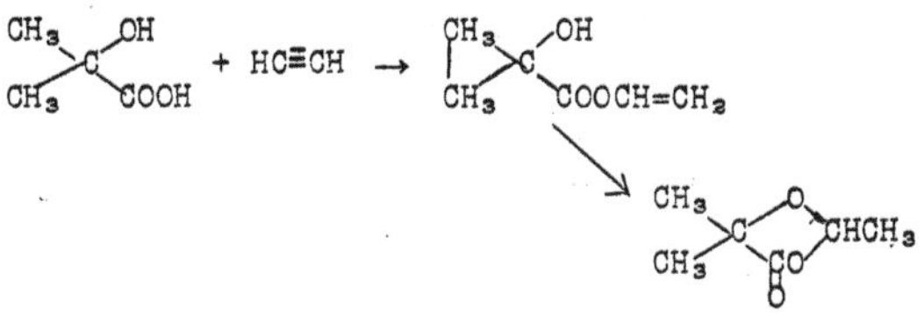

Preparation of Vinyl Esters

In the presence of a BF_3 and HgO catalyst, carboxylic acids add to acetylene and alkyl acetylenes to form vinyl esters. Vinyl acetate is prepared with an 80% yield by adding acetylene to acetic acid containing about 1% HgO and 1% $(CH_3COOH)_2{\cdot}BF_3$ (14). This reaction is preferable to other methods of preparation of vinyl esters in that there are no side products.

Preparation of Ketones and β-Diketones (15)

BF_3 may be substituted for $AlCl_3$ in the synthesis of ketones from acid anhydrides and aromatic hydrocarbons. In general, however, the yields are smaller than those obtained with $AlCl_3$. For the synthesis of β-diketones, BF_3 is unsurpassed. Acetophenone reacts with acetic anhydride to form benzoyl acetone in an 85% yield. Acetone, diethylketone, cyclohexanone and tetralone all react readily to form β-diketones. In every reaction involving acetic anhydride and BF_3, some acetyl acetone is formed through a self-condensation of the anhydride.

[illegible] (TP)

[illegible]

Condensations with Active Hydrogen Compounds

The preparation of benzoylacetone from acetophenone in the presence of BF_3 involves a reaction which is analogous to the Claisen condensation. Hauser (16) has demonstrated the possibility of effecting analogs of the Perkin, Knoevelagel and Michael condensations in the presence of BF_3. The process has also been used successfully in alkylating acetoacetic ester with benzyl chloride, isopropyl alcohol, diisopropyl ether, isopropylacetate, tert-butyl alcohol, and tert-butyl ethyl ether, (16, 17). The use of BF_3 for alkylations of this kind is of special value, because of the well-known difficulties in the introduction of benzyl and tert-alkyl groups into active methylene compounds by the common methods in which a base is employed.

Alkylation of Aromatic Compounds.

It is possible to use BF_3 and some of its coordination compounds ($BF_3 \cdot 2H_2O$, $BF_3 \cdot (CH_3COOH)_2$) in the alkylation of aromatic hydrocarbons with olefins, alcohols, ethers, and esters. In the presence of BF_3, benzene and propylene react to give a 98% yield of p-diisopropylbenzene and a 2% yield of o-diisopropylbenzene (18); with $AlCl_3$ the meta derivative is preponderant. With normal alcohols and ethers, only secondary alkyl groups are introduced. For this reason it has been suggested that in many cases an olefin is formed as an intermediate. Phenols may also be alkylated with olefins and alcohols (19). A mixture of mono- and poly-alkyl phenols and their alkyl ethers is usually formed; the relative proportions of the products depend to a great extent on the reaction conditions.

Rearrangement of Aryl-Alkyl Ethers.

The fact is well-known that if certain allyl aryl ethers are heated to a high temperature, they are transformed into isomeric substituted phenols (Claisen Rearrangement). If an alkyl group is substituted for the allyl group, the presence of inorganic substances is necessary for the rearrangement to take place. By the use of BF_3 the time necessary for this process has been materially shortened; the products are the same as those obtained when $ZnCl_2$ and HCl or H_2SO_4 and HOAc are used (20). If there are no substituents in the positions ortho to the ether group, the alkyl group will enter an ortho position; if an ortho position is filled, the alkyl group will enter the para position.

By allowing the temperature to rise during the addition of BF_3, a number of products are formed (21); isopropylphenyl ether gives a mixture of phenol, 2-isopropylphenol, 2,4-diisopropylphenol, 2,4,6-triisopropylphenol, 2-isopropylphenyl isopropyl ether, 2,4-diisopropylphenyl isopropyl ether, and 2,4,6-triisopropylphenyl isopropyl ether.

Cleavage of Ethers

In the presence of BF_3 and organic acids, acid chlorides or

acid anhydrides aliphatic ethers react to form esters (22). High temperatures are necessary for this ether cleavage. The yields are seldom above 50%.

Fries Rearrangement

BF_3 has been used with great success in effecting the Fries rearrangement. It allows the reactions to be carried out at a lower temperature than $AlCl_3$, and also differs from it in that para hydroxy ketones are formed preferentially. Auwers (23) has prepared p-hydroxypropiophenone in an 88% yield from phenylpropionate.

Polymerization and Isomerization

There are numerous references in the patent literature to the use of BF_3 for the polymerization of all kinds of olefins, diolefins, and substituted olefines to form synthetic rubbers, plastics, fuels and lubricants. Butyl rubber, for example, is formed by the polymerization below 32° F of a mixture of 95% isobutylene and 5% butadiene. BF_3 is also used as a catalyst for isomerization and cyclization. Price (24) was able to convert cis-stilbene to trans-stilbene in a 92.3% yield. Low molecular weight paraffins may be isomerized with BF_3. Thus isobutane may be prepared from n-butane and isopentane from n-pentane.

Sulfonation and Nitration

BF_3 is an effective promoter and dehydrating agent for sulfonation and nitration reactions (25). When water is a product of a reaction, BF_3 may drive the reaction to completion by combining with the water produced to form $BF_3 \cdot H_2O$. At the completion of the reaction enough water may be added to form $BF_3 \cdot 2H_2O$, which can then be separated from the reaction products by distillation.

Bibliography

1. Kästner, Angew. Chem., 54, 273 (1941)
2. Meerwein, Ann., 455, 227 (1927)
3. Hinton and Nieuwland, J. Am. Chem. Soc., 54, 2017 (1932)
4. Dorris, Sowa and Nieuwland, ibid., 56, 2689 (1934)
5. Sowa and Nieuwland, ibid., 55, 5052 (1933)
6. McKenna and Sowa, ibid., 60, 124 (1938)
7. Sowa, ibid., 60, 654 (1938)
8. Meerwein and Burneleit, J. prakt. chem., 141, 139 (1934)
9. Loder (to E. I. duPont deNemours and Co.) U. S. 2,158,031 (May 9, 1938)
10. Loder (to E. I. duPont deNemours and Co.) U. S. 2,217,650 (October 8, 1940)
11. Sowa and Nieuwland, J. Am. Chem. Soc., 59, 1202 (1937)
12. Nieuwland and Coworkers, ibid., 52, 2892 (1930), 52, 1015 (1930)

13. Nieuwland and Coworkers, ibid., 56, 1130 (1934), 56, 1384 (1934)
14. I. G. Farbenindustrie, D.R.P., 582,544
15. Meerwein and Vossen, J. prakt. Chem., 141, 149 (1934)
16. Breslow and Hauser, J. Am. Chem. Soc., 62, 2385 (1940); 62, 2389 (1940); 62, 2611 (1940).
17. Hauser and Adams, ibid., 64, 728 (1942)
18. Slanina, Sowa and Nieuwland, ibid., 57, 1547 (1935)
19. Sowa, Hinton and Nieuwland, ibid., 54, 3694 (1932)
20. Sowa, Hinton and Nieuwland, ibid., 54, 2019 (1932)
21. Sowa, Hinton and Nieuwland, ibid., 55, 3402 (1933)
22. Hennion, Hinton and Nieuwland, ibid., 55, 2859 (1933)
23. Auwers, Pötz and Noll, Ann., 535, 228 (1938)
24. Price and Meister, J. Am. Chem. Soc., 61, 1595 (1939)
25. Thomas, Anzilotti and Hennion, Ind. Eng. Chem., 32, 408 (1940)

Reported by E. W. Maynert
November 11, 1942

REDUCTION OF MULTIPLE CARBON-CARBON BONDS

The methods used for the reduction of multiple carbon-carbon bonds may be divided into three large groups: chemical, electrolytic and catalytic. The object of this seminar is to give certain generalizations governing each type of reduction together with specific illustrations and mechanisms of reactions. The term multiple as it appears in the title is taken to mean double and triple bonds, although the subject of dienes, polyenes, enynes and diynes is also covered.

I. Chemical Reduction

The reducing agents in the chemical group include all the dissolving metal combinations, however most of the work has been done with the alkali-metals and aluminum amalgam.

A. Theories of Chemical Reduction.

Any mechanism proposed must account for the bimolecular products, for the mixture of 1,4- and 1,2-dihydro products from conjugated dienes, for the trans reduction of acetylenes, and for the influence of activating groups.

1. Baeyer's theory (1892) attributed reduction to the action of 'nascent' hydrogen. While this theory held sway for many years, it is no longer considered the true mechanism, since it has been shown that in many cases there is no correlation between hydrogen production and reducing activity.

2. Willstatter's theory (1928) proposed that reduction occurs by the addition of metallic sodium to the double bond followed by solvent hydrolysis. This theory was limited, as Huckel pointed out, since it is reasonable to assume the same mechanism of reduction for all dissolving metals, and on such a basis, it would be difficult to explain the action of a divalent metal like calcium.

3. Burton and Ingold (1929) set forth the ionic theory postulating that the double bond polarizes in the reaction medium. This explains the activating action of certain groups - like aryl or carbonyl - which are strongly electron attracting. The polarized molecule adds a proton from solution forming a positive fragment which can stabilize itself by acquiring two electrons from the metal surface (or reducing salt), followed by another proton. If the fragment acquires one electron instead of two, a free radical will be formed which can stabilize itself by dimerization. This also explains 1,4- and 1,2-dihydro products, since the intermediate positive fragment can undergo allylic rearrangement to a tautomeric form. If it is assumed that a sodium ion can be added in place of a proton when the reduction is carried out in liquid ammonia, then this mechanism will also explain the di-sodium naphthalene formed as an intermediate in the reduction of naphthalene.

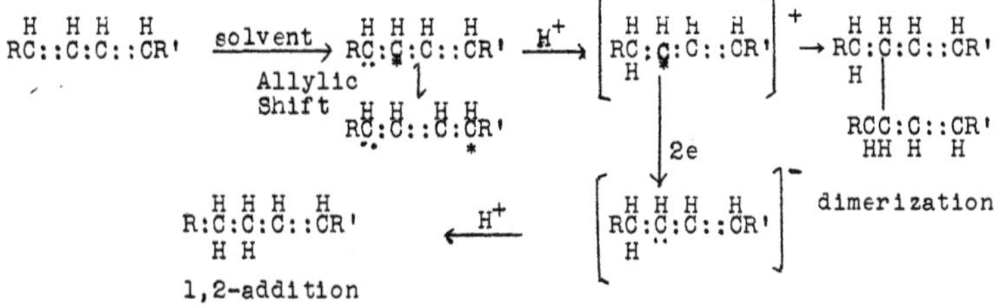

B. Reduction of Olefins.

Simple aliphatic olefins are not reduced by dissolving metal combinations, but the presence of one aryl group adjacent to an ethylenic bond activates it sufficiently to permit reduction by alkali metals. The reductions may be carried out by three general methods:

(1) Treatment of an olefin with an alkali metal in an inert medium, such as ether, followed by hydrolysis of the organo-alkali compound. In general, aryl olefins with two alkyl groups, or an alkyl group and hydrogen on the same carbon will not be reduced.

From a practical point of view, it has been found that lithium frequently adds more rapidly than sodium. Reaction may also be facilitated by the use of dioxane, or ether previously treated with P_2O_5, or the use of ethylene glycol di-methyl ether. Dimerization can often be avoided by the use of indene or ethylaniline which replace the sodium by hydrogen as quickly as the metal-adduct forms. Internal dimerization may occur to give a cyclic structure.

(2) Treatment of the olefin with a solution of sodium in liquid ammonia. This method is the strongest of the three and will reduce $(C_6H_5)_2C{=}CHCH_3$ which is not attacked by sodium in ether.

(3) Sodium or sodium-amalgam may be used in alcoholic or aqueous media.

Fuson and his co-workers found that 1,2-diarylacetylenes and olefins could be reduced at room temperature by benzene and aluminum chloride forming diphenylethanes. The mechanism is not merely the addition of hydrogen, for substituted aryl groups are replaced by phenyl-, and tri-phenyl ethylene is converted to dibenzyl.

$$p\text{-}Cl\text{-}C_6H_5CH{=}CHC_6H_5pCl \xrightarrow[\text{benzene}]{AlCl_3} C_6H_5CH_2CH_2C_6H_5$$

$$(C_6H_5)_2C{=}CHC_6H_5 \xrightarrow[\text{benzene}]{AlCl_3} C_6H_5CH_2CH_2C_6H_5$$

C. Reduction of Acetylenes.

In contrast to simple olefins, acetylenes without adjacent aryl or carbonyl groups can be reduced by chemical agents.

Campbell and Eby have shown that mono and di-alkylacetylenes can be reduced by sodium in liquid ammonia to the corresponding olefin - and here the reaction would stop - in good yield, and that the olefins are of known structure and in a high state of purity--uncontaminated by any saturated hydrocarbons.

D. Reduction of Polyunsaturated Hydrocarbons.

A conjugated system of double bonds is reducible by dissolving metal conbinations and in this respect resembles the acetylenes and aryl olefins. In many cases such a system is even more reactive than the last two classes.

The course of the reduction of conjugated hydrocarbons, according to Thiele's theory, should always originate at the ends of the conjugated chain, but recent work indicates that this is not always so. In unsymmetrical molecules, according to the theory of Burton and Ingold, varying amounts of terminal and non-terminal addition will take place. A seemingly similar example would be the reduction

$$RCH{=}CH{-}CH{=}CH{-}Y + 2e + H^{+} \rightarrow$$

$$R\dot{C}H{=}CH{-}\bar{C}H{-}CH_2Y \xrightarrow{H^{+}} RCH{=}CH{-}CH_2{-}CH_2Y \quad \text{1,2-addition}$$

$$\updownarrow$$

$$R\bar{C}H{-}CH{=}CH{-}CH_2Y \xrightarrow{H^{+}} RCH_2{-}CH{=}CH{-}CH_2Y \quad \text{1,4-addition}$$

of cinnamalfluorene with sodium amalgam and then again with aluminum amalgam.

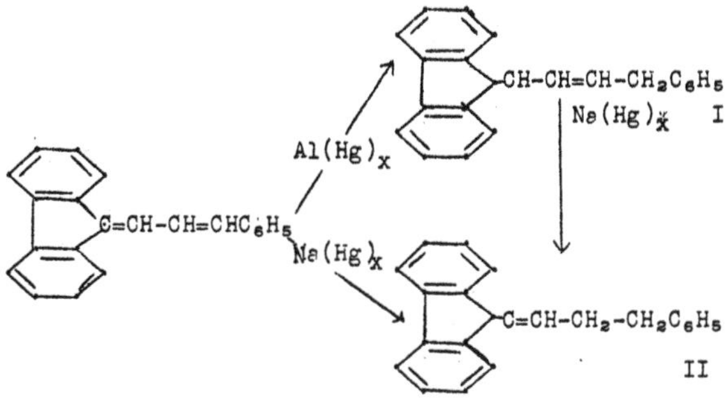
C=CH-CH=CHC6H5
Al(Hg)x
Na(Hg)x
CH-CH=CH-CH2C6H5
Na(Hg)x
I
C=CH-CH2-CH2C6H5
II

The mechanism in both cases is probably the same-normal 1,4-addition, but the less stable form I is rearranged by the sodium amalgam to the more stable conjugated form.

Dupont and his coworkers, using Raman spectra data, have shown two cases in which terminal addition does not occur exclusively.

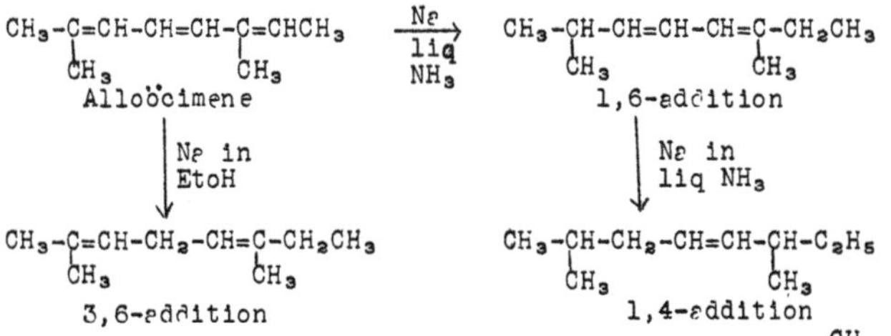

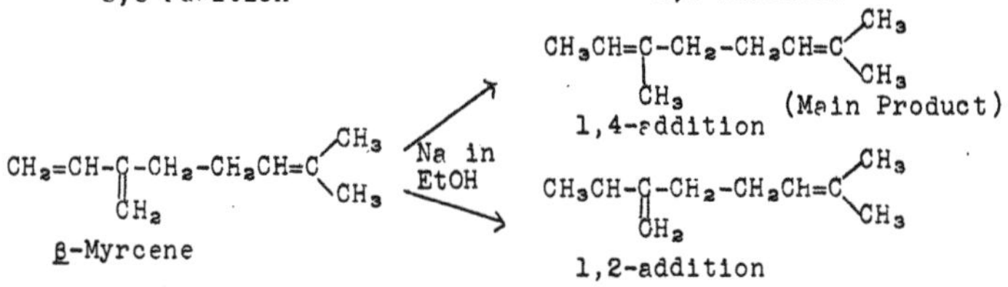

An interesting reaction is one observed by Fischer and Wiedmann in which polyene alcohols, acids, or aldehydes are reduced by certain yeasts, the double bond adjacent to the alcohol group being the only one attacked. The reduction of enynes, diynes, allenes and

$$CH_3CH{=}CHCH{=}CHCH{=}CHCH_2OH \rightarrow CH_3CH{=}CHCH{=}CHCH_2CH_2CH_2OH$$

crossconjugated systems has also been accomplished chemically.

E. Reduction of Aromatic Hydrocarbons.

Benzene and its simple homologs are not, in general, reduced by alkali metals. Lithium adds to polynuclear compounds with increasing ease, in the order-naphthalene, phenanthrene, anthracene. The addition of hydrogen or metals tends to occur in the 1,4-positions.

F. Reduction of α,β-Unsaturated Carbonyl Compounds.

These reactions probably proceed by 1,4-addition to the C=C-C=O system. This is substantiated by the stable enol formed by the zinc and acetic acid reduction of β-phenylbenzalacetomesitylene which was accomplished by Kohler and Thompson.

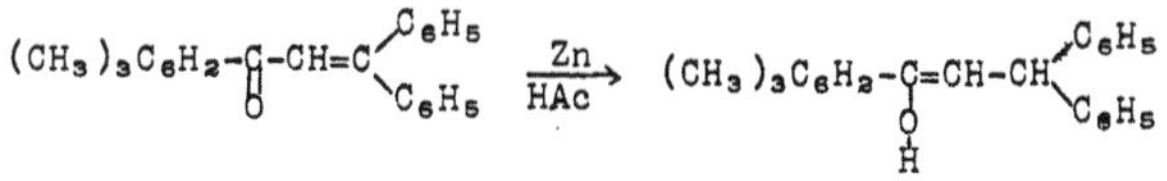

G. Reduction of Polyene Acids, Aldehydes and Ketones.

These reactions proceed by the following mechanism.

$$RC=C-C=C-C=O \xrightarrow[2e]{H^+} R\overset{..}{C}-C-C=C-OH \xrightarrow{H^+} RCH \cdot C=C-CH-C=O$$

I   III

$$RC=C-\overset{..}{C}-C=C-OH \xrightarrow{H^+} RC=C-CH-CH-C=O$$

II   IV

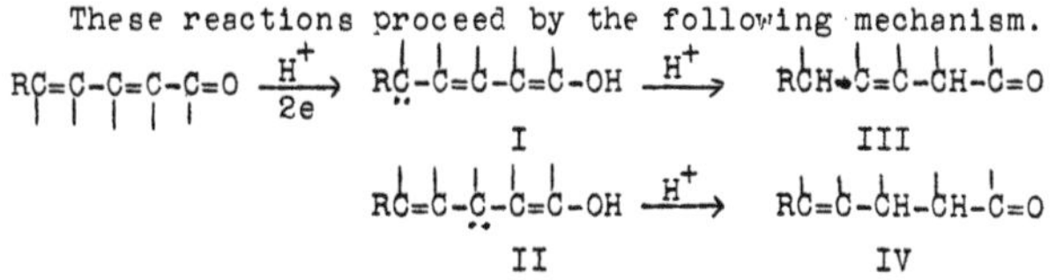

If R is electron attracting I should be the more stable of the two tautomeric forms and III would be the dihydro product. If R is alkyl or hydrogen both III and IV would result in varying amounts. In a crossed conjugated system the doubly conjugated group is reduced.

$$C_6H_5CH=\overset{\overset{O}{\|}}{C}-OH \xrightarrow{Na(Hg)_x} C_6H_5CH_2-CHCOOH$$
$$\overset{|}{CH}=CHC_6H_5 \qquad\qquad \overset{|}{CH}=CHC_6H_5$$

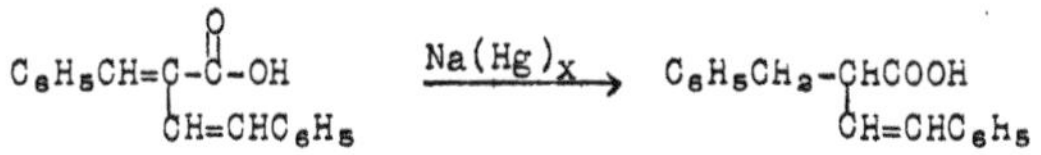

a-Styrylcinnamic Acid

H. The Stereochemical Course of Reduction.

Since chemical reduction of acetylenes and olefins occurs by a stepwise mechanism, the intermediate fragment can adjust to a position of minimum potential energy before the second stage of addition occurs. On this basis, the product of the reduction of acetylenes should be trans, and similarly, as we shall see later, catalytic hydrogenation, where the hydrogen adds simultaneously, should give cis addition products. Campbell and Eby showed that

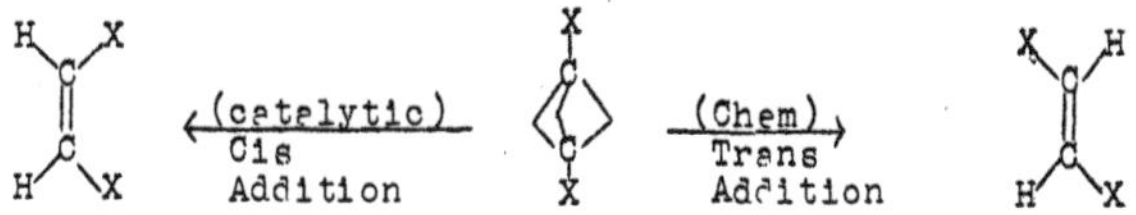

trans reduction of di-alkylacetylenes always occurred, and that this was not a rearrangement of the cis form since the cis-dialkylole-fins were stable.

I. Reactions of Calcium Ammonia (Ca(NH₃)₄).

This reagent is made by passing dry ammonia over calcium at 0°, 15° and 30° and has been studied by Kazanskii and his co-workers.

Ca + 4NH₃ → Ca(NH₃)₄ + Benzene → Ca(NH₂)₂ + 2NH₃ + cyclohexene

+ 1,3-cyclohexadiene

Toluene yielded tetrahydrotoluene and naphthalene yielded tetralin. At higher temperatures calcium ammonia becomes unstable losing

its efficiency.

The hydrocarbons are allowed to react at room temperature for twenty-four hours. Calcium ammonia isomerizes various diolefins and aromatic hydrocarbons. These reactions are all carried out at 0°. Ammonia complexes of lithium, strontium and barium were also prepared. These showed a similar action but lowered activity.

II. Electrolytic Reduction

In electrolytic reduction of organic compounds, there are many variables to be considered; therefore, the results are not as susceptible to generalization as are those of chemical reduction and catalytic hydrogenation.

Cathodes used may be divided into two large classes (1) Those of low overvoltage: In this class belong cathodes of iron, nickel, platinum, and palladium. Electrolytic reduction at these cathodes is probably a catalytic reduction, (2) Those of high overvoltage: In this class belong the cathodes of copper, lead, mercury, zinc, tin and cadmium. Reductions at these cathodes resemble reductions by chemical agents. Wilson has postulated that the Burton-Ingold theory of reduction by dissolving metals can be applied to electrolytic reduction at high overvoltage cathodes.

A. Reduction of Unsaturated Hydrocarbons.

Little work has been reported on electrolytic reduction of ethylenic hydrocarbons, but unsaturated acids have been studied in which the double bond is sufficiently far removed from the carboxyl group so that there is little influence of such a group on the bond. In view of the analogies presented above, such unsaturated acids should be reduced at cathodes of spongy nickel or platinum and not at high-overvoltage cathodes.

Acetylenes are reduced at cathodes of class 1, and the course of the reaction is very similar to catalytic hydrogenation. When acetylene is reduced at a platinized platinum cathode in acid or basic medium, by adjusting the current density, the reaction can be controlled to give principally ethane or principally ethylene. At spongy nickel cathodes, dialkyl acetylenes are reduced to the cis olefins.

Polynuclear hydrocarbons, which can be reduced both catalytically and by dissolving metals, are capable of reduction at cathodes of either class.

B. Reduction of Unsaturated Acids, Aldehydes and Ketones.

α,β-Unsaturated acids are comparatively easily reduced to saturated acids by electrodes of either class; occasionally the acid group is attacked.

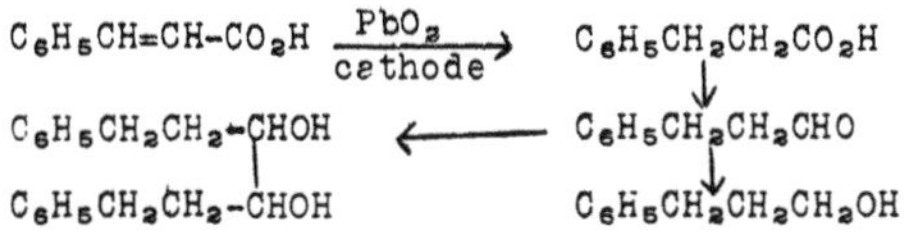

$$C_6H_5CH=CH-CO_2H \xrightarrow[\text{cathode}]{PbO_2} C_6H_5CH_2CH_2CO_2H$$

$$C_6H_5CH_2CH_2-\underset{|}{C}HOH \longleftarrow C_6H_5CH_2CH_2CHO$$

$$C_6H_5CH_2CH_2-\underset{|}{C}HOH \qquad C_6H_5CH_2CH_2CH_2OH$$

Reduction of benzoic acid at a mercury or lead cathode shows close resemblance to reduction of this acid by sodium amalgam.

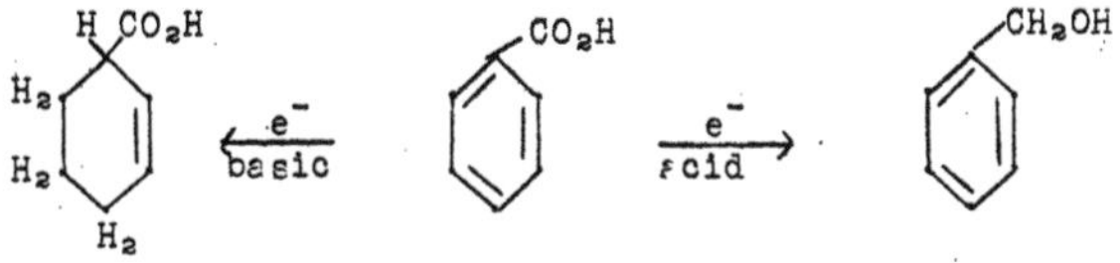

With unsaturated aldehydes or ketones, electrolytic reduction is usually accomplished at a lead or copper cathode. The reaction does not always stop at the saturated carbonyl compound.

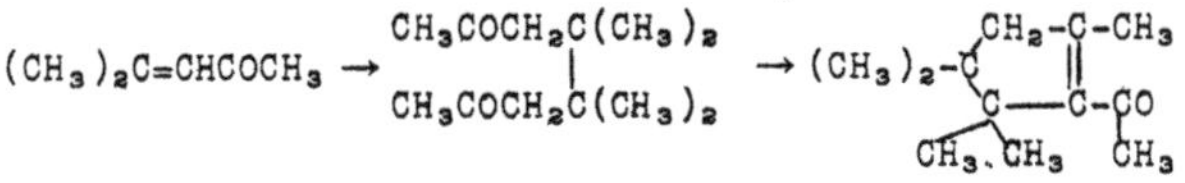

III. Catalytic Hydrogenation

Hydrogen can be added in the presence of a catalyst to practically any carbon to carbon double bond, the ease of addition varying widely with the position of unsaturation and nature of the adjacent groups. The choice of catalyst and its activity, the solvent used, as well as the temperature and pressure are all important variables in determining the speed of reaction and the products to be formed.

A. Hydrogenation of Olefins.

A wide variety of catalysts and experimental conditions can be used for hydrogenating olefins. Palladium, platinum and Raney nickel catalysts will saturate most olefins at room temperature and 2 to 3 atmospheres pressure. At higher temperatures, other catalysts may be effective, but since polymerization and cracking may occur, reductions under mild conditions are favored. In contrast to chemical methods, simple olefins are usually hydrogenated catalytically more easily than conjugated olefins or α-β-unsaturated compounds.

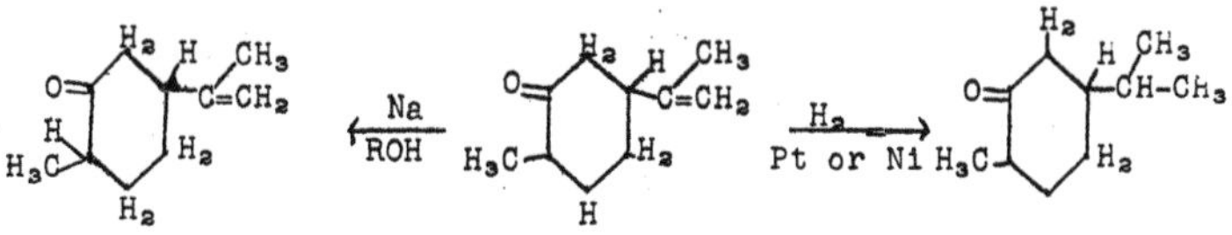

Due to the many variables, study of the effect of structure and substitution on the rate of the hydrogenation is difficult. However, certain generalities may be made. With platinum black and Raney nickel, the order of decreasing ease of hydrogenation is:

$$RCH{=}CH_2 > RCH{=}CHR' > R_2C{=}CH_2 > R_2C{=}CHR' > R_2C{=}CR'_2$$

In general, the substitution of aryl groups retards the reaction, but with palladium as catalyst the reverse is true.

The effect of the geometrical configuration of the double bond has received some attention, but the results obtained are not general and in many cases change with the conditions used.

B. Selective Hydrogenation.

The answer to the question of which unsaturated group will be saturated first when there are two or more in the molecule depends on the nature of the groups, catalyst used, and experimental conditions. In general, the aliphatic ethylene group is so easily hydrogenated that when it is present in a molecule together with an aryl nucleus or a carboxyl group, it is almost always reduced first. However, if the aryl nucleus is activated or the conditions are strenuous, they may be saturated simultaneously. Other exceptions may occur due to a difference in catalyst action. Sauer and Adkins have shown that butyl oleate in the presence of zinc chromite at 300°C undergoes selective hydrogenation of the carboxyl group, yielding octadecenol.

With unsaturated aldehydes or ketones, selective hydrogenation of the ethylenic group is more difficult and in many cases, the product obtained is the saturated alcohol or even the hydrocarbon. Adams and co-workers found that the platinum oxide-platinum black catalyst, in the presence of small amounts of ferrous ion and zinc acetate, would give preferential hydrogenation of the carbonyl group in many α,β-unsaturated aldehydes.

$$(CH_3)_2C{=}CHCH_2CH_2\underset{\displaystyle CH_3}{\underset{|}{C}}{=}CH{-}CHO \xrightarrow{H_2} (CH_3)_2C{=}CHCH_2CH_2\underset{\displaystyle CH_3}{\underset{|}{C}}{=}CHCH_2OH$$

C. Hydrogenation of Acetylenes.

The acetylenes are very readily hydrogenated under mild conditions and a wide variety of catalysts. Ordinarily there are few side reactions. With tertiary acetylenic carbinols some cleavage to a ketone may take place.

$$R_1\overset{\displaystyle R_2}{\overset{|}{\underset{\displaystyle OH}{\underset{|}{C}}}}{-}C{\equiv}CH \rightarrow R_1{-}\overset{\displaystyle R_2}{\overset{|}{C}}{=}O + HC{\equiv}CH$$

With substituted acetylenes, the course of hydrogenation depends mainly on the catalyst used. With platinum black and

palladium black, the reaction is non-selective and can not be
stopped at the olefin stage. Colloidal palladium, however, gives
selective reduction, and if the reaction is stopped when one molar
equivalent of hydrogen has been added, the product is largely the
olefin. This selectivity is not due to the fact that acetylenes
are reduced more rapidly than olefins for this is not always true.
The selective action of ~~palladium~~ colloidal palladium is more
probably due to selective adsorption of acetylene on the catalyst.
Raney nickel shows a similar selectivity and Raney iron shows a
more marked selectivity than both.

Recently much work has been done to clear up the stereochemical course of the catalytic hydrogenation of acetylenes.
Bourgeul and his students have shown that the olefin obtained by
hydrogenating disubstituted acetylenes over colloidal palladium
is almost always the <u>cis</u> form. This is also true for Raney nickel.
Bourgeul suggested that in any such hydrogenation the primary
product is the <u>cis</u> form and that this may then undergo stereochemical conversion to the <u>trans</u> form in certain cases. From
studies of ethylene hydrogenation and the catalytic exchange
reaction between deuterium and ethylene, Farkas and Farkas have
proposed that low temperature catalytic hydrogenation occurs by
simultaneous addition of the two hydrogen atoms from one molecule to
the unsaturated bond. Catalytic hydrogenation of disubstituted
acetylenes should thus always yield <u>cis</u>-olefins; these should form
<u>meso</u> saturated derivatives. If the <u>trans</u> ethylenic compounds are
hydrogenated, the product will be the racemic saturated compounds.

$$XC{\equiv}CX \xrightarrow{\text{H}_2} \begin{matrix} X\ X \\ C=C \\ H\ H \end{matrix} \underline{cis}$$

$$\begin{matrix} X\ X \\ HC-CH \\ Y\ Y \\ \underline{meso} \end{matrix} \xleftarrow{\text{H}_2} \begin{matrix} X\ X \\ C=C \\ Y\ Y \\ \underline{cis} \end{matrix} \quad and \quad \begin{matrix} X\ Y \\ C=C \\ Y\ X \end{matrix} \xrightarrow{\text{H}_2} \begin{matrix} X\ Y \\ HC-CH \\ Y\ X \\ recemic \end{matrix}$$

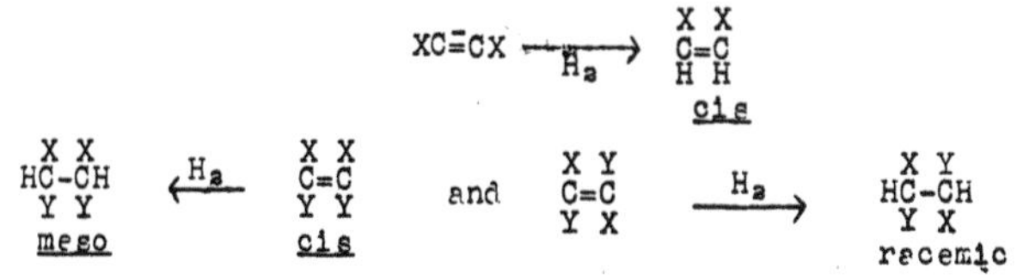

D. Hydrogenation of Conjugated Systems.

In general conjugated systems are fairly easily reduced
catalytically.

Recent work has shown that, contrary to older belief, 1,4-
addition does take place in certain cases and that this is true for
both C-O and C-C conjugations. In compounds with C-C conjugation,
there is a strong tendency for both double bonds to saturate at
about the same rate. Lebedev and Yakubchik found that with purely
aliphatic diene hydrocarbons reduced over platinum black, dihydro
compounds could be isolated at the half-reduction point. Raney
nickel gives more stepwise reduction than does platinum black.

The addition of hydrogen in the presence of palladium or Raney
nickel to the system $\overset{4\ 3\ 2\ 1}{C=C-C=C-CO_2H}$ tends to occur in the 3,4-positions,
although some 1,4-addition is obtained.

In cases where an acetylenic bond is in conjugation with an ethylenic bond, it is frequently possible to obtain selective hydrogenation of the former with formation of the corresponding diene. Palladium and nickel catalysts may be used.

$$HC\equiv C\underset{H}{-C}=CH_2 \xrightarrow[Pd]{H_2} H_2C=\underset{H}{C}-\underset{H}{C}=CH_2$$

D. Hydrogenation of Aromatic Hydrocarbons.

Catalytic hydrogenation of the double bonds of an aromatic ring is much more difficult than with open chain unsaturations, and, therefore, requires more strenuous conditions of temperature and pressure, and a longer reaction time.

Benzene and its homologs can be reduced to the cyclohexane derivatives at low pressures and temperatures by use of catalysts such as platinum black, colloidal palladium and platinum, and nickel black. Reduced nickel, Raney nickel, and nickel on kieselguhr are effective at higher temperatures and pressures. The hydrogenation can not be stopped at any intermediate stage.

The effect of substituents in the ring on the ease of hydrogenation is not clear-cut. Phenols undergo hydrogenation more easily than the hydrocarbons, while ethers do not. Amine groups and carbethoxy groups retard the saturation over most catalysts. With alkyl substituted compounds, the relative ease of reaction depends chiefly on the catalyst.

Few side reactions occur, but in some cases hydrogenolysis may occur, or the reaction product may undergo secondary reactions.

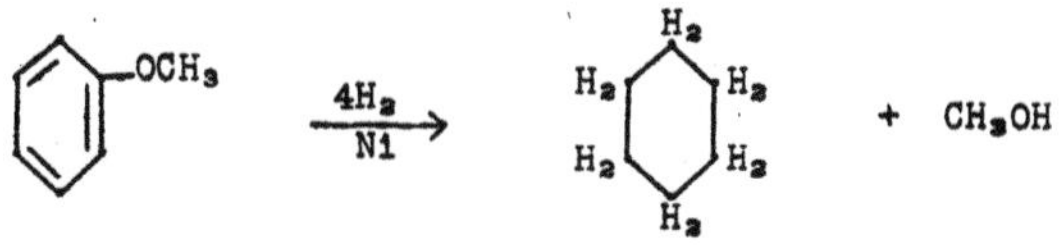
OCH₃
4H₂
Ni
H₂
H₂
H₂
H₂
H₂
H₂
+ CH₃OH

In contrast to benzene, the polynuclear hydrocarbons are hydrogenated stepwise, and the intermediate compounds may frequently be isolated. Naphthalene can be hydrogenated to tetralin or decalin depending on the conditions used. In most cases both are formed and in some cases either product may be made to predominate by changing the temperature and pressure conditions.

Naphthols are more easily hydrogenated than naphthalene. With α-naphthol, either ring may be preferentially saturated. Raney nickel at 150°C hydrogenates the unsubstituted ring, and with copper chromite the substituted ring is attacked.

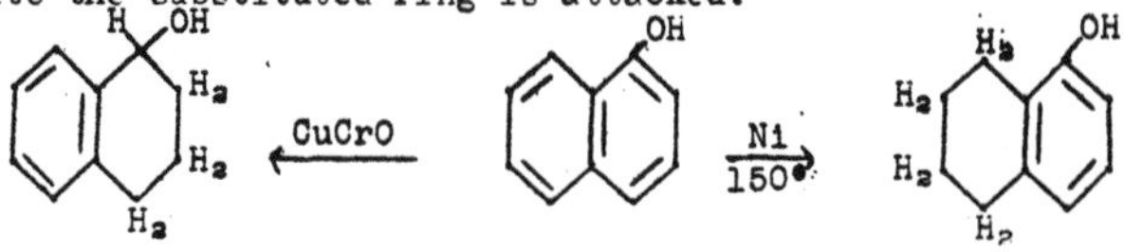
H OH
H₂
H₂
H₂
CuCrO
OH
Ni
150°
OH
H₂
H₂
H₂
H₂

The hydrogenation of anthracene can be done in stages, the reaction usually proceeding to the octahydro stage. Schroeter considered the following stepwise reduction to take place with nickel as a catalyst:

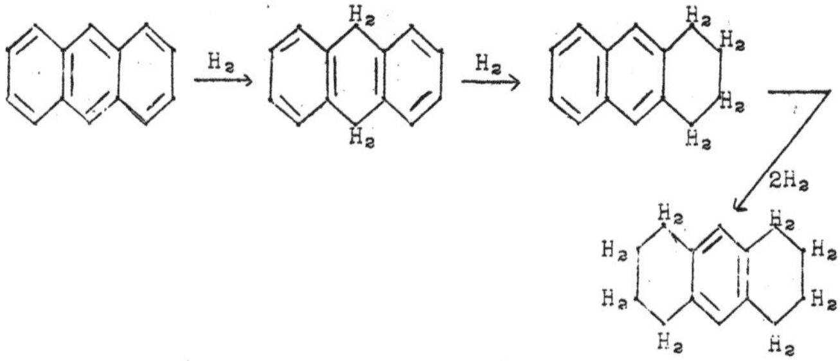
H2
H2
H2
H2
H2
H2
H2
2H2
H2
H2
H2
H2
H2
H2
H2
H2

Other investigators, however, do not consider the 9-10 dihydro compound as a necessary intermediate. They have suggested that two simultaneous processes occur:

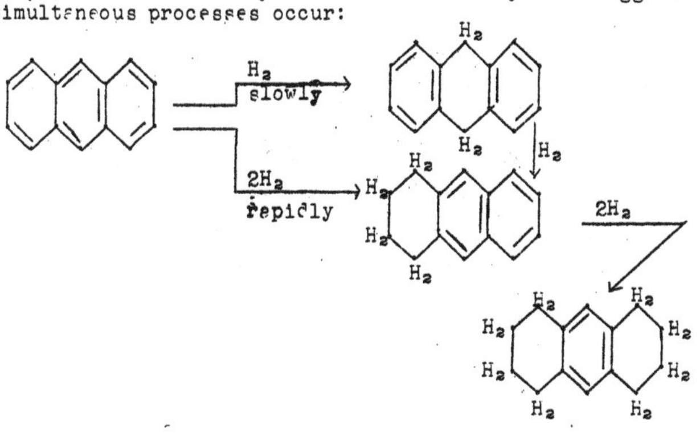
H2
slowly
H2
H2
H2
2H2
rapidly
H2
H2
H2
H2
2H2
H2
H2
H2
H2
H2
H2
H2
H2

Phenanthrene can also be hydrogenated stepwise, the first attack being at the 9,10 positions. The aliphatic nature of this double bond is shown by the fact that it can be saturated using copper chromite as catalyst. The hydrogenation tends to stop at the octahydro stage, but by a suitable choice of conditions other hydrogenation products may be made to predominate. With Raney nickel at 110° or palladium black at room temperatures, the tetrahydro phenanthrene is formed. Dodecahydrophenanthrene can be obtained if Raney nickel is used at 200°C.

<u>Bibliography</u>

Adkins, Reactions of Hydrogen with Organic Compounds over Chromium Oxide and Nickel Catalysts, Univ. of Wis. Press, Madison, Wis., (1930).
Alexander and Fuson, J. Am. Chem. Soc., <u>58</u>, 1745 (1936).
Burton and Ingold, J. Chem. Soc., 2022 (1929).
Bourgeul, Bull. soc. chim. [4] <u>45</u>, 1067 (1929).
Campbell and Campbell, Chem. Reviews, <u>31</u>, 77 (1942) (includes 366 references).
Campbell and Eby, J. Am. Chem. Soc., <u>63</u>, 2683 (1941).
Ellis, Hydrogenation of Organic Substances, 3rd edition (1930).
Greenlee and Fernelius, J. Am. Chem. Soc., <u>64</u>, 2505 (1942).
Jeannes and Adams, J. Am. Chem. Soc., <u>59</u>, 2608 (1937).
Kazanskii and Tatevoysan, Chem. Abstracts, <u>34</u>, 4731.
Kern, Shriner and Adams, J. Am. Chem. Soc., <u>47</u>, 1147 (1925).
Paal and Hartman, Ber., <u>43</u>, 248 (1910).
Robinson, Reduction of Multiple Carbon-Carbon Bonds, Organic Seminar 1st semester (1941-1942).
Shaw and Thompson, J. Am. Chem. Soc., <u>64</u>, 363-6 (1942).

Reported by Sidney Melamed and John M. Stewart
November 18, 1942.

THE ADDITION OF β,γ-UNSATURATED ALCOHOLS TO THE ACTIVE METHYLENE GROUP

The addition of β,γ-unsaturated alcohols to the active methylene group, although appearing somewhat similar to a Michael reaction in regard to the products formed, cannot be formally classified as such.

When this reaction was first investigated by Carroll, it was in connection with the reversibility of the acetoacetic ester condensation and included saturated as well as β,γ-unsaturated alcohols. To introduce the discussion of this particular reaction, it is necessary to mention first, the course of the reaction in regard to saturated alcohols.

It might be expected that the reversible nature of the acetoacetic ester condensation might be of value in preparing various esters of acetic acid; that is,

$$CH_3-\overset{O}{\overset{\|}{C}}-CH_2-\overset{O}{\overset{\|}{C}}-O-Et + ROH \overset{OEt^-}{\rightleftharpoons} \underset{I}{CH_3-\overset{O}{\overset{\|}{C}}-O-R} + CH_3-\overset{O}{\overset{\|}{C}}-O-Et$$

This reaction, however, does not usually take place since simple alcoholysis occurs:

$$CH_3-\overset{O}{\overset{\|}{C}}-CH_2-\overset{O}{\overset{\|}{C}}-O-Et + ROH \rightleftharpoons EtOH + CH_3-\overset{O}{\overset{\|}{C}}-CH_2-\overset{O}{\overset{\|}{C}}-O-R$$

By raising the temperature, however, carbon dioxide and some of I is obtained. If equimolar quantities of the two reactants are used, a mixture of the original alcohol (ROH) and the ester (I) is obtained. Other products obtained in small amounts are acetone, substituted acetones, and the olefin of the original alcohol. Hence it can be seen that the products obtained from ethyl acetoacetate and an alcohol are

$$EtOH,\ CO_2,\ CH_3-\overset{O}{\overset{\|}{C}}-O-R,\ CH_3-\overset{O}{\overset{\|}{C}}-CH_3,\ CH_3-\overset{O}{\overset{\|}{C}}-CH_2-R \text{ or } R'$$

if rearrangement occurs and the olefin from ROH. Table I shows the mole per cent of ester, ketone, and olefin formed. In none of these saturated alcohols, is any noteworthy addition experienced to give a substituted ketone.

The first β,γ-unsaturated alcohols that Carroll studied were linalöol and geraniol. If γ or what will be referred to as normal addition occurs, the following would take place:

$$\text{A.}\quad R-\overset{CH_3}{\underset{OH}{\overset{|}{\underset{|}{C}}}}-CH=CH_2 + CH_3-\overset{O}{\overset{\|}{C}}-CH_2-\overset{O}{\overset{\|}{C}}-OEt \rightarrow \underset{II}{\overset{CH_3}{\underset{R}{\overset{|}{\underset{|}{C}}}}=CH-CH_2-\overset{O}{\overset{\|}{C}}-CH_3}$$

linalöol $R=(CH_3)_2C=CH-CH_2-CH_2-$

B. 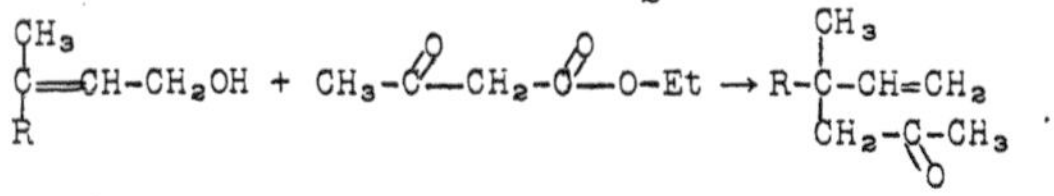
$$\underset{R}{\overset{CH_3}{C}}=CH-CH_2OH + CH_3-\overset{O}{C}-CH_2-\overset{O}{C}-O-Et \rightarrow R-\underset{CH_2-\underset{O}{C}-CH_3}{\overset{CH_3}{C}}-CH=CH_2$$

geraniol   III

However, no trace of III could be found with geraniol and only a
small amount of II was obtained along with large amounts of geranyl
acetate. Linalool gave no geranyl acetate but only the ketone (II)
and a small amount of an unidentified alcohol. The identity of
the ketone in reaction A was established by its synthesis from
geranyl chloride and acetoacetic ester and mixed melting point on
the semicarbazones.

Carroll has also studied the action of other β,γ-unsaturated
alcohols and these results have been summarized in Table II.
For example, cinnamyl alcohol reacts with acetoacetic ester in the
following manner:

$$C_6H_5-CH=CH-CH_2OH + CH_3-\overset{O}{C}-CH_2-\overset{O}{C}-O-Et \rightarrow C_6H_5-\underset{CH=CH_2}{\overset{CH_2-\overset{O}{C}-CH_3}{CH}}$$

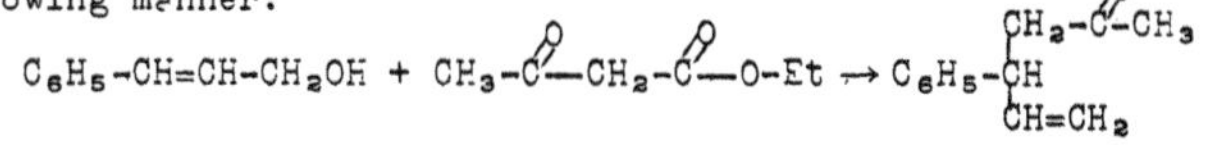

Phenyl vinyl carbinol also reacts in a normal fashion.

$$C_6H_5-\overset{OH}{CH}-CH=CH_2 + CH_3-\overset{O}{C}-CH_2-\overset{O}{C}OEt \rightarrow C_6H_5-CH=CH-CH_2-CH_2-\overset{O}{C}-CH_3$$

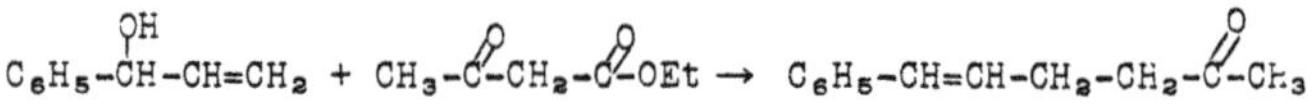

The constitution of the ketone from cinnamyl alcohol was proven
by oxidation to α-phenyllevulinic acid, which was prepared from
ethyl α-bromophenyl acetate and ethyl acetoacetate.

As seen previously in the linalool-geraniol pair, addition may
take place in two ways to the active methylene group: normal or
γ-addition and α-addition.

$$\overset{\gamma}{C}=CH-\overset{\alpha}{C}-OH \left< \begin{array}{l} \overset{\alpha}{\longrightarrow} CH_3-\overset{O}{C}-CH_2-\overset{\alpha}{C}-CH=\overset{\alpha}{C} \\ \searrow \overset{\gamma}{C}=CH-\overset{\alpha}{C}-CH_2-\overset{O}{C}-CH_3 \end{array} \right.$$

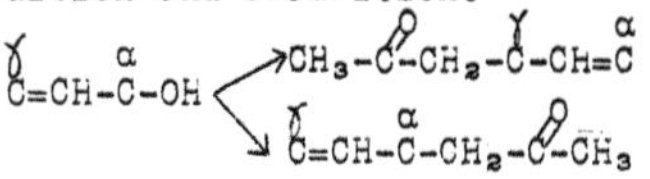

Another example is the condensation of ethyl malonate with cinnamyl
alcohol or phenyl vinyl carbinol. Cinnamylacetic acid is obtained
from both reactions and could only be formed by γ-addition to
cinnamyl alcohol, α-addition to phenyl vinyl carbinol.

The occurrence of α-addition introduces complications in the
mechanism of the reaction. The experimental conditions do not favor
aniontropic change.

$$\overset{\gamma}{C}=CH-\overset{\alpha}{C}-OH \rightleftharpoons \underset{OH}{\overset{\gamma}{C}}-CH=\overset{\alpha}{C}$$

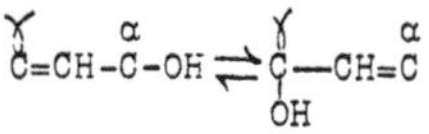

Linalool and geraniol show no interconversion under similar conditions.

The α-addition may be represented by an ionic mechanism.

$$[\overset{\gamma}{C}=CH-\overset{\alpha}{C}]^{+} + [CH_3-\overset{O}{C}-CH-\overset{O}{C}-O-Et]^{-} \rightarrow CH_3-\overset{O}{C}-\underset{COOEt}{\overset{\overset{\alpha}{C}-CH=\overset{\gamma}{C}}{\underset{CH}{|}}}$$

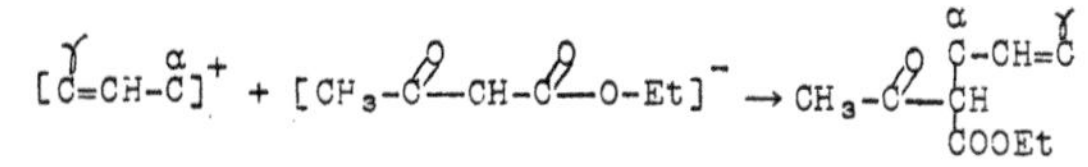

However, in order to explain γ-addition it is necessary to postulate an α → γ shift. This mechanism fails to explain the behavior of the linalool-geraniol pair as the γ-addition product is formed at a much lower temperature than the α. Also the shift should yield mixtures of α- and γ-addition products in all cases, but the cinnamyl alcohol-phenyl vinyl carbinol pair gave one compound only, the γ-addition product in each case.

A different mechanism is therefore involved in γ-addition. Evidently, γ-addition takes place before removal of the hydroxyl group.

$$\overset{\gamma}{C}=CH-\overset{\alpha}{C}-OH + CH_3-\overset{O}{C}-CH_2-\overset{O}{C}-O-Et \rightarrow CH_3-\overset{O}{C}-\underset{COOEt}{\overset{\overset{\gamma}{C}-CH_2-\overset{\alpha}{C}-OH}{\underset{CH}{|}}}$$

$$CH_3-\overset{O}{C}-CH_2-\overset{\gamma}{C}-CH=\overset{\alpha}{C}$$

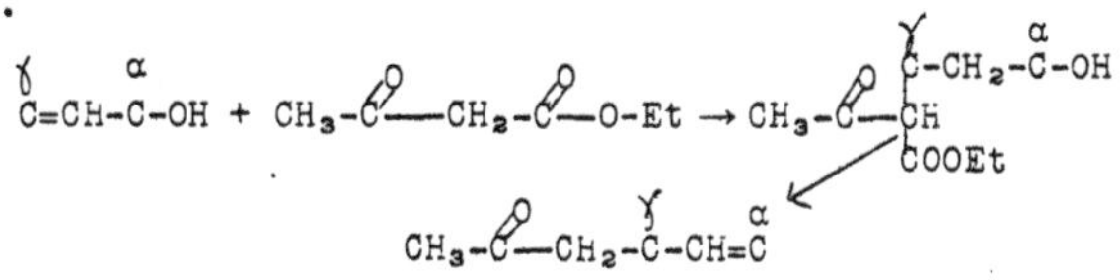

Similar explanations may be given of the results of Bergman and Corte who caused ethyl-O-cinnamyl-β-oxycrotonate to rearrange by heating to 260°.

$$CH_3-\underset{O-CH_2-CH=CH-C_6H_5}{\overset{C=CH-COOEt}{|}}$$

$$CH_3-\overset{O}{C}-\underset{CH_2-CH=CH-C_6H_5}{\overset{CH-COOEt}{|}}$$

$$CH_3-\overset{O}{C}-\underset{\overset{CH-CH=CH_2}{|}}{\overset{CH-COOEt}{|}}$$
$$C_6H_5$$

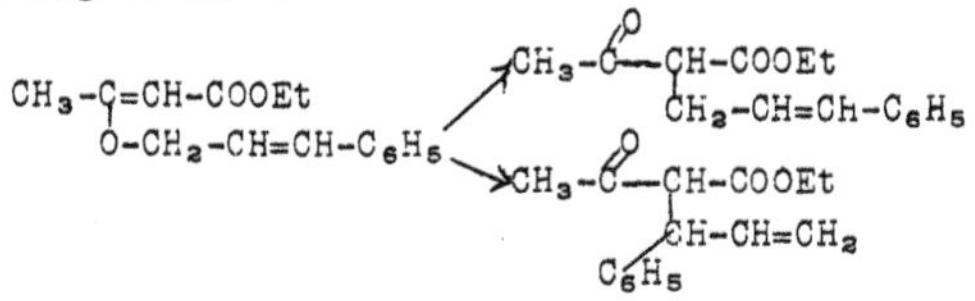

γ-addition product

In the analogous rearrangement of the phenyl allyl ethers, the normal reaction corresponds to γ-addition.

Evidently, α-addition only occurs when the alcohol ionizes easily. Triphenyl carbinol and ethyl malonate on prolonged refluxing give small quantities of the α-addition product β,β,β-triphenyl propionic acid. Benzyl alcohol and butyl malonate, on prolonged refluxing gave a small quantity of the α-addition product β-phenyl-propionic acid.

TABLES

Table I. Ethyl acetoacetate and saturated alcohols

Exp.		Ester	Ketone	Olefin
1.	ethyl alcohol	2	0	0
2.	butyl alcohol	4	0	0
3.	heptyl alcohol	14	0	0
4.	benzyl alcohol	46	1	0
5.	phenyl ethyl alcohol	42	0	3
6.	anisyl alcohol	55	0	I
7.	phenyl methyl carbinol	11	1	11
8.	phenyl methyl ethyl carbinol	0	0	83

Table II. Ethyl acetoacetate and unsaturated alcohols

		Ester	Ketone	Olefin
9.	allyl alcohol	6	1	0
10.	methallyl alcohol	9	2	0
11.	crotyl alcohol	21	8	0
12.	methyl vinyl carbinol	12	12	0
13.	geraniol	16	(abnormal) 12	trace
14.	linalool	10	41	trace
15.	cinnamyl alcohol	16	33	0
16.	phenyl vinyl carbinol	5	75	0
17.	isopulegone	12	2	5

Table III. Ethyl butylacetoacetate and unsaturated alcohols.

		Ester	Ketone	Olefin
18.	methallyl alcohol	49	1	0
19.	cinnamyl alcohol	40	15	0
20.	linalool	10	27	trace

Table IV. Ethyl malonate and unsaturated alcohols

		Unsaturated acid.
21.	allyl alcohol	6 per cent
22.	crotyl alcohol	13 per cent
23.	cinnamyl alcohol	66 per cent (abnormal)
24.	phenyl vinyl carbinol	51 per cent

Bibliography

Arndt, Ber., 69, 2373 (1936).
Bergmann and Corte, J. Chem. Soc., 59, 2856 (1935).
Carroll, J. Chem. Soc., 704, 1266 (1940); 507 (1941).
Tuier and Kilburn, J. Am. Chem. Soc., 59, 2856 (1937).
Fosse, Compt. rend., 145, 196, 1292 (1907).

Reported by C. G. Overberger
November 25, 1942

CARBONYL BRIDGE COMPOUNDS

By a carbonyl bridge compound is meant any cyclic substance in which two non-adjacent atoms are joined by a carbonyl group. Most of these compounds give the expected ketone reactions. In this report a special group of compounds will be discussed in which both bonds to the carbonyl group are β,γ-to an ethylenic linkage, i.e., (I)

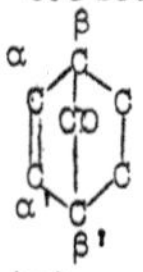

β
α
CO
α'
β'

(I)

Japp and later Gray noted that a "bimolecular compound" was formed when anhydracetonebenzil and its homologs were treated with acidic dehydrating agents. These compounds had a very peculiar property. They evolved one mole of carbon monoxide quantitatively on heating. Although neither worker attempted to assign structures to the molecules, they had probably prepared the first carbonyl bridge compounds.

Some years later Allen and coworkers investigated more thoroughly these interesting compounds. From their study has grown the work that is reported here. It should also be mentioned that Zincke and his associates recognized similar compounds in their work with chlorinated cyclopentenone.

When Allen began to study Japp's bimolecular compound, he proposed a series of reactions that might have resulted in its formation. First, a cyclopentadienone (II) could be formed by dehydration. This could then enter into a diene synthesis with itself to give III. This structure (III) was suggested after a long series of reactions that finally gave *o*-terphenyl. The existence of cyclopentadienone as an intermediate was shown by dehydrating anhydracetonebenzil with an excess of maleic anhydride.

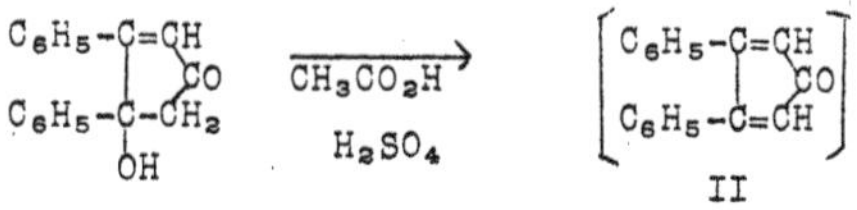

CH_3CO_2H
H_2SO_4

II

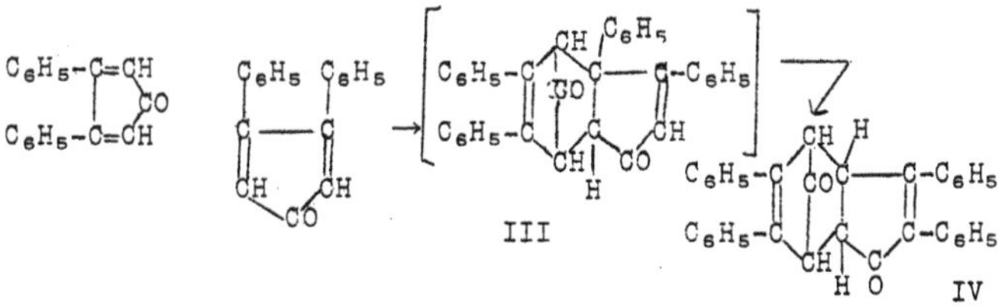

III

IV

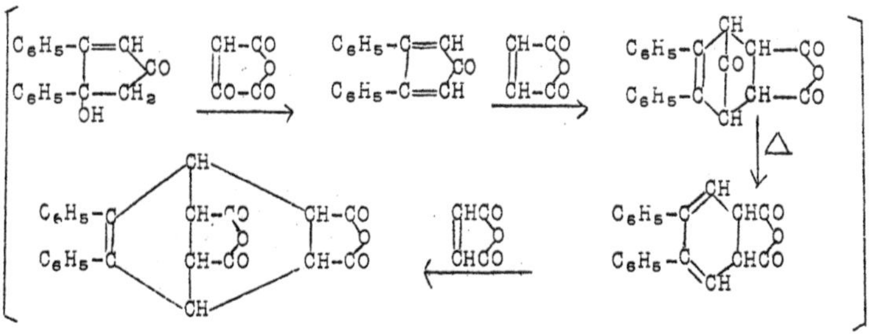

One correction to the originally assigned structure (III) for the bimolecular compound has been made. The phenyl group is not on the atom common to the two rings but undergoes an allylic rearrangement to give IV. Whenever there is a hydrogen present to exchange with the phenyl group, this rearrangement occurs.

Preparation

In general carbonyl bridge compounds are prepared by the diene synthesis. The reaction given above for the dehydration of anhydracetonebenzil is typical. The methylated homolog of III reacts as if it were dissociated into its diene component (V). Use has been made of this fact in preparing a large number of compounds in which the bridge contains the only carbonyl group in the molecule.

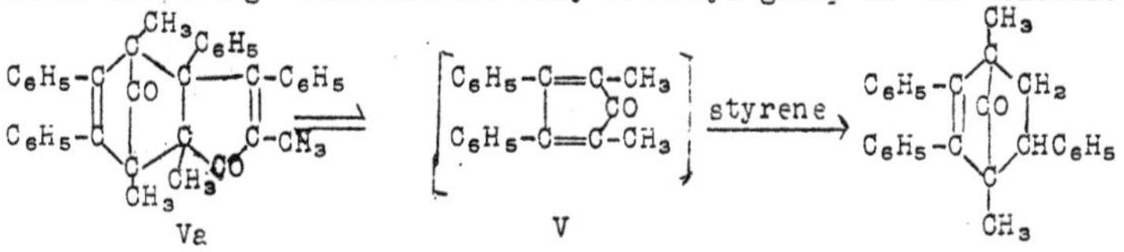

Styrene, maleic anhydride, and methyl fumarate are examples of the variety of unsaturated compounds that have been successfully added to it.

Pyrolysis

All carbonyl bridge compounds in which both ends of the bridge are joined β, γ to an ethylene linkage lose carbon monoxide on heating. This is in accord with Schmidt's double bond rule which states that when unsaturated molecules are decomposed by heating, cleavage takes place at the saturated bond β, γ to the unsaturated linkage.

Alkaline Reagents

When IV is heated with alcoholic potassium hydroxide, the bridge is destroyed and a carboxylic acid is formed. If sodium

methylate or ethylate are used in place of potassium hydroxide the corresponding esters of the carboxylic acid are formed.

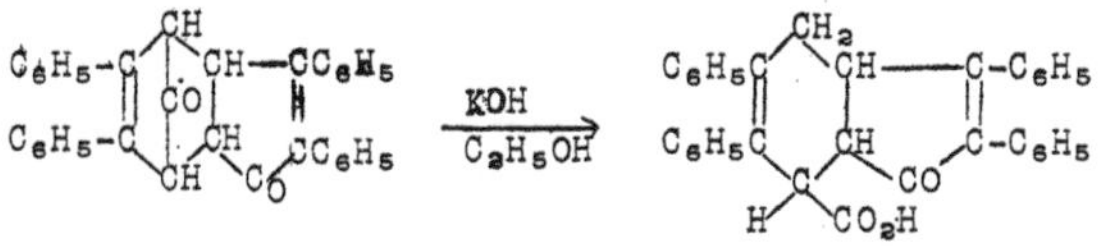
KOH
C_2H_5OH

Grignard Reagent

In the Grignard machine compound IV gives an unexpected result. One mole of gas is evolved showing the presence of one active hydrogen, and one instead of the expected two additions is observed. The product resulting after hydrolysis loses carbon monoxide on heating, indicating that the carbonyl bridge is protected during the reaction. Enolization involving the carbonyl bridge would account for these facts. This view is further supported by the observation that the tribromo derivative of IV, in which enolization is impossible, gives the expected two Grignard additions.

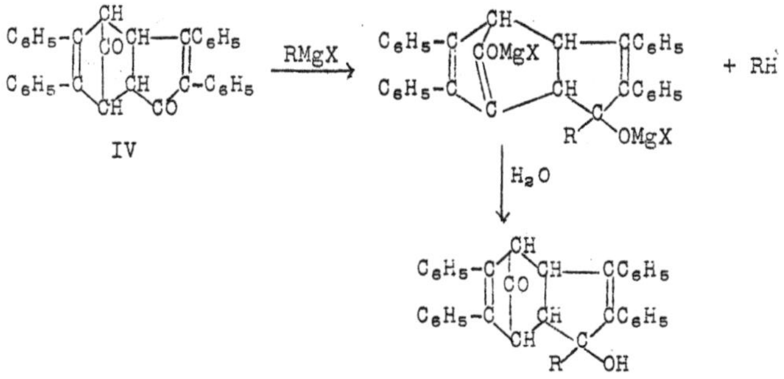
RMgX
+ RH
H_2O

IV

Alkaline Peroxide

When compound IV is treated with alkaline peroxide a new substance stable at room temperature is formed. Analysis shows four additional oxygen atoms. No attempt has been made to assign a formula to this molecule. When dissolved in acetic acid, the peroxide quantitatively regenerates an isomer of IV. This isomer shows no reaction with sodium peroxide. Both isomers give identical derivatives except for the products obtained on pyrolysis. The methylated homolog of IV gives no peroxide, suggesting that

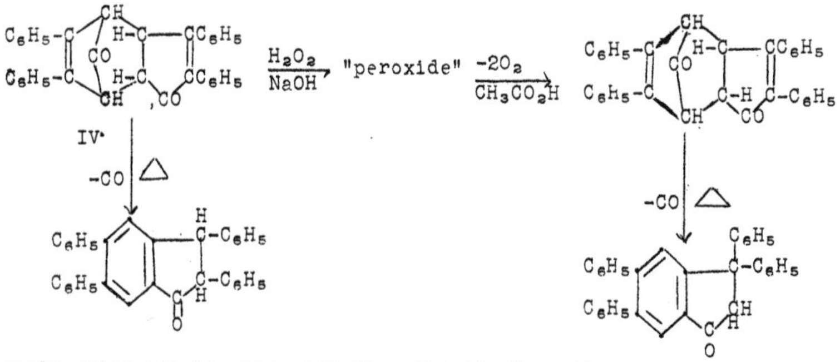

IV
H₂O₂
NaOH
"peroxide"
-2O₂
CH₃CO₂H
-CO
-CO

enolization may be necessary for peroxide formation.

Bibliography

Japp and Burton, J. Chem. Soc., 51, 420 (1887).
Gray, ibid., 95, 2131 (1909).
Zincke and Meyer, Ann., 367, 1 (1909).
Zincke and Pfaffendorf, Ann., 394, 3 (1912).
Allen and Spanagel, J. Am. Chem. Soc., 55, 3773 (1933).
Allen, Bell and Van Allen, ibid., 62, 665 (1940).
Allen and Van Allen, ibid., 64, 1260 (1942).
Allen and Gates, ibid., 64, 2120, 2123, 2439 (1942).
Allen and Spanagel, Can. J. Res., 8, 414 (1933).
Allen and Sheps, ibid., 11, 171 (1934).
Allen and Rudoff, ibid., B15, 327 (1937).

Reported by S. Parmerter
November 25, 1942

MECHANISM AND USES OF WURTZ-TYPE REACTIONS

Mechanism

Because of the variety of products formed in any given instance of the Wurtz or Fittig reaction, there has long been a controversy concerning the intermediate steps occurring between the mixing of the organic halide with the alkali metal and the separation of the products. First, the evidence supporting both sides of the question of mechanism will be introduced; then the applications and experimental conditions will be considered.

Two principal routes are proposed for the reaction: (a) the removal of the halogen atom by the alkali metal with the production of a free alkyl or aryl radical; (b) the formation of an organometallic intermediate. Sometimes a combination of the two is postulated.

The basis for the free radical theory lies mainly in the products found and in a kinetic study of the reaction. The products usually resulting from the reaction of an alkyl halide RX are: (a) a saturated hydrocarbon RR; (b) a saturated hydrocarbon RH; (c) an unsaturated hydrocarbon R-H; and (d) polymers. It should be noted that disproportionation has frequently been considered a property of (and often a proof of the existence of) free radicals. At first, Schlubach and Goes combined the free radical hypothesis with the older organometallic theory:

$$RX + Na \rightarrow R^{\cdot} + NaX$$
$$2R^{\cdot} \rightarrow RR$$
$$2R^{\cdot} \rightarrow RH + R\text{-}H$$
$$R^{\cdot} + Na \rightarrow RNa$$
$$RX + RNa \rightarrow RR + NaX$$
$$RX + RNa \rightarrow RH + R\text{-}H + NaX$$

Bachmann and Clarke boiled chlorobenzene with sodium and obtained these products, which can readily be accounted for on a free radical basis: benzene, biphenyl, 2-phenyl biphenyl, triphenylene, 2,2'-diphenyl biphenyl, and traces of terphenyl. From n-heptyl bromide at 150° they obtained, in addition to the usual C_7 and C_{14} hydrocarbons, $C_{21}H_{44}$, which they attributed to disproportionation to $C_6H_{13}CH=$. From chlorobenzene and excess toluene they isolated benzene, diphenylmethane, and 4-methyl biphenyl, but no biphenyl. All of these products, they maintained, could best be accounted for by assuming an intermediate free radical. Whitmore (16), umtil recently, gave the same reasoning for the production from neopentyl chloride and sodium of neopentane, 1,1-dimethylpropane, and the expected tetramethylhexane.

In the study of the interaction of lithium butyl with a variety of halides, Marvel observed in most instances a pronounced color, which appeared and later faded; this he believed was the sign of a free radical. Wurtz reactions in general show a color, usually blue, at the outset.

On the basis of kinetic studies of the reaction between ethyl iodide and sodium, which showed that the reaction was first order with respect to iodide concentration, Richards also advocated the free radical mechanism. He assumed a high stationary concentration of radicals at the sodium surface; these radicals reacted by disproportionation to give ethane and ethylene, or with the (less concentrated) ethyl iodide to give butane.

Proponents of the organometallic intermediate theory have perhaps more direct evidence to support their views. The reaction can be stopped at the intermediate stage, either by using low temperatures and excess sodium and separating the resulting sodium alkyl, or by passing carbon dioxide through the mixture to trap the sodium alkyl as the sodium salt of the carboxylic acid (2).

Fuchs and Metzl obtained an amorphous product (among others) from 1,3,5-tribromobenzene and potassium. This product contained oxygen and bromine; it was split by hydrogen iodide to give some resorcinol. They assumed this to be a polymeric ether resulting from air oxidation of the metallic intermediate to a phenol salt, followed by Williamson-type coupling. The transient dark blue color which is commonly observed during a Wurtz reaction they attributed to the organometallic compound.

Goldschmidt and Schon supported this theory by pointing out the higher yields obtained in the Wurtz-Fittig modification as compared to either of the others. This is commonly explained by the fact that these reactions:

$$ArX + 2Na \rightarrow ArNa + NaX$$
$$ArNa + RX \rightarrow ArR + NaX$$

go more readily than any of the others which are possible in theory. A free radical mechanism would call for correspondingly high yields of ArAr and RR.

Morton at first believed (3) that free radical formation accompanied the formation of the organometallic intermediate, but he has now concluded that no free radical is present either before the formation of the intermediate or during its reaction with the halide. The former part of this conclusion (13) is based upon the fact that if the reaction (at low temperature) is halted at the sodium alkyl stage and the mixture is poured over solid carbon dioxide, only one carboxylic acid is formed, and that nearly quantitatively; thus, no disproportionation or migration of the odd electron has taken place. The latter part of the conclusion (14) is based upon the contention that both disporportionation and polymerization products are predictable on an organometallic basis. The alkyl halide loses a halide ion, the sodium alkyl loses a sodium ion; the resulting positive and negative alkyl ions can couple, or the positive ion can lose a proton to give an olefin while the negative ion becomes a paraffin. If the positive ion is an aryl, it can lose a proton from the ortho position without double bond formation, and the resulting "zwitterion" can produce triphenylene and other polymers. If this reasoning is true, the olefin resulting from RNa - R'X should always

be R'-H; experimental evidence shows that the ratio of R'-H to R-H is usually quite large, although because of interchange of radicals there is always some formation of R-H. Whitemore (17) agrees that where organometallic intermediates take an important part, this mechanism is the plausible one, and he has showed that from neopentyl chloride and sodium propyl the ratio of dimethylcyclopropane to propene is 15:1. Also, with sodium ethyl and hexyl chloride the ratio of hexene to ethylene is 23:1.

Morton points out (14) that it is not reasonable to suppose that a positive and a negative alkyl ion would first exchange an electron and then react with one another as free radicals. Also, he has calculated that the free energy change is more negative for free radical coupling than for disproportionation, to the extent that the former should practically exclude the latter. Finally, he explains the first order reaction rate of ethyl iodide (13) as being the result of the formation of a layer on the surface of the metal through which the iodide must diffuse in order to react with the metal.

The existence of the organometallic intermediate can scarcely be questioned. Whether the possibility of free radicals at one point or another has been disproved is another matter. Certainly they are present in the vapor-phase reactions of Saffer at 320°; the high temperature production of $C_{21}H_{44}$ by Bachmann and Clarke hints at their presence. It seems conceivable that the extent to which free radicals take part is dependent upon the temperature and perhaps other conditions.

The Organo-Metallic Intermediate

The existence of the sodium intermediate has been postulated by many investigators (32, 23, 33, 34). Morton studied the formation of alkyl sodium extensively, and was able to prepare many of them in good yields. In general, this is accomplished by slowly adding alkyl chloride with much stirring to extremely finely divided sodium at low temperatures. Factors to be considered are:

1. Temperature needed for the reaction.
2. Thermal stability of sodium alkyl.
3. Absence of a protective sodium chloride coating on the sodium particles. Secondary halides favor a hard protective coating; primary halides do not.
4. An unreactive carbon-halogen bond so that alkyl halide can't react too rapidly with alkyl sodium--primary alkyl halides and chlorobenzene, for example. Moderate to poor results are obtained with secondary and tertiary chlorides and with alkyl bromides and iodides (12).

In studying the relative stability of various sodium alkyls Morton obtained decreasing yields as follows (2):

Halide	% Yield of the Na Intermediate
n-amyl	57
n-hexyl	25
chlorobenzene	16.8
n-butyl	13.8
t-amyl	1.0
2-chloro-3-methylbutane	trace
n-propyl	trace

Since then by means of improved high-speed stirring and specially creased flasks, and with better temperature control the yields have all been greatly improved--a 72% yield of n-amyl sodium recently has been reported (11, 13).

Carrying the Wurtz reactions out in a step-wise manner has found direct application where the exothermic nature of the first step is harmful to the stability of the product desired. Schumb (31) found that by allowing the energy released in the formation of phenyl sodium to be dissipated before adding further reactants, he was able to get good yields of disilanes, whereas in a one-step Wurtz only monosilanes could be obtained.

Sodium alkyls may also be prepared from the corresponding mercury dialkyls. Schlenk, 1917, (35) was the first to use this method. Since then Gilman (24), and Whitmore (17,18) have both made good use of this method. Whitmore obtained propyl sodium in 88% yield:

$$Hg(C_3H_7)_2 + 2Na \rightarrow 2C_3H_7Na$$

Factors Involved in Organo-Metallic Interchange

Schorigin (23) in 1908 first noticed the tendency of alkyl sodium to undergo an interchange with its solvent. Control of this has led to simplified methods of obtaining other intermediates and the use of more economical reactants. Morton has prepared many aryl sodium intermediates by means of an interchange with n-amyl sodium (8). In trial runs he obtained the following interchanges:

Solvent	% Interchange	Compound Formed
Benzene	78	Phenyl Sodium
Toluene	87	Benzyl Sodium
Dimethyl Aniline	18	o-Sodium dimethylaniline
Anisole	20	o-Sodium anisole
Diphenyl methane	14	Sodium diphenylmethane
Fluorene	18.5	9-Sodium fluorene

In all cases pentane was the other product. The interchange is a rather slow reaction often facilitated by a temperature higher than that possible for the formation of the sodium alkyl (8). Comparing toluene, benzene, and n-alkyl sodium compounds, the tendency toward formation by interchange is in the decreasing order: benzyl, phenyl, n-amyl, butyl, and propyl sodium.

Where a choice of sodium intermediates is possible benzyl sodium is by far the most desirable. It is the most stable of the aryl derivatives studied and leads to the most consistent results. The strong tendency for sodium derivatives of toluene to assume this form is clearly brought out by the "lateral metallation" studies of Gilman (26). Thus:

o-sodium toluene —refluxing→ benzyl sodium
m-sodium toluene —refluxing→ benzyl sodium
p-sodium toluene —refluxing→ benzyl sodium

Benzyl sodium may be prepared in 87% yield from chlorobenzene, toluene, and sodium.

Morton has also made studies of exchange reaction of amyl sodium with biphenyl (9), naphthalene, and naphthalene analogs (15). Decreasing aromaticity of the ring systems resulted in greatly increasing the ease of metallation, leading to polymetallation and heterogenity of products. In consequence the reactions are of little importance. Benzene can be dimetallated to a small degree giving mixtures of meta and para isomers (6, 27).

Factors Involved in the Coupling of the Na Intermediate

The introduction of another halide into the reaction mixture containing the sodium intermediate may have two simple consequences: the two may couple or dimerize. It is not at once apparent how such a side reaction as that below may take place:

$$2\ RNa + 2R'X \rightarrow R\text{-}R + R' - R' + 2NaX$$

However, if we visualise an intermediate step in which there is a metal-halogen interchange, we can then account for the dimeric products. Morton has suggested this, (14) and shown that the following reaction takes place:

$$C_5H_{11}Na + CH_3I \rightarrow CH_3Na + C_5H_{11}I\ (47\%)$$

Morton has studied the tendencies of various halides to couple with n-amyl sodium and to form dimers by measuring the yield of the Wurtz product and decene (4, 14).

Halide Used	% Coupled Product	% Decane
Methyl Iodide	0.0	58.4
n-Hexyl iodide	32.6	39.8
Methyl bromide	13.1	44.4
Ethyl iodide	2.1	[illegible]
n-Butyl iodide	17.8	[illegible]
Ethyl bromide	24.2	[illegible]
n-Butyl bromide	32.0	[illegible]
n-Hexyl bromide	45.6	15.2
n-Hexyl chloride	13.4	19.8
i-Butyl chloride	5.5	11.4

n-Butyl chloride	8.8	6.4
Ethyl Chloride	30.2	9.0
Methyl Chloride	43.2	6.3
Benzyl Chloride	8.1	---
Dichloromethane	0.0	20.0
Ethylene chloride	0.0	18.0

Iodides appear to favor the metal-halogen interchange before coupling. Methyl iodide acts preponderantly in this manner. Chlorides, in general, seem to favor desired coupling.

This is only a partial picture of the success of the Wurtz reaction, and does not show the tendency to disproportionate and to polymerize, inevitable side reactions leading to undesirable products. Whitmore (17) found the following reaction to take place:

$$C_2H_5Na + n\text{-}C_6H_{13}Cl \rightarrow C_4H_9CH{=}CH_2 + C_2H_6$$

46% 52%

$$\searrow n\text{-}C_8H_{18} + NaCl$$

40%

The picture is clearer with aryl compounds. In general coupling predominates. Thus:

$$C_6H_5Na + CH_3I \rightarrow C_6H_5CH_3 \text{ and not } C_6H_5\text{-}C_6H_5$$

Dimethyl and diethyl sulfates are convenient coupling agents in like reactions. Phenyl sodium can be partially dimerized by treatment with iodine (4):

$$C_6H_5Na + I_2 \rightarrow [C_6H_5I] + C_6H_5Na \rightarrow C_6H_5\text{-}C_6H_5 \; (12.5\%)$$

Benzyl sodium is recommended in all Wurtz-Fittig reactions where the nature of the products permits (5). The coupling of an aliphatic side-chain to a benzene nucleus may be successfully carried out with benzyl sodium, while the same product from phenyl sodium is obtained in a comparatively unsatisfactory yield.

Some work has been done on coupling with polyhalogen compounds, though no reactions in which the Wurtz was carried out in a completely step-wise manner were found. Van Alphen studied the one-step coupling of dibromoparaffins, and bromobenzene with sodium. He obtained α,ω-diphenyl paraffins in yields varying from 40-70%. The reaction is complicated by the cyclization of small paraffins. Ethylene bromide was quite anomalous in that it "cyclized" to ethylene exclusively. 1,3-Dibromopropane also shows this tendency, though considerable 1,3-diphenylpropane may be obtained. (28)

Dihalides are theoretically capable of polymerization through normal coupling reactions. Such is the case where the distance between the halogens is increased. Carothers (29) succeeded in synthesizing the largest pure paraffin hydrocarbon known, $C_{70}H_{142}$, by the following series simultaneously carried out in ether solution:

$$Br(CH_2)_{10}Br + 2Na \rightarrow Br(CH_2)_{10}Na + NaBr$$

$$Br(CH_2)_{10}Na + Br(CH_2)_{10}Br \rightarrow Br(CH_2)_{20}Br, \text{ etc.}$$

The reactions of 3,5-dibromotoluene with sodium have been studied by Fuchs and Metzl. They obtained an analogous polymerization.

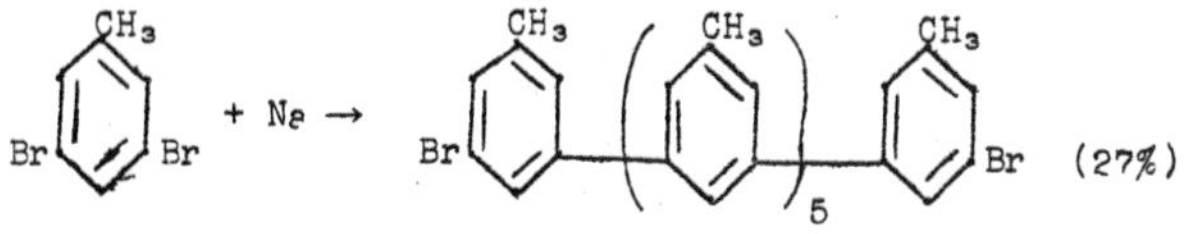

Whitmore found it possible to cyclize certain monohalides. He has prepared 1,1-dimethylcyclopropane in 75% yield by treating neopentyl chloride with sodium propyl (17):

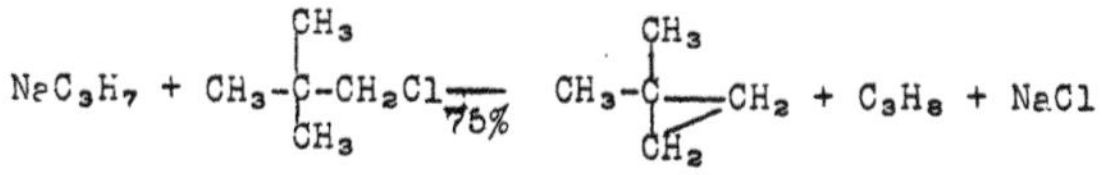

Recognition of the practicability of stopping the Wurtz reaction in the sodium alkyl stage has enormously widened the scope of the reaction. Many of the modifications are close parallels to the Grignard reactions. For example, it is possible to prepare the corresponding carboxylic acids by treating the sodium intermediate with carbon dioxide. This is of special interest in that malonic acids may be obtained in this manner in yields varying from 20-80%.

Bibliography

1. Morton, et al, J. Am. Chem. Soc., 58, 754; (2) 1697; (3) 2599 (1936); (4) 59, 2387 (1937); (5) 60, 1429; (6) 1924 (1938); (7) 62, 120 (8) 123; (9) 126; (10) 129 (1940); (11) 63, 327 (1941); (12) 64, 2239; (13) 2240; (14) 2242 (15) 2250 (1942).
16. Whitmore, et al, J. Am. Chem. Soc., 63, 124 (1941); (17) 64, 1783 (1942); (18) 51, 1491 (1929).
19. Schlubach and Goes, Ber., 55B, 2889 (1922)
20. Bachman and Clark, J. Am. Chem. Soc., 49, 2089 (1927).
21. Richards, Trans. Faraday Soc., 36, 956 (1940).
22. Saffer and Davis, J. Am. Chem. Soc., 64, 2039 (1942).
23. Schorigin, Ber., 41, 2711 (1908).
24. Gilman, et al, J. Am. Chem. Soc., 55, 2893 (1933); (25) 62, 1301 (1940); (26) 62, 1514 (1940); (27) 58, 2074 (1936).
28. Van Alphen, Rec. trav. chim., 59, 580 (1940).
29. Carothers, Hill and Kirby, J. Am. Chem. Soc., 52, 5279 (1930).
30. Schumb, et al, ibid., 60, 2486 (1938); (31) 63, 93 (1941).
32. Acree, Am. Chem. J., 29, 588 (1903).
33. Goldschmidt and Schon, Ber., 59, 2717 (1926).
34. Ziegler, Ann., 479, 135 (1930).
35. Schlenk and Holtz, Ber., 50, 262 (1917).
36. Fuchs and Metzl, ibid., 55, 738 (1922).
37. Marvel, Hager, and Coffman, J. Am. Chem. Soc., 49, 2323 (1927).

Reported by George E. Inskeep and Louis F. Reuter
December 2, 1942

THE ACTION OF DIAZOMETHANE UPON ACYCLIC SUGAR DERIVATIVES

Hans Meyer discovered that an aldehyde could be converted to the corresponding methyl ketone by the action of diazomethane. Later Schlotterbeck investigated the reaction much more thoroughly. However, the reaction of diazomethane with an aldehyde may yield products other than the simple methyl ketone. A general mechanism for this process has been postulated by Arndt.

$$R-C\overset{O}{\underset{H}{\diagdown}} \xrightarrow{CH_2N_2} R-\overset{O}{\underset{\;}{C}}H-CH_2-N\equiv N- \xrightarrow{-N_2} R-\overset{O}{\underset{\;}{C}}H-CH_2- \longrightarrow \begin{cases} \longrightarrow R-C\overset{O}{\diagup}CH_3 \\ \longrightarrow R-\overset{O}{C}H-CH_2 \\ \longrightarrow H-C\overset{O}{\diagup}CH_2-R \end{cases}$$

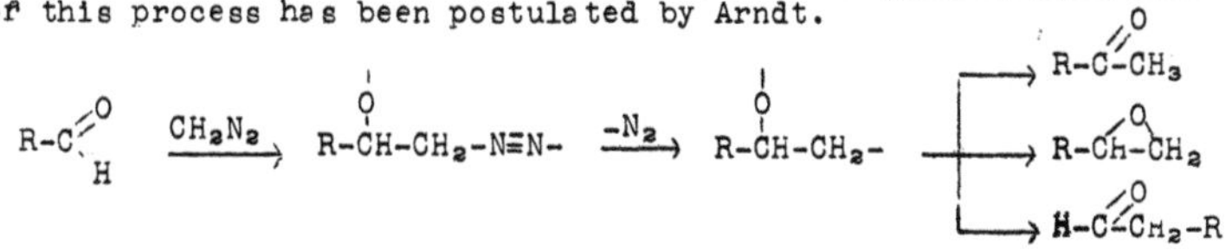

As to which one of these three reactions is to dominate depends upon the nature of "R" and in part on catalytic influences. Brigl and his co-workers made use of this reaction in attempting to prove the structure of aldehydo-d-glucose 3,4,5,6-tetrabenzoate. A crystalline product was obtained which exhibited the analysis expected for a methyl ketone but for which no complete proof of structure was offered.

Clibbens and Nierenstein stated that an analogous reaction occurred to a predominating extent when certain acid chlorides were employed. α-chloroacetophenone was obtained from benzoyl chloride. However, Arndt and his co-workers showed that such a reaction may follow a different course and obtained diazomethyl ketones from a number of acid chlorides.

In the sugar field, Götzi and Reichstein obtained a sirupy diazomethyl ketone by the action of diazomethane upon diethylidene-l-xylonyl chloride. Iwadare obtained a sirupy diazomethyl ketone from isopropylidene-d-glyceroyl chloride and diazomethane.

Wolfrom and his co-workers have studied the action of diazomethane upon aldehydo-d-arabinose-tetraacetate (III) and observed that a crystalline product was obtained. It was characterized as the methyl ketone and designated 1-desoxy-d-fructose tetraacetate (IV).

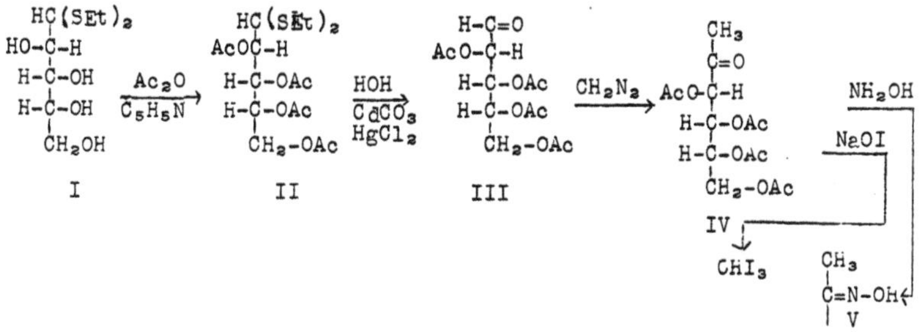

Upon analysis, it gave the correct empirical formula, formed a crystalline oxime and gave positive iodoform, Seliwanoff (ketose), and Fehling reduction reactions. Wolfrom first studied the action of diazomethane upon the l-form; but since a d-derivative is of more immediate interest than is the l-variety, he began a study of the d-enantiomorph.

Upon treating d-gluconyl chloride pentaacetate (VI) with diazomethane, a crystalline product resulted which showed Fehling reduction and which was characterized by analysis as the diazomethyl ketone. It was designated 1-diazo-1-desoxy-keto-d-glucoheptulose pentaacetate (VII). Evidence for the formation of a chlorine-containing compound, probably the chlormethyl ketone, was also obtained. Similarly 1-diazo-1-desoxy-keto-d-fructose tetraacetate was synthesized from d-arabonyl chloride tetraacetate.

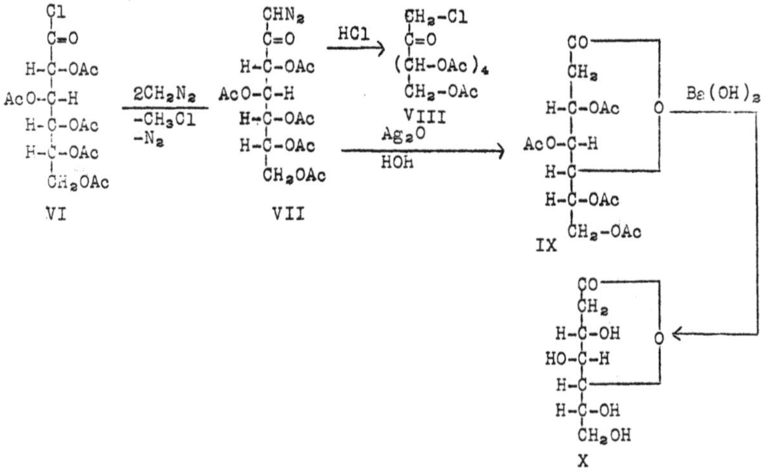
Cl
C=O
H-C-OAc
AcO-C-H
H-C-OAc
H-C-OAc
CH₂OAc
VI
2CH₂N₂
-CH₃Cl
-N₂
CHN₂
C=O
H-C-OAc
AcO-C-H
H-C-OAc
H-C-OAc
CH₂OAc
VII
HCl
CH₂-Cl
C=O
(CH-OAc)₄
CH₂-OAc
VIII
Ag₂O
HOH
CO
CH₂
H-C-OAc
AcO-C-H
H-C
H-C-OAc
CH₂-OAc
IX
O
Ba(OH)₂
CO
CH₂
H-C-OH
HO-C-H
H-C
H-C-OH
CH₂OH
X
O

Treatment of VII with ethereal solutions of dry hydrogen chloride and dry hydrogen bromide produced 1-chloro (VIII) and 1-bromo-keto-d-glucoheptulose pentaacetate, respectively. Similar treatment of 1-diazo-1-desoxy-keto-d-fructose tetraacetate produced 1-chloro and 1-bromo-keto-d-fructose tetraacetate. The chloro derivative (VIII) was the uncharacterized compound isolated in crude form from the mother liquors of the original preparations of the diazomethyl ketone (VII). It is believed that this by-product is formed by the action on the diazomethyl ketone of hydrogen chloride formed in the reaction mixture by traces of moisture. When the reaction was carried out under nearly anhydrous conditions, only traces of the chloro derivative were obtained.

Wolfrom and his co-workers have applied the Wolff rearrangement to the acetylated aldonic acids. A suspension of VII in hot water was treated with catalytic amounts of silver oxide, followed by silver ion removal with hydrogen sulfide. Concentration of the solution yielded 2-desoxy-d-glucoheptono-δ-lactone tetraacetate (IX) of specific rotation 39.5° ($CHCl_3$). According to Lane and Wallis, this rearrangement may take place with racemization provided that the terminal carbon of the migrating group contains a hydrogen atom. The compound must be of the general type $\left(\begin{smallmatrix}R \\ R'\end{smallmatrix}\!>\!\underset{H}{C}-\overset{O}{C}-CHN_2\right)$.

This prerequisite structure is present in VII and thus a mixture (not necessarily of equimolecular amounts) of 2-desoxy-d-glucoheptonic acid (or lactone) and 2-desoxy-d-mannoheptonic acid (or lactone) would be predicted. However, only one product was isolated (70% yield). This substance was designated as 2-desoxy-d-glucoheptono-lactone. Saponification of IX with barium hydroxide, followed by removal of barium ions with sulfuric acid and concentration, yielded crystalline 2-desoxy-d-glucoheptonolactone (X) of specific rotation +20° (HOH). The analytical data definitely indicated a lactone structure for IX and X. If, according to Hudson's lactone rule, the rotations of these compounds are due mainly to the position of the lactone ring, the ring closure in IX and X must be on the right (in Fischer projection formula). Thus IX and X must be delta lactones since carbon four in their projection formulas is on the left.

The delta lactones of normal sugar acids are rather unstable and are hydrolyzed rapidly. Thus, additional evidence for the 1,5-ring in IX lies in the fact that it was found to be rapidly titratable to a stable end point within one and one-half minutes from the time of solution in acetone-water. X was too slowly soluble to permit rapid titration but was titratable to a stable end-point within four and one-half to five minutes.

An abnormally high rate of hydrolysis has been found for the glycopyranosides of the 2-desoxy sugars in comparison with rate exhibited by the normal sugar pyranosides. Contrary to expectation, IX exhibited no mutarotation in either methanol or aqueous acetone. X was too slowly soluble in water for early polarimetric readings in that medium; however, no mutarotation was observed after an initial rotation of +20° at twenty minutes. Attempts were made to follow lactonization of the free acid by the general method of Levene and Simms. An initial reading at three minutes gave a specific rotation of +20°. There was no observed mutarotation over a period of eighty hours.

Explanation of the anomalous optical behaviour of IX and X may be due to an effective increase in the rate of hydrolysis due to the influence of the 2-desoxy carbon. Again, perhaps there is a very close similarity in rotation between the lactone and its acid, in the event of which Hudson's lactone rule would not hold. The expectation would be that in X, gamma lactone formation would be slow enough to be readily observable and an appreciable quantity of the gamma lactone in the equilibrium mixture should yield a more negative

value. As more information is obtained concerning the little known group of 2-desoxy sugar acids, it is not unlikely that their behavior will be found to differ markedly from that of the normal sugar acids.

Keto-d-fructose pentaacetate in absolute chloroform solution containing a trace of methanol, yielded the ethylene oxide derivative (XI, 75% yield) on treatment with diazomethane. Saponification with barium methylate followed by carbonation yielded XII. Neither of the compounds showed coloration on heating with a methanol solution of potassium hydroxide but reduced Tollens' reagent (pyridine soln. of XI).

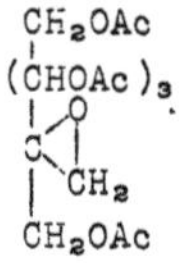

XI

M.P. = 86 - 87°

$[\alpha]^{24°}_{d} = + 32°$

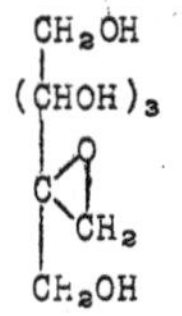

XII

M.P. = 136°

Staudinger has shown that ethyl diazoacetate can be acylated.

$$2\ EtOOC\text{-}CHN_2 + COCl_2 \rightarrow EtOOC\text{-}CN_2\text{-}C(=O)\text{-}Cl + EtOOC\text{-}CH_2\text{-}Cl + N_2$$

Similarly, Bradley and Robinson described the reaction of a diazomethyl ketone with acetic acid to produce an acetoxymethyl ketone.

It was deemed feasible to apply this reaction to the synthesis of ketoses in their open chain or keto-acetate structure. The reaction was first established for the known fructose derivative. The diazomethyl ketone obtained from d-arabonyl chloride tetraacetate and diazomethane was found to react smoothly with glacial acetic acid to produce the open chain or keto-form of d-fructose pentaacetate, a derivative that had been synthesized by Hudson and Brauns by direct acetylation procedure and whose open-chain structure is established. The reaction of VII, 1-diazo-1-desoxy-keto-d-glucoheptulose pentaacetate, with acetic acid produced keto-d-glucoheptulose hexacetate (XIII), isomeric with and convertible (by saponification and reacetylation) to the one known cyclic hexaacetate of d-glucoheptulose.

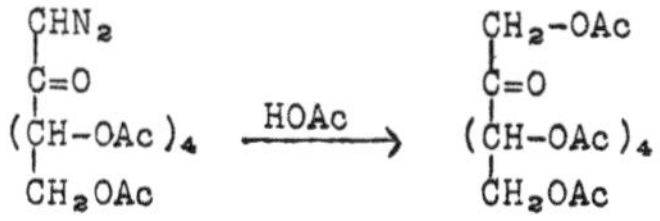
CHN₂
C=O
(CH-OAc)₄
CH₂OAc
HOAc
CH₂-OAc
C=O
(CH-OAc)₄
CH₂OAc

VII XIII

Utilizing the acetolysis procedure of Tambor and DuBois, 1-bromo-1-desoxy-d-glucoheptulose pentaacetate was converted into XIII in somewhat lower yield than from the diazomethyl ketone (VII) and acetic acid. This new acyclic derivative (XIII) exhibited no mutarotation in either chloroform or methanol solution. Its ultraviolet absorption spectrum revealed an absorption maximum at 2830Å which corresponds closely to the value reported for keto-d-fructose pentaacetate and keto-l-sorbose pentaacetate. Since it has been shown that the cyclic acetates do not exhibit an absorption maximum in the region of that for the ketonic carbonyl group, these data prove that no ring closure has occurred during the series of reactions leading to the formation of the new derivatives and confirm the assignment of the acyclic or keto structure to XIII.

This series of reactions provides a new general method for the synthesis of acyclic ketose acetates or for the transformation of an aldose to the next higher ketose. Wolfrom then applied these reactions to a dibasic sugar acid, mucyl dichloride tetraacetate (XIV). From XIV the bisdiazomethyl ketone XV was obtained. Treatment of XV with hydrogen chloride yielded the dichloro derivative and treatment with acetic acid produced the well-crystallized diketose acetate, XVI, designated 1,8-dihydroxymucyldimethane hexaacetate. This is an open chain acetate of a diketose, a new type of derivative in the sugar field.

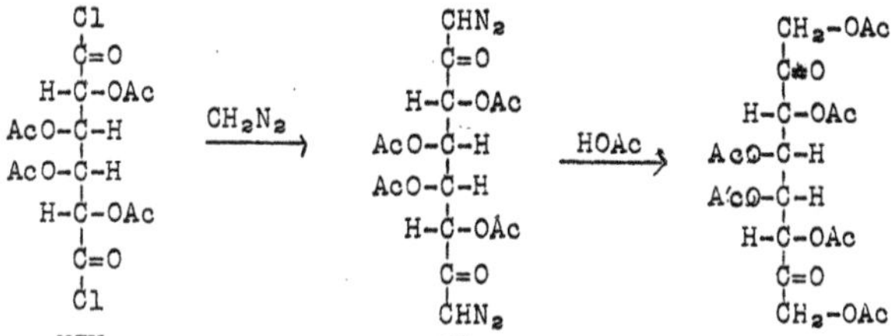
CH2N2
HOAc

XIV XV XVI

Bibliography

Meyer, Monatsch, 26, 1295 (1905).
Schlotterbeck, Ber., 40, 479 (1907), 42, 2559 (1909).
Arndt and Eistert, Ber., 68, 196 (1935).
Brigl, Muhlschlegel, and Schinle, Ber., 64, 2921 (1931).
Clibbens and Nierenstein, J. Chem. Soc., 107, 1491 (1915).
Arndt, Eistert, and Partale, Ber., 60, 1364 (1927).
Arndt and Amende, Ber., 61, 1122 (1928).
Gätzi and Reichstein, Helv. Chim. Acta, 21, 186 (1938).
Iwadare, Bull. Chem. Soc. Japan, 14, 131 (1939).
Wolfrom, Weisblat, Zophy, and Waisbrot, J. Am. Chem. Soc., 63, 201 (1941)
Wolfrom, et al, ibid., 63, 632 (1941); 64, 2329 (1942); 64, 1701 (1942)
Lane and Wallis, J. Org. Chem., 6, 443 (1941).
Levine and Simm, J. Biol. Chem., 65, 31 (1925).
Bradley and Robinson, J. Chem. Soc., 1310 (1928).
Staudinger, et al, Ber., 49, 1973, 1978 (1916).

Reported by Louis D. Scott
December 2, 1942

THE STRUCTURE OF BIOTIN

du Vigneaud, et al, Cornell University Medical College

Biotin is the yeast-growth factor originally isolated by Kögl from egg yolk. In 1940 du Vigneaud's group, working in collaboration with György and Rose of Western Reserve University, showed Kögl's substance to be identical with vitamin H. Vitamin H had been recognized as a factor capable of preventing the injury to experimental animals which can be brought about by feeding large amounts of raw egg-white. Biotin is found in yeast, liver, milk, and various other foods, and is now recognized as a member of the vitamin B complex. It is believed to play an important, though as yet unelucidated, role in animal and plant nutrition.

The synthesis of biotin has not yet been reported, but du Vigneaud and his group have presented evidence that biotin has the structure shown in formula I below. The data and considerations from which this structure was deduced are outlined briefly in this abstract.

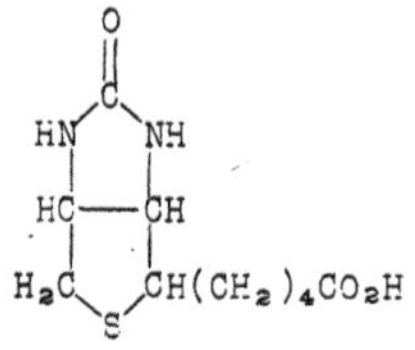

I

Biotin was isolated in the form of a methyl ester from liver, extracts and from milk concentates by a chromatographic procedure. The ester was finally obtained as a pure compound witn the empirical formula $C_{11}H_{18}O_3N_2S$. Free biotin was obtained from the ester by saponification in cold alkali, and was found to have the empirical formula $C_{10}H_{16}O_3N_2S$. It had a neutral equivalent of 244, the value expected for a monocarboxylic acid of this formula, and the proper titration curve for such an acid. The biological activity of these pure preparations was very high, and consistent activity values were found using both the yeast-growth method and the vitamin H assay method carried out with rats.

The first feature of the biotin structure to be elucidated was the cyclic urea portion of the molecule. It was found that treatment of biotin with strong barium hydroxide for twenty hours at 140° gave a diamino acid of formula $C_9H_{18}O_2N_2S$ in high yield. The loss of one carbon and one oxygen atom and the addition of two hydrogen atoms in the hydrolysis suggested a urea structure, and resynthesis of biotin in 98% yield by treatment of the diamino acid with phosgene confirmed this surmise.

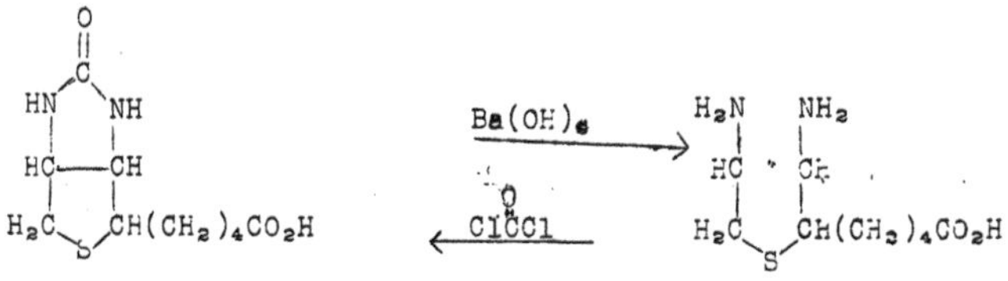
HN
NH
HC
CH
H₂C
CH(CH₂)₄CO₂H
S
Ba(OH)₂
ClCCl
H₂N
NH₂
HC
CH
H₂C
CH(CH₂)₄CO₂H
S

With the establishment of the presence of a cyclic urea structure in addition to a carboxyl group, all of the oxygen and nitrogen atoms in the molecule were accounted for, but the character of the sulfur atom had not been determined. It became apparent that the sulfur was present in a thio ether structure when it was found that treatment of biotin with hydrogen peroxide yielded a sulfone, $C_{10}H_{16}O_5N_2S$.

The ratio of hydrogen to carbon, together with the absence of unsaturation in the molecule, led to the conclusion that some sort of bicyclic structure was present. Oxidative degradation of biotin provided a further clue to the structure. Treatment with either permanganate or nitric acid yielded adipic acid.

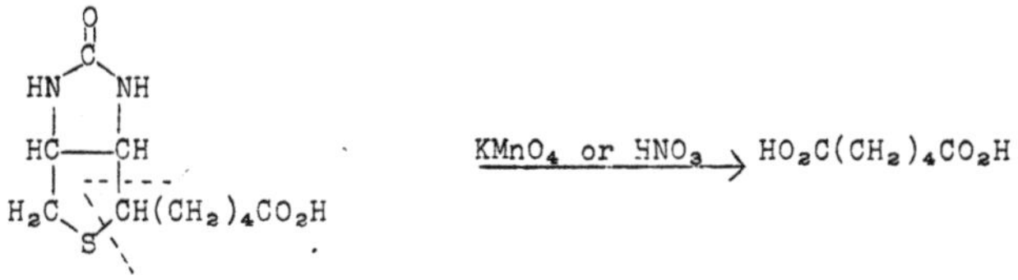
HN
NH
HC
CH
H_2C
$CH(CH_2)_4CO_2H$
S
$KMnO_4$ or HNO_3
$HO_2C(CH_2)_4CO_2H$

I

To make it possible to decide whether one of the carboxyl groups of the adipic acid was the original carboxyl group of biotin, this group was subjected to a Curtis degradation. (R represents the bicyclic portion of the molecule in the following equations.)

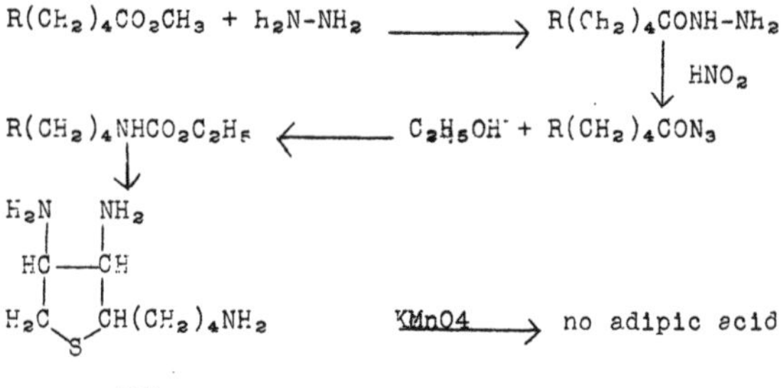
$R(CH_2)_4CO_2CH_3$ + H_2N-NH_2
$R(CH_2)_4CONH-NH_2$
HNO_2
$R(CH_2)_4NHCO_2C_2H_5$
C_2H_5OH + $R(CH_2)_4CON_3$
H_2N
NH_2
HC
CH
H_2C
$CH(CH_2)_4NH_2$
S
KMnO4
no adipic acid

III

Failure to obtain adipic acid by oxidation of the triamine (III) indicated that the adipic acid resulting from the oxidation of biotin contained the original carboxyl group, and hence resulted from oxidation of the side chain holding the carboxyl group, not entirely from oxidative cleavage of a cyclic structure. It was considered that the most reasonable interpretation of this result was that bio- contained a n-valeric acid side chain which was split off together with one carbon from the ring during oxidation. It was thought

possible, however, that the adipic acid might have resulted from the decarboxylation of a malonic or α -substituted β -keto acid initially formed. The possibility of a butyric acid side chain was therefore not excluded.

That the cyclic urea ring was five membered was established by the formation of a quinoxaline when the diaminocarboxylic acid was treated with phenanthrenequinone, a reagent which is known to form a ring structure only with 1,2-diamines.

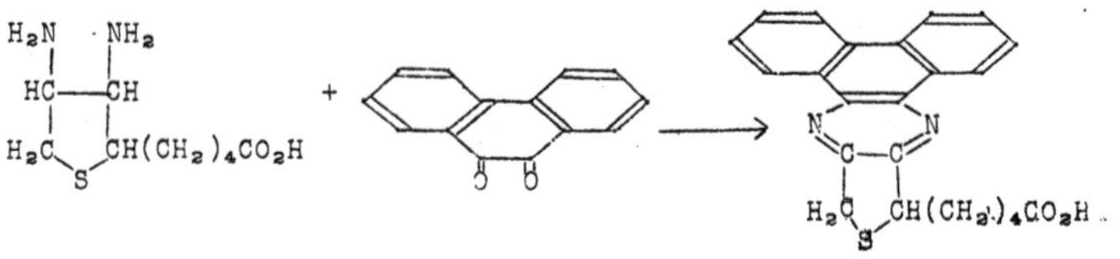

H₂N NH₂
HC—CH
H₂C CH(CH₂)₄CO₂H
S
+
N N
C—C
H₂C CH(CH₂)₄CO₂H
S

II

The product was shown to be a quinoxaline, not a dihydroquinoxaline, by comparison of its ultraviolet absorption spectrum with those of the dibenzoquinoxaline and dibenzodihydroquinoxaline derivatives of 3-4-diaminotetrahydrothiophene, the compounds shown in the formulas (V and VI) below.

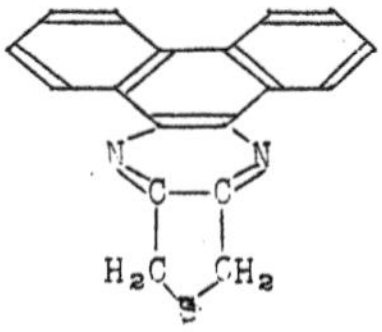

N N
C—C
H₂C CH₂

V

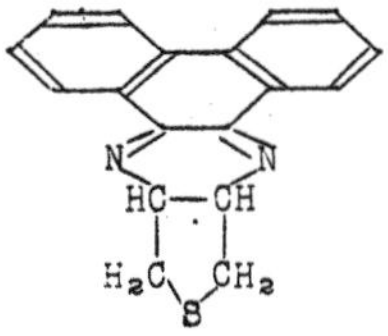

N N
HC—CH
H₂C CH₂
S

VI

The absorption spectrum of the product from the diamino acid (II) was almost identical with that observed for compound V and was quite unlike that for compound VI. Formation of the oxidized product indicated that both carbons carrying amino groups must also carry hydrogen atoms. From this conclusion it followed that if these carbons were assumed to be part of a ring structure, neither could be the point of attachment of the side chain previously shown to exist.

At this point in the work the accumulation of data was thought to permit only the two possible structures shown in formulas I and VII.

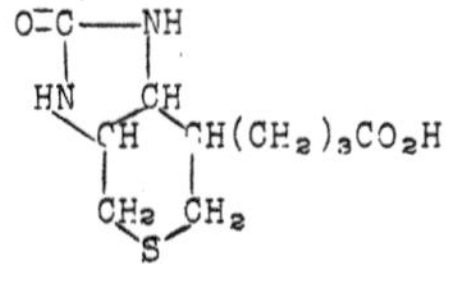

VII

The problem of deciding between these structures was attacked in two ways. One approach to the problem was afforded by the discovery by Mozingo of a method by which the sulfur atom of organic sulfides can be removed and replaced by two hydrogen atoms. Treatment of biotin methyl ester with Raney nickel according to Mozingo's procedure resulted in the formation of the "desthiobiotin"ester (VIII), which was hydrolyzed to give a diaminocarboxylic acid (IX). Oxidation of this acid with alkaline periodate yielded pimelic acid, as expected from formula I, not α-methyl adipic acid as expected from formula VII. Confirmatory evidence in favor of formula I was supplied by Folkers and Harris of the Merck Laboratories, who synthesized the diaminocarboxylic acid IX. Comparison of the racemic synthetic product with the partly racemized material obtained from natural biotin was accomplished by preparation of quinoxaline derivatives (X), which had no asymmetric carbon atoms, and proved to be identical.

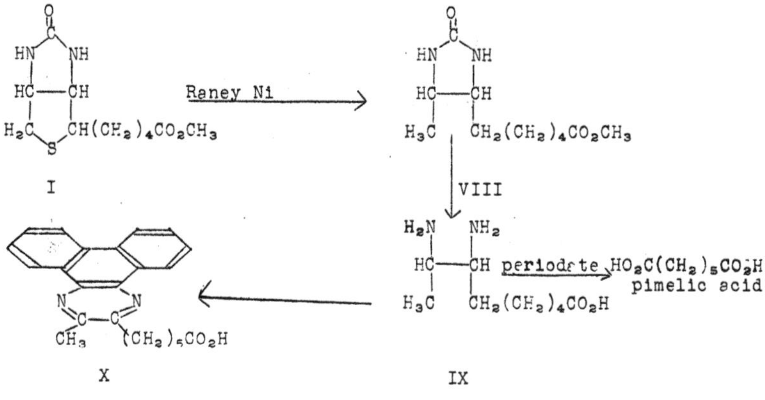

Raney Ni
I
VIII
periodate
$HO_2C(CH_2)_5CO_2H$
pimelic acid
X
IX

By a second approach it was possible to establish unequivocally the reduced thiophene structure of biotin. Conversion of the diaminocarboxylic acid II to δ-(α-thienyl)-valeric acid XI was accomplished by the process indicated below.

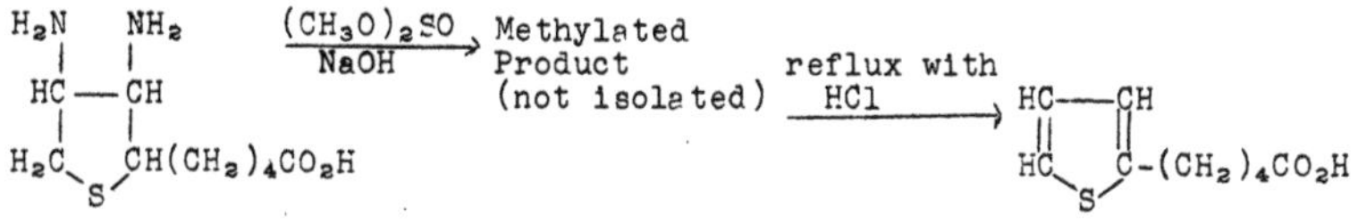

II XI

Compound XI was prepared according to the scheme shown below, a method analogous to that used by Fieser and Kennelly in the preparation of γ-(α-thienyl)-butyric acid.

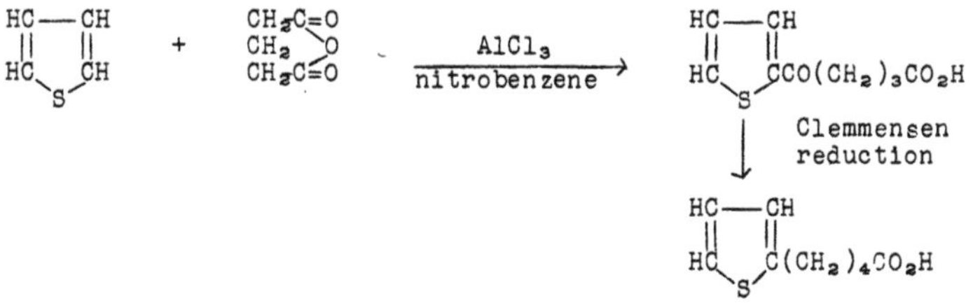

XI

Bibliography

du Vigneaud, Science, 96, 455, (1942). (Lists twenty-two of the more important references, including papers now in press.)
György, Melville, Burk and du Vigneaud, Science, 91, 243, (1940)
du Vigneaud, Melville, György and Rose, Science, 92, 62, (1940)
György, Rose, Hofmann, Melville and du Vigneaud, Science 92, 609, (1940)
du Vigneaud, Melville, Folkers, Wolf, Mozingo, Keresztesy, and Harris, J. Biol. Chem., (in press)
Kögl and Tönnis, Z Physiol. Chem., 242, 43, (1936)
Fieser and Kennelly, J. Am. Chem. Soc., 57, 1611, (1935)

Reported by P. L. Southwick
December 9, 1942

THE STRUCTURE OF AGAR

For many years agar-agar, the gelatinous extract of certain East Indian sea-weeds, has been well known, but the elucidation of its structure has been hindered by its physical properties and also by the fact that the commercial varieties are not homogeneous since structural changes take place to a certain extent during its purification. The principal features of the structure of this compound were determined in 1937 by E. G. V. Percival and J. C. Somerville who studied the hydrolysis products of methylated agar.

These investigators identified the chief product of the hydrolysis of methylated agar as 2,4,6-trimethyl-d-galactopyranose, and from this result they concluded that agar consisted for the most part of d-galactose units joined by 1,3-glycosidic linkages. The next step toward clarification of the structure was the observation made by N. W. Pirie and confirmed by Hands and Peat that a derivative of l-galactose was present among the hydrolysis products of methylated agar, thus indicating that l-galactose was a definite part of the polysaccharide.

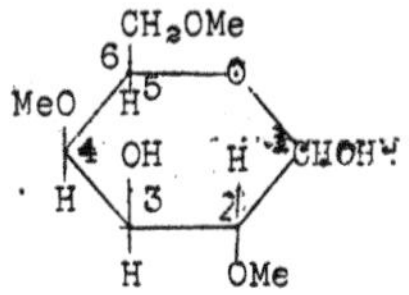
CH₂OMe
6
5
MeO
H
4
OH
H
H
3
2
H
OMe

The derivative of l-galactose which was found also contained a 3, 6-anhydro bridge which might have existed originally in the polysaccharide molecule or which might have been formed during its purification or methylation. The fact that commerical agar always has sulfuric acid residues present in organic combination was established by Neuberg and Ohle, and since it is well known that these are easily hydrolyzed by acid or alkali, an anhydro ring might be expected if an hydroxyl group is available for its formation. The conclusion that the hydrolysis of a sugar sulfuric ester yields an anhydro ring was drawn from the analogous reaction of the hydrolysis of a sugar toluenesulfonic ester, and Percival and Soutar have shown that the alkaline hydrolysis of α-methyl galactoside sulfate gives an anhydro-methylhexoside which Duff and Percival identified as 3,6-anhydro-α-methyl galactoside.

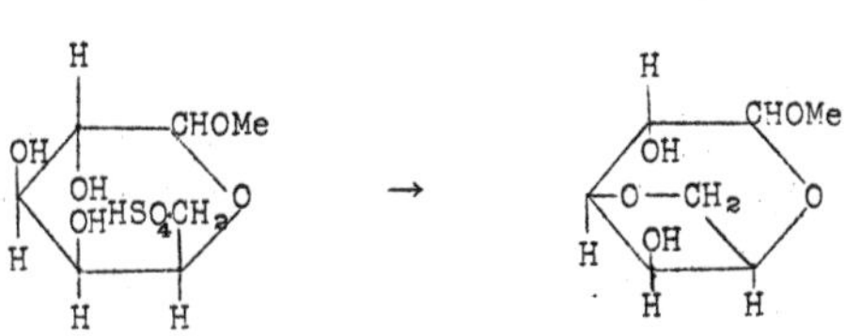
H
CHOMe
OH
OH
OHHSO₄CH₂
O
H
H
H
→
H
CHOMe
OH
O—CH₂
O
H
OH
H
H

From the various bits of evidence collected and from the results of analysis it was concluded that agar is composed entirely of galactose units arranged in chains. Each unit chain consists of approximately nine d-galactopyranose residues combined by 1, 3-glycosidic linkages, the chain being terminated at the reducing end by one residue of l-galactopyranose, the link in this case

involving C_4. Also, the primary hydroxyl group on carbon atom six of each l-galactose residue is esterified with sulfuric acid. Thus agar is a polymer with the following structure:

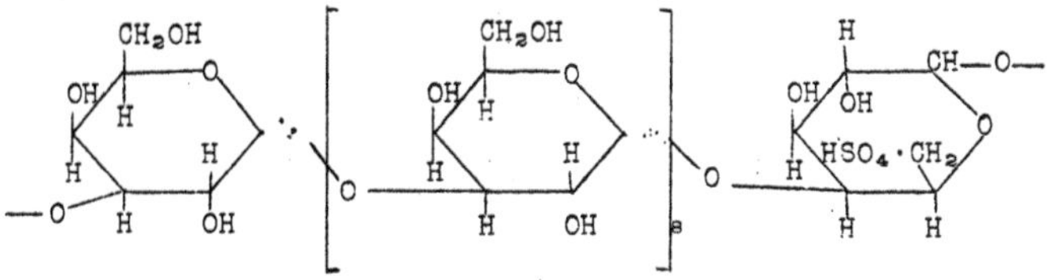

If this structure represents not the repeating unit but the whole molecule of the polysaccharide, then methylated agar would contain tetramethyl d-galactopyranose as the non-reducing end group, but this compound has never been detected among the hydrolysis products of the methylated compound.

The presence of the sulfuric ester in the l-galactose residue has already been mentioned. According to the above structure the theoretical percentage of sulfur in agar would be 1.8 per cent whereas values given in the literature vary from 0.5 to 1.5 per cent. However, this discrepancy can be easily explained by the assumption that some of the sulfate hydrolyzes during the commercial purification of the compound, and it has been found that chlorine is frequently employed for bleaching purposes. Thus some of the unit chains in commercial agar would be expected to have the following structure:

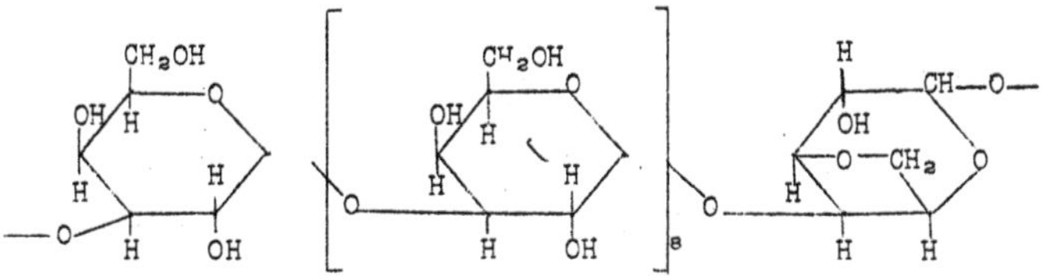

Since neither the acetylated nor the methylated agar was found to contain sulfur, it was assumed that during those processes the anhydro ring was formed. Thus when agar is methylated in alkaline solution, the product will be:

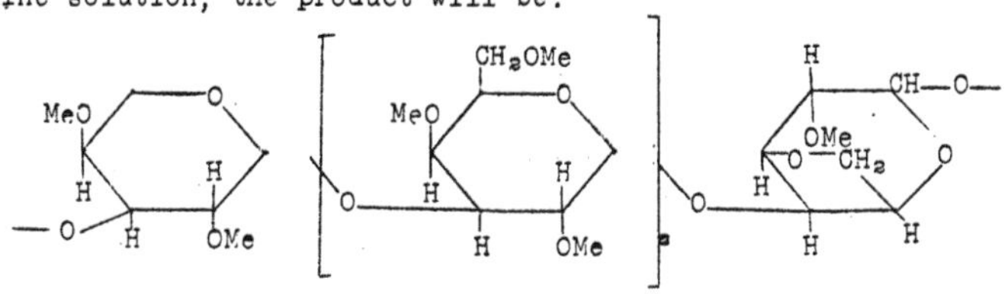

The methanolysis of this product will give 2,4,6-trimethyl-d-methyl galactoside (I) and 2-methyl-3,6-anhydro-1-methyl galactoside (II):

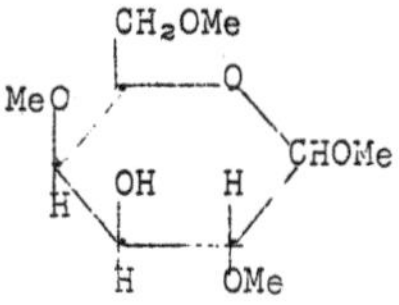

I

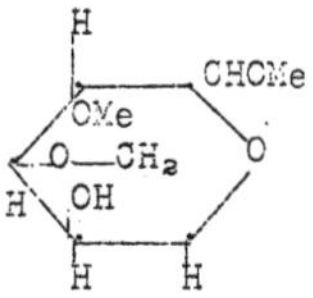

II

Although the momomethyl anhydro-1-methyl galactoside (II) was not isolated, further methylation of the higher boiling fraction yielded crystalline 2,4-dimethyl-3,6-anhydro-β-1-methyl galactoside, and the actual increase in methoxyl content corresponded with the theoretical conversion of the monomethyl to the dimethyl compound.

The linkage of the 1-galactose unit to the remainder of the chain at position C_4 was assigned after the following study. Positions 3 and 6 were excluded because of the anhydro bridge. Exhaustive methylation of commercial agar followed by fractional precipitation yielded two fractions, A and B, the first of which was dissolved in dilute sulfuric acid and subjected to dialysis for a long period of time. This process of purification gave a product which was strongly acid and which, when treated with diazomethane, gave an ester with an increased methoxyl content. The amount of this increase together with the determination of the saponification equivalent indicated the presence of one carboxyl group in every ten hexose units. Haworth, Jackson, and Smith showed that the presence of both a hydrofuranol (anhydro) and a pyranose ring in a hexose makes it very sensitive to acid, the effect of the acid being that of opening the pyranose ring with the subsequent formation of an aldehydo hexose. If this were the case, the aldehyde group might be easily oxidized to a carboxyl during the dialysis. The following reactions represent the changes so far.

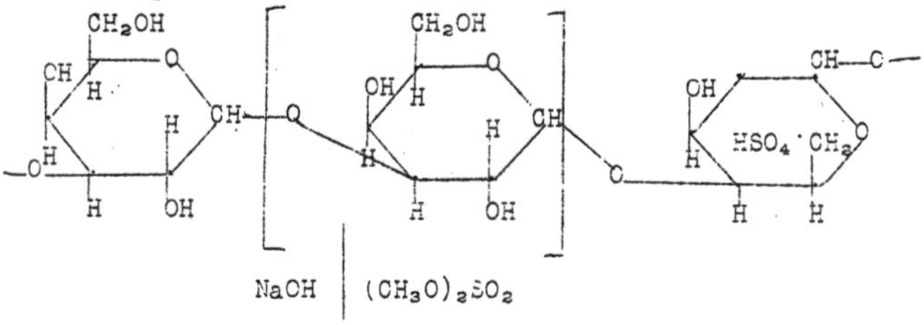
NaOH
$(CH_3O)_2SO_2$

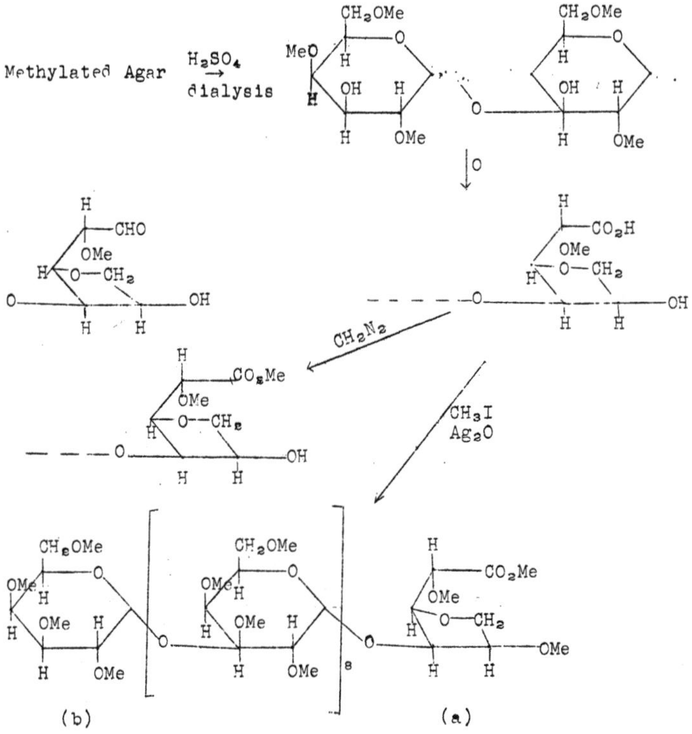

Methylated Agar
H_2SO_4
dialysis
CH_2N_2
CH_3I
Ag_2O
(b)
(a)

Treatment of the acid with methyl iodide and silver oxide would then not only esterify the carboxyl but also methylate the hydroxyls as shown (C_5 of the <u>l</u>-galactose unit and C_3 of the first <u>d</u>-galactose unit in the unit chain). Complete hydrolysis of this ester would then yield principally 2,4,6-trimethyl-<u>d</u>-galactose plus equimolar amounts of tetramethyl-<u>d</u>-galactose (from b) and 2,5-dimethyl-3,6-anhydro-<u>l</u>-galactonic acid (from a). These were the products which were actually obtained, the acid being isolated as the amide.

The positions of the methyl groups on this amide should locate the position of attachment of the l-galactose residue to the remainder of the chain. The following structures show the three possibilities:

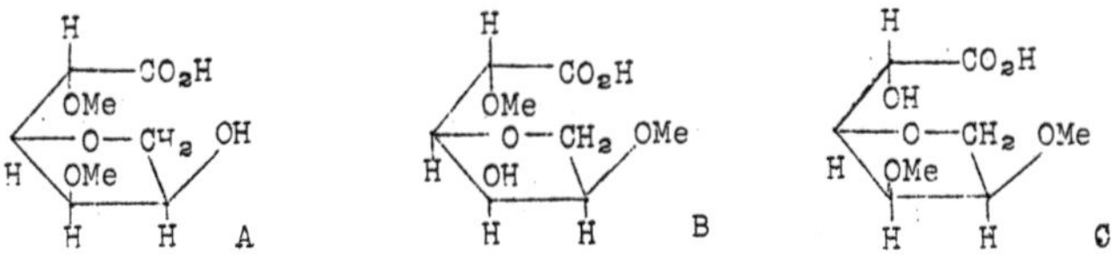
H
CO_2H
OMe
O—CH_2
OH
H
OMe
H H A
H
CO_2H
OMe
O—CH_2
OMe
H
OH
H H B
H
CO_2H
OH
O—CH_2
OMe
H
OMe
H H C

Structure C was eliminated by a negative Weerman test for α-hydroxy amides and structure A was eliminated by comparison of the melting points of the two amides in question. Thus the isolation of 2,5-dimethyl-3,6-anhydro-l-galactonic acid definitely established the linkage of the l-galactose residue as being a 1,4-glycosidic linkage.

Bibliography

Peat, S., Ann. Rep. Prog. Chem. 153 (1942).
Jones and Peat, J. Chem. Soc. 225 (1942).
Percival and Somerville, ibid., 1615 (1937).
Pirie, N. W., Biochem. J., 30, 369 (1936).
Hands and Peat, Chem. and Ind., 937 (1938).
Neuberg and Schweitzer, Monatsh., 71, 46 (1937).
Percival and Soutar, J. Chem. Soc., 1475 (1940).
Duff and Percival, ibid., 830 (1941)
Haworth, Jackson, and Smith, ibid., 625 (1940).

Reported by A. B. Spradling
December 9, 1942

SOME RECENT INVESTIGATIONS OF THE GRIGNARD REACTION

As the Grignard reaction has been used for ever more complicated molecules, the number of abnormal reactions has increased and with it, interest to classify and learn to predict these types of abnormal reactions. Two such reactions occur when an acyl chloride, preferably branched, reacts with a branched Grignard reagent. When a sterically hindered ketone reacts with a Grignard reagent, there are three "abnormal" reactions, two of which may be considered as reduction and enolization. In the former there are two important possibilities.

(expected reaction)

$$RCOCl + R'MgX \rightarrow RCOR' \xrightarrow{R'MgX} R{-}C(OH)(R'){-}R'$$

(abnormal reactions)

$$RCOCl + 2R'MgX \rightarrow RCH_2OH$$
$$\rightarrow RCHOHR'$$

A great deal of work has been done on such reactions by Whitmore and his coworkers. There have been several mechanisms postulated.

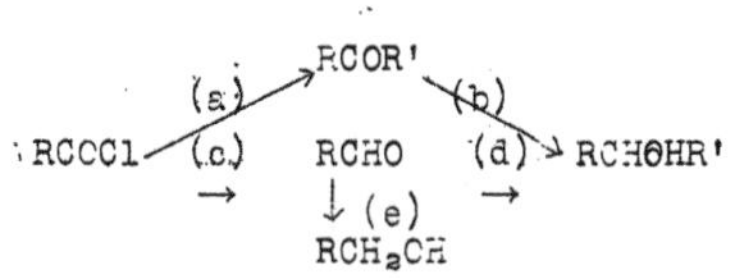

Reaction (a) has been observed as a very general one, for instance in the reaction of acetyl chloride and t-butylmagnesium chloride to form pinacolone.

Reaction (b) is entirely possible, and many such instances have been recorded in the literature. For example, methyl isopropyl ketone is reduced to the secondary alcohol by t-butylmagnesium chloride.

However, the secondary alcohol RCHOHR', does not seem to be produced through steps (a) and (b), since the yields from acid chlorides are higher than those from the ketones. The conversion of RCHO to RCHOHR' is a normal and expected reaction except for highly branched reagents.

Although reaction through the aldehyde had been surmised early, the intermediate aldehyde was not isolated until the reaction of methyl-t-butylneopentylacetyl chloride and t-amylmagnesium chloride was investigated. Methyl-t-butylneopentylacetaldehyde was the product of the reaction.

Some of the first work done by Whitmore specifically on the above reaction was done on the reaction of t-butylmagnesium chloride with n-butyryl chloride, isobutyryl chloride and trimethylacetyl chloride, the results of which are summarized in the following table.

Acid Chloride	% Addition	% Reduction
pivalyl (a)	32.4	8.00
pivalyl (b)	1.5	94.0
iso-butyryl	63.0	20.0
n-butyryl	71.0	9.0

(a) Run at -10°C. in excess acid chloride.
(b) Run at 40°C. in excess Grignard reagent.

In each case it has been found that there are two moles of the unsaturated hydrocarbon corresponding to R' for each mole of RCH_2OH and one for each mole of RCHOHR' formed. If there is any excess acyl chloride, the secondary alcohol formed is usually found as an ester of that acyl chloride.

When t-butylacetyl chloride was added to t-butylmagnesium chloride, there was a 71% yield of t-butylneopentyl carbinol and 1% of neopentyl carbinol. When a slight excess of the above Grignard reagent was added to t-butylacetyl chloride, there was found 51% t-butylneopentyl ketone and 17% t-butyl acetate.

n-Butylmagnesium chloride and trimethylacetyl (pivalyl) chloride produced 27% neopentyl alcohol and 69% n-butyl t-butyl carbinol. This yield compared with above figures, using t-butylmagnesium chloride, illustrates how branching of the Grignard agent seems to increase reduction to the primary alcohol. That this branching is not absolutely essential is illustrated by the reaction of n-butylmagnesium chloride with the following compounds to give as side products primary and secondary alcohols:

CH_3COCl → 8% CH_3CH_2OH

$CH_3CO_2C_2H_5$ → 3% $CH_3CHOHC_4H_9$

CH_3CHO → 18% CH_3CH_2CH

The action of n-propyl-, n-butyl- and n-amylmagnesium bromides on t-butylacetyl chloride produced the tertiary carbinols and the olefins resulting from partial dehydration, also the secondary alcohols in yields of 24%, 20.5% and 19.3% respectively. No neopentyl carbinol could be found, thus being analogous to the very low yield using the more powerful reducing agent, t-butylmagnesium chloride.

To compare with the action of t-butylmagnesium chloride in excess with pivalyl chloride, the following Grignard reagents have been used:

Grignard Reagent	$\%(CH_3)_3CCH_2OH$	$\%(CH_3)_3CCHOHR$
Ethyl	none	67
n-propyl	20	76
isopropyl	23	53
n-butyl	28	71
isobutyl	61	26
n-amyl	20	75
isoamyl	15	71

It has been reported that in the addition of ethyl-, n-propyl-, n-butyl- and n-amylmagnesium chlorides to an excess of t-butylacetyl chloride, the primary product was a ketone, in yields of 51%, 37%, 34% and 29% respectively. Reduction of these ketones by the Grignard reagent gave the secondary carbinols in yields of 7%, 20%, 23% and 21%. Thus, the larger the primary Gringnard, the less ketone formed and in general the more reduction takes place. No neopentyl carbinol was found in the above reaction, greatly in contrast to the approximate yields of 20% given by n-propyl-, n-butyl-, and n-amyl Grignard reagents with trimethylacetyl chloride. This contrast brings out the general rule that branching on the α carbon atom of the acyl chloride leads to greater reduction to primary alcohols.

It is a well known fact that the reaction of a Grignard reagent with a ketone can go one of four ways - by addition (the normal path), or by reduction, enolization or condensation. The last named reaction has been eliminated from the discussion by the nature of the ketones used.

normal addition $$R-\overset{O}{\overset{\|}{C}}-CH_2R' + R''MgX \rightarrow R-\underset{R''}{\overset{OH}{C}}-CH_2R'$$

reduction $$R-\overset{O}{\overset{\|}{C}}-CH_2R' + R''MgX \rightarrow R-CHOHCH_2R'$$

+ an unsaturated derivative of R"

enolization $$R-\overset{O}{\overset{\|}{C}}-CH_2R' + R''MgX \rightarrow R-\underset{OMgX}{C}=CHR' + R''H$$

It was formerly thought by many that this enolization was primarily a function of hindrance in the ketone. Due to more extensive study, this theory has been invalidated by Whitmore

and George. Thus:

Reaction of Diisopropyl Ketone with Grignard Reagents.

Grignard reagent	% Enolization	% Reduction	% Addition
MeMgBr	0	0	95
EtMgBr	2	21	77
n-PrMgBr	2	60	36
i-PrMgBr	29	65	0
i-BuMgBr	11	78	8
NpMgCl (Np - Neopentyl)	90	0	4

It can be seen that as the hindrance increased, the enolization and reduction increased. The amount of reduction is directly correlated to the primary, secondary or tertiary character of the hydrogen atoms attached to the β carbon atom of the Grignard reagent. Thus the ethyl reagent with three primary hydrogen atoms ranks lowest of those that react at all. The isopropyl ranks above *n*-propyl because of more hindrance and more primary hydrogen atoms than the propyl reagent. Since a β hydrogen is essential to reduction, the neopentylmagnesium chloride produces no reduction.

Considerable investigation has been done on mesityl ketones and those hindered by aromatic groups. Whitmore has worked on sterically hindered aliphatic ketones. These ketones sometimes are difficult to reduce and resist oximation, alkali cleavage and the haloform reaction. A highly hindered aliphatic ketone was prepared by the following reaction:

$$\underset{(R)}{(CH_3)_3CCH_2}-\overset{\displaystyle CH_3}{\underset{\displaystyle C(CH_3)_3}{C}}-COCl + CH_3MgBr \rightarrow RCOCH_3$$

It has been mentioned before that the above acyl chloride reduces to the aldehyde and primary alcohol with *t*-butylmagnesium chloride. The methyl ketone formed in the equation above was 94% enolated in a Grignard machine. Such enolates react as true Grignard reagents to give diketones with acyl chlorides, β-keto acids with carbon dioxide and β-ketols with aldehydes.

$$RCOCH_2MgBr + CH_3COCl \rightarrow MgClBr + RCOCH_2COCH_3$$

$$RCOCH_2MgBr + CO_2 \rightarrow RCOCH_2COOH$$

$$RCO\underset{\displaystyle CH_3}{CH}MgBr + C_6H_5CHO \rightarrow RCOCH(CH_3)CH(OH)C_6H_5$$

1) $Np_2CHCOCl + CH_3MgBr \rightarrow Np_2CHCOCH_3$
I → II

2) $II + CH_3MgBr \rightarrow Np_2CHCOCH\ MgBr + CH_4$
III

3) $III + I \rightarrow Np_2CHCOCH_2COCHNp_2$

Reaction 1) above was favored by excess Grignard reagent. The ketone II was easily reduced by hydrogenation but not by highly branched Grignard reagents, giving an enolate instead. Such reactions many times go to give reduction. We see that the dineopentyl carbinyl group exerts a steric influence similar to the mesityl group. The phenyl, o-tolyl, and p-tolyl dineopentyl carbinyl ketones add methyl Grignard reagent with no trace of enolization, to give tertiary alcohols. This must mean that the tertiary hydrogen α to the carbonyl in the above mentioned ketones is inactive, a unique situation.

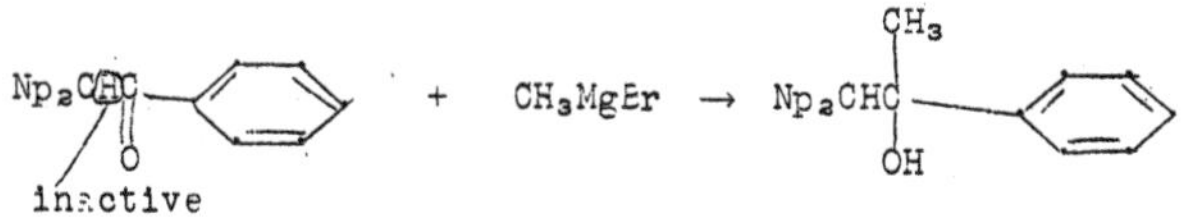

Np₂CHC
O
inactive
+ CH₃MgBr →
CH₃
Np₂CHC
OH

$Np_2CHC(=O)CH_3 + CH_3MgBr \rightarrow Np_2CHC(OMgBR) = CH + CH_4$

The only compound studied so far in which the tertiary hydrogen of the dineopentyl carbinyl group can be forced to react as an enol hydrogen is 2,2-bis-(dineopentylacetyl)-propane, $(Np_2CHCO)_2$-$C(CH_3)_2$. This compound will liberate 0.63 equivalent of methane, indicating enolization, and will add 1.36 equivalents of the methyl Grignard reagent.

$Et_3CCOCl + MeMgBr \rightarrow Et_3CCOCH_3 + MgBrCl.$

This ketone thus formed gives 94% enolate and no addition, thus being the lowest molecular weight ketone which will only form the enolate with the methyl Grignard reagent. It is almost as sterically hindered as acetomesitylene, a very interesting fact, remembering that the corresponding methyl compound, pinacolone exhibits only slightly sterically hindered carbonyl reactions. This shows a great difference between the steric effect of the methyl and ethyl groups.

Compound	% Enolization	% Addition
Methyl isopropyl ketone	0	100
Ethyl isopropyl·ketone	0	100
Methyl t-butyl ketone	5	86
Ethyl t-butyl ketone	9	86
Pentamethyl acetone	0	49
Methyl pincolyl ketone	48	47
Ethyl pincolyl ketone	62	33
Methyl s-butyl ketone	32	
Propyl s-butyl ketone	53	40
2,2, Dimethyl-4-ethyl-3-hexanone	5	19
2,2,4,6,6-Pentamethyl,3,5-heptadione	27/2	129/2

From these measurements, it can be seen that the amount of addition and enolization is controlled by the steric influence of the groups about the carbonyl. The fact that pentamethyl acetone gave only 49% addition, but no enolization recalls the failure of phenyl dineopentyl carbinyl ketone to give an enolate. The effect of the ethyl group is shown to be much greater than that of the methyl group. It is also apparent that in several cases steric hindrance in carbonyl compounds may retard or prevent either or both enolization or addition.

Bibliography

Whitmore, Rec.Trav.Chim., 57, 563(1938)

Greenwood, Whitmore, and Crooks, J. Am.Chem.Soc., 60,2028,(1938)

Whitmore and Heyd, J.Am.Chem.Soc., 60,2030(1938)

Whitmore,Popkin,Whitaker,Mattil and Zech,J.Am.Chem.Soc.,60,2458, 2462(1938)

Whitmore, Meyer,Pedlow, and Popkin,J. Am.Chem.Soc.,60,2788(1938)

Whitmore, Whitaker, Mattil and Popkin,J.Am.Chem.Soc.,60,2790(1938)

Whitmore, and Wheeler, J.Am.Chem.Soc.,60,2899(1938)

Whitmore, Whitaker, Mosher, Brevik, Wheeler, Miner, Sutherland, Wagner, Clapper, Lewis, Lux and Popkin, J.Am.Chem.Soc., 63,643(1941)

Whitmore and Lester, J.Am.Chem.Soc.,64,1247,1251(1942)

Whitmore and George, J.Am.Chem.Soc.,64,1239,(1942)

Whitmore and Randall, J.Am.Chem.Soc.,64,1242(1942)

Whitmore and Lewis, J.Am.Chem.Soc.,64,1618(1942)

Whitmore and Block, J.Am.Chem.Soc.,64,1619(1942)

Reported by H. J. Sampson, Jr.
December 9, 1942

ACETYLENIC ETHERS

Acetylenic compounds in which the triple-bond carbon is attached directly to oxygen are of interest because they are derivatives of an "yne-ol" system related to aldoketenes.

$$-C\equiv C-OH \rightarrow -\overset{H}{C}=C=O$$

Acetylenic ethers might be expected to show considerable reactivity since such ketone derivatives as ketene acetals are unusually reactive and enol ethers possess an active double bond and are easily hydrolyzed. One might also expect these compounds to exhibit some of the peculiarities of halogenated acetylenes since in both classes the carbon-carbon triple bond is attached directly to an atom containing unshared pairs of electrons.

Although such compounds have been mentioned several times in the early literature and have been postulated as intermediates in organic reactions as recently as 1937, few compounds of this type have been isolated. In 1903 Slimmer prepared phenoxyacetylene, an unstable oil which became a black, viscous mass in a few hours. He confirmed its structure by analysis and by analyses of the silver and copper derivatives. Slimmer also reported a stable compound which he thought to be benzoylphenoxyacetylene, but this has recently been shown to be an altogether different compound. A systematic investigation of acetylenic ethers was not begun, however, until 1940 when Jacobs and his co-workers became interested in these compounds.

Preparation

Phenoxyacetylene.--Jacobs and his co-workers obtained phenoxyacetylene by the method of Slimmer:

$$C_6H_5OK + BrCH{=}CHBr \xrightarrow{\text{abs. } CH_3OH} \underset{(35\text{-}45\%\ \text{yield})}{C_6H_5OCH{=}CHBr} \xrightarrow{KOH} \underset{(60\text{-}80\%\ \text{yield})}{C_6H_5OC\equiv CH}$$

The pure product, a white solid of m.p. -37 to -36°C. and b.p. 43-44°C./10 mm., turns pink on melting and darkens rapidly at room temperature. It can be stored without decomposition at low temperatures. The decomposition at room temperature is not accompanied by absorption of oxygen and is apparently not accelerated by light; polymerization seems to be the reaction taking place. Heating above 100°C. in a sealed tube causes it to explode violently leaving a charcoal-like residue. The structure of phenoxyacetylene was confirmed by catalytic hydrogenation to phenetole, the preparation of a diiodide and dibromide, and hydrolysis with concentrated sulfuric acid to phenol, phenolsulfonic acids, and acetic acid. Substituted phenoxyacetylenes are listed under reactions of phenoxyacetylene.

Ethoxy- and Butoxy-acetylene.--Two alkoxyacetylenes have been synthesized by Jacobs and his co-workers according to the following series of reactions:

$$CH_2BrCH(OR)_2 \rightarrow CHBr_2CH(OR)_2 \xrightarrow[C_2H_5OH]{Zn} BrCH{=}CHOR \xrightarrow{KOH} HC{\equiv}COR$$

If R equals C_2H_5-, 50-55% yield
If R equals C_4H_9-, 34-56% yield

The elimination of hydrogen bromide from the bromoalkoxyethylene corresponds to the final step in the synthesis of phenoxyacetylene. Ethoxyacetylene (b.p. 27.5-28.5°C./300 mm.) and butoxyacetylene (b.p. 50.5°C./110 mm.) are colorless evil-smelling liquids. Their structures were proved by analysis, molecular weight determinations and hydrogenation to the corresponding saturated ethers. Observed molecular refractions compared favorably with calculated values. When heated in a sealed tube to around 100°C. these compounds also decomposed explosively with the formation of a black solid.

Reactions of Phenoxyacetylene

Metallic Derivatives.--Phenoxyacetylene forms explosive silver and copper derivatives. They are sufficiently stable, however, to permit analysis. Jacobs and his co-workers have shown that sodium phenoxyacetylide reacts with benzoyl chloride to give phenyl benzoate instead of benzoylphenoxyacetylene as reported by Slimmer.

Phenoxyethynylmagnesium Bromide.--A further attempt to obtain benzoylphenoxyacetylene was made by treating phenoxyethynylmagnesium bromide with benzoyl chloride and bromide, but the principal products were only tar and phenyl benzoate. The reaction with acetyl chloride or acetic anhydride also resulted in much tar, but with the latter a small amount of phenyl acetate was obtained.

Phenoxyethynylmagnesium bromide has been found to react in two ways: (1) the metal is replaced as it is in most Grignard reactions; and (2) the carbon atoms of the acetylenic linkage are lost, the final product being phenol (or phenyl benzoate). In Table 1 it can be seen that the reaction took both courses with most reactants.

Table 1

The Reactions of Phenoxyethynylmagnesium Bromide

Reagent	Product	Yield %	Yield of phenol %
Ethyl p-toluene sulfonate	C_6H_5-O-C≡C-C_2H_5	15	74
Butyl p-toluene sulfonate	C_6H_5-O-C≡C-C_4H_9	52	38
Acetone	C_6H_5-O-C≡C-C(OH)$(CH_3)_2$	63	15
Allyl bromide	Recovered phenoxyacetylene	61	11

Benzoyl chloride	$C_6H_5CO_2C_6H_5$	38	10
Acetaldehyde	$C_6H_5-O-C\equiv C-CH(OH)CH_3$	28	9
Water	$C_6H_5OC\equiv CH$	80	3
Carbon dioxide		0	2
Benzoyl bromide	$C_6H_5CO_2C_6H_5$	26	2

The formation of phenyl benzoate in the reaction of metallic derivatives of phenoxyacetylene with benzoyl halides and the formation of phenol in other reactions of these derivatives are believed to be closely related. Although the system $C_6H_5OC\equiv C$-Metal is unique, it does have certain analogies with the metallic derivatives of enols and with Grignard reagents of allylic halides. Reactions of these compounds have been explained on the basis of mechanisms involving ionization and chelation or, if non-ionic, the existence of two metallic derivatives differing in the position of the metal or a single metallic derivative capable of reacting in two ways. It is not possible to write for $C_6H_5\overset{(2)}{O}-C\equiv\overset{(1)}{C}-MgBr$ two structures differing in the position of attachment of the metal, nor is it possible to write for the ion $[C_6H_5\overset{(2)}{O}-C\equiv\overset{(1)}{C}-]^-$ a second form in which the charge resides on the oxygen (2), without cleavage. The system is quite unstable, however, and with many reactants there is a cleavage resulting in separation of the phenoxyl and $-C\equiv C-$. If the metal-carbon bond is covalent, the electron pair must certainly be unequally distributed and the carbon must be relatively negative as in the ionic form. This would leave the electron density high at (1), but due to the polarizability of the triple bond a distribution of the effect over the system would take place and could, under the influence of the various reactants, result in a high electron density at (2) followed by loss of phenoxide ion or a reaction in which the phenoxyl becomes incorporated in another molecule. According to this idea, reactions occur when the electron density becomes critically high and the different results with different reactants can be explained by this polarizing influence. Furthermore, it seems certain that the phenoxyl remains attached to the $-C\equiv C-$ until phenoxyethynylmagnesium bromide enters into reaction, for hydrolysis with water gave a high yield of phenoxyacetylene and very little phenol. It was also found that one equivalent of methane was produced from phenoxyacetylene by the Zerewitinoff procedure; no additional methylmagnesium iodide was used in any cleavage reaction and the gas produced contained no unsaturated hydrocarbons.

Alkylphenoxyacetylenes.--These compounds are stable liquids showing the usual reactions of the triple bond. Butylphenoxyacetylene gave a liquid dibromide and was readily hydrogenated to hexyl phenyl ether. Hydration of this acetylene with dilute hydrochloric acid in the presence of mercuric acetate yielded phenyl caproate. No evidence was found for the formation of 1-phenoxyhexanone-2 which would have resulted if water had added in the

reverse manner. This differs from the behavior of halogenated acetylenes such as 1-bromoheptyne-1 which yields 1-bromoheptanone-2.

Polymerization.--Phenoxyacetylene polymerizes rapidly to give a black solid as the final product. It apparently proceeds in two steps, the first leading to a red substance believed to contain a conjugated linear polyene structure. This polyene may then change over to a cross-linking system, possibly by a Diels-Alder reaction. Polymers obtained at high temperatures may be separated into polyene and cross-linked polymer by chromatographic adsorption.

These polymers lose 50% of the phenoxyl groups as phenol on heating. This occurs after cross-linking and is believed to be due to the aromatization of the cyclohexene rings formed during the process. The remaining phenoxyl groups are substituents on the aromatic system. It is interesting that the rate of polymerization is not influenced by benzoyl peroxide, phenyl-α-naphthylamine, p-toluenesulfonic acid, or other substances. Intermediate polymers are thermosetting plastics and heat treatment or long continued polymerization produces infusible, insoluble, brittle substances which are undoubtedly highly cross-linked.

Halogen Derivatives.--Although acetylenic ethers might be expected to exhibit many of the peculiarities of halogenated acetylenes, few close similarities have been observed. One of these is the great tendency toward polymerization shown in both series by those members having an acetylenic hydrogen (H-C≡COR and H-C≡CX); compounds in which this hydrogen is replaced by an alkyl group are relatively stable. Halogenated phenoxyacetylenes have been investigated in order to compare them with dihalogenated acetylenes, XC≡CX, which are likewise characterized by ready polymerization.

All attempts to prepare iodophenoxyacetylene resulted in the formation of triiodophenoxyethylene and liquids which rapidly polymerized. Bromophenoxyacetylene, $C_6H_5OC{\equiv}CBr$, was obtained by the action of potassium hypobromite in alkaline solution on phenoxyacetylene. It polymerized very readily and therefore could not be purified successfully. It could be distilled only at very low pressures. Addition of bromine gave tribromophenoxyethylene and hydration with dilute hydrochloric acid in the presence of mercuric acetate yielded phenyl bromoacetate.

Reactions of Alkoxyacetylenes

Metallic Derivatives.--Although phenoxyacetylene gave a white silver salt, the alkoxyacetylenes gave only black precipitates probably containing metallic silver. These showed less tendency to explode. Treatment of the freshly precipitated derivative of ethoxy- or butoxy-acetylene with dilute nitric or sulfuric acid yielded a small amount of ethyl or butyl acetate. Ethoxyacetylene formed a white mercury derivative which rapidly turned brown.

No attempt has yet been made to synthesize such substituted acetylenic ethers as R–C≡C–O–Alkyl by the use of sodium or bromomagnesium derivatives of alkoxyacetylenes. Ethoxyacetylene, however, gave one mole of methane by the Zerewitinoff procedure and no additional reagent was used in any other manner.

Polymerization.--The alkoxyacetylenes polymerize much less rapidly than phenoxyacetylene, but they remain completely colorless only when sealed in glass and kept at -80° C. Carefully purified material developed a yellow color in a few minutes when exposed at room temperature.

Hydrolysis.--The dilute acid hydrolysis of ethoxyacetylene to ethylacetate was too rapid to measure. Butoxyacetylene was hydrolyzed rapidly by dilute acids and was completely converted to butylacetate and polymer by refluxing with distilled water.

Addition of Alcohol.--Ethoxyacetylene did not react with ethanol under reflux conditions. At 0° C., however, in the presence of boron trifluoride-mercuric oxide catalyst the products were ethyl acetate and diethyl ether. Ethyl orthoacetate yielded the same products under similar conditions.

Uses

The acetylenic ethers have been tested as anesthetics. Ethoxyacetylene, interesting because of its similarity to diethyl and ethyl vinyl ethers, was found to have some anesthetic action, but was quite poisonous even at low concentrations. Phenoxyacetylene had no anesthetic action but was quite poisonous.

Bibliography

Sabanejeff and Dworkowitsch, Ann., 216, 279 (1883).
Nef, ibid., 298, 337 (1897).
Lawrie, Am. Chem. J., 36, 487 (1906).
Grignard and Perrichon, Ann. chim., [10], 5, 5 (1926).
Reihlen, Friedolsheim, and Oswald, Ann., 465, 84 (1928).
Rhinesmith, The action of the Grignard reagent on esters of propiolic acid. A paper presented before the Organic Section of the American Chemical Society at the Rochester meeting, 1937.
Slimmer, Ber., 36, 289 (1903).
Jacobs, Cramer, and Weiss, J. Am. Chem. Soc., 62, 1849 (1940).
Jacobs, Cramer, and Hanson, ibid., 64, 223 (1942).
Jacobs and Whitcher, ibid., 64, 2635 (1942).
Tuttle and Jacobs, The Polymerization of phenoxyacetylene. A paper presented before the Organic Section of the American Chemical Society at the Buffalo meeting, 1942.

Reported by F. W. Spangler
December 9, 1942

REACTIONS OF THE TERPHENYLS

Whereas the three isomeric terphenyls have been recognized for many years, only recently has any systematic work been undertaken relative to determining the position taken by entering groups during halogenation, nitration and Friedel-Crafts reactions. This paper deals mainly with such substitution reactions and the proof of structure of the products. A brief discussion of general methods of preparation and the systems of numbering positions on the rings is also given.

Preparation.—A supply of the isomeric terphenyls is now available as a by product in the commercial preparation of biphenyl. The best laboratory method consists of treatment of the N-nitroso-acetylbiphenyl amine with benzene.

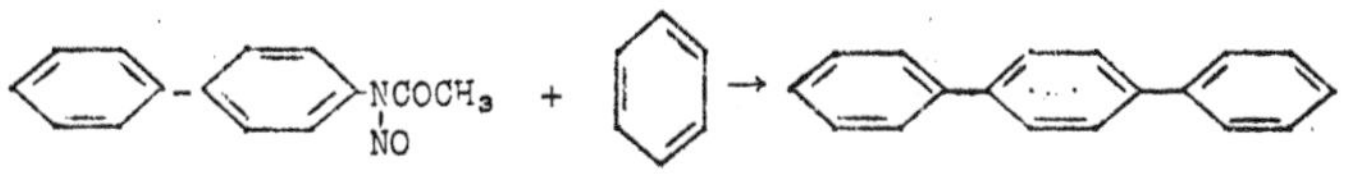
NCOCH₃
NO
+
→

Treatment of biphenyl with cyclohexene in the presence of aluminum chloride yields 4-cyclohexylbiphenyl which may be dehydrogenated with selenium to give p-terphenyl.

An interesting case of rearrangement is observed when 1,2-dibromocyclohexane is caused to react with benzene in the presence of aluminum chloride. The product consists mainly of 1,4-diphenylcyclohexane with a small amount of the 1,3-isomer but none of the 1,2-isomer. These isomers can be dehydrogenated with selenium to the corresponding terphenyl.

Nomenclature.—The hydrocarbons will be designated as ortho, meta or para depending upon the position of the phenyl groups with respect to the central ring and the numbering of the positions in the ring will be as follows:

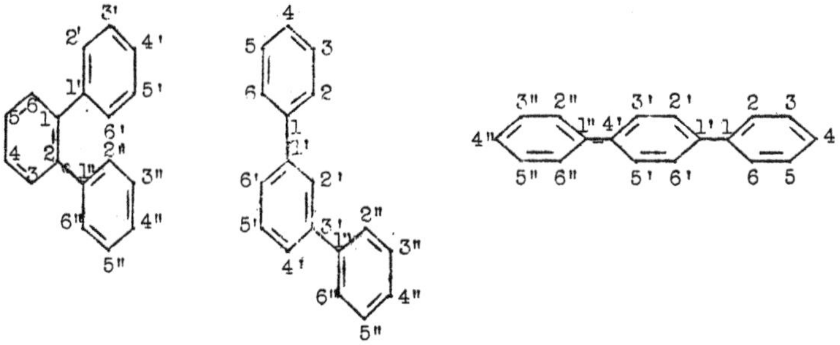

<u>ortho-Terphenyl</u>.--Allen and Pingert have observed an interesting rearrangement involving a phenyl group. If o-terphenyl is refluxed in benzene solution in the presence of aluminum chloride, the resulting product is found to be p-terphenyl. ~~Pure m-terphenyl will yield p-terphenyl.~~ Pure m-terphenyl will yield p-terphenyl under the same conditions.

In view of the above transformations, it would be expected that difficulty would be encountered in preparation of o-terphenyl by means of the Friedel-Crafts reaction. However, a number of such derivatives have been prepared. The yields, however, are usually quite low. The benzoyl and the acetyl derivatives were prepared and substitution was found to take place in th 4' position. Proof of structure of the ketones was established by synthesis using the Ullman procedure. For example 4'-acetyl-o-terphenyl was prepared in the following fashion:

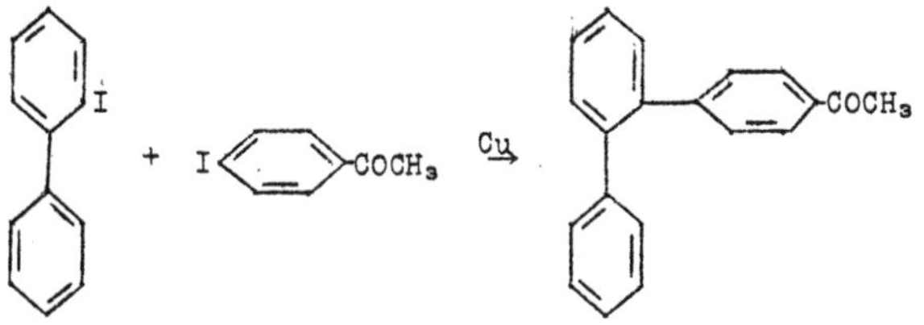
I
+
I
COCH3
Cu
COCH3

Bromination of o-terphenyl using a variety of conditions always resulted in the dibromo product, 4', 4"-dibromo-o-terphenyl, since oxidation yielded only p-bromobenzoic acid. Further bromination led to 4',4",4,5-tetrabromo-o-terphenyl. Forced treatment with bromine produced 3,6,10,11-tetrabromotriphenylene.

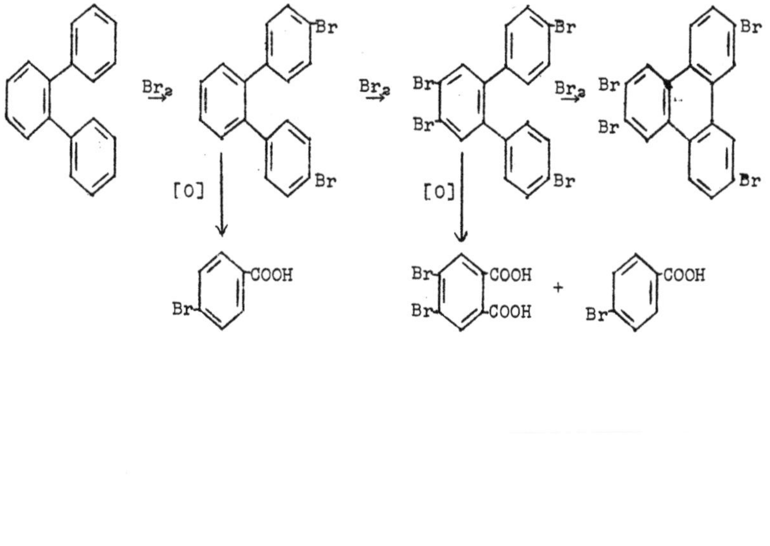
Br2
Br
Br
[O]
COOH
Br
Br2
Br
Br
Br
Br
[O]
Br
COOH
Br
COOH
+
COOH
Br
Br2
Br
Br
Br
Br

Mononitration yields 4'-nitro-o-terphenyl and dinitration yields a mixture of 4',4" and 2',4'-dinitro-o-terphenyl. These structures were also proven by oxidation to the corresponidng acids.

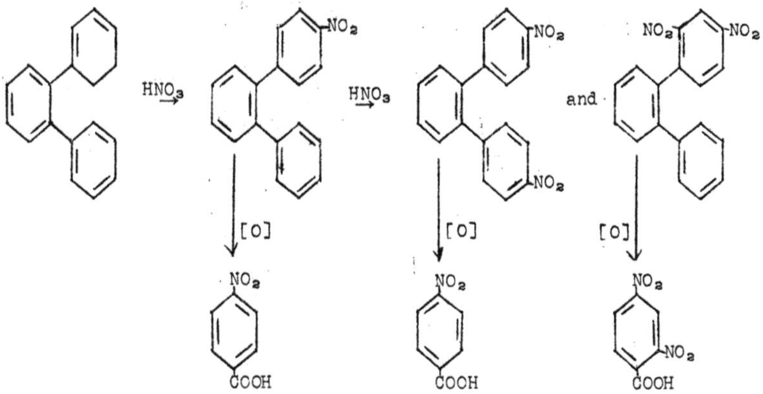
NO₂
HNO₃
HNO₃
and
[O]
[O]
[O]
COOH
COOH
COOH

Reduction of the 4',4"-dinitro-o-terphenyl converted it to the corresponding diamine. When the diamine was diazotized and treated with alcohol, the o-terphenyl was obtained indicating that no rearrangements had occurred during the process. Treatment of the diazotized diamine with cuprous bromide produced a dibromo compound identical with that obtained by direct bromination.

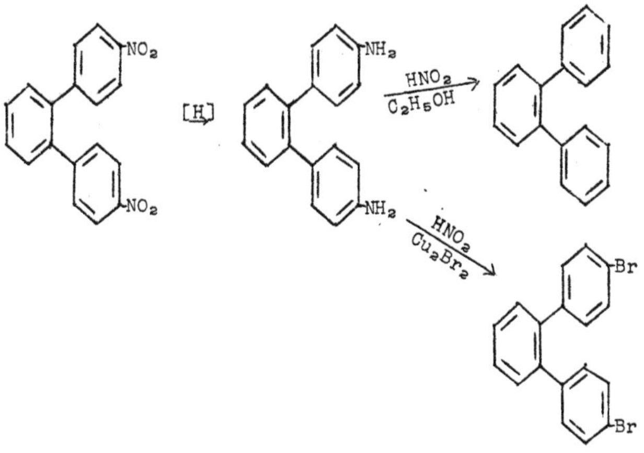
NO₂
NO₂
[H]
NH₂
NH₂
HNO₂
C₂H₅OH
HNO₂
Cu₂Br₂
Br
Br

meta-Terphenyl.--From the nitration of m-terphenyl Wardner and Lowy ~~obtained an oily~~ mononitration product. Oxidation gave a nitrobiphenylcarboxylic acid which was unreported. Later, France, Heilbron and Hey obtained the same acid in the same manner and also synthesized it thus establishing the structure of the mononitro product.

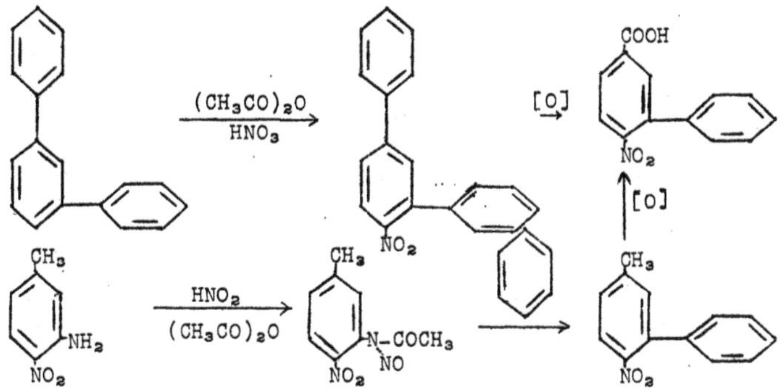

$(CH_3CO)_2O$
HNO_3
[O]
COOH
NO_2
[O]
CH_3
NH_2
NO_2
HNO_2
$(CH_3CO)_2O$
CH_3
$N-COCH_3$
NO
NO_2
CH_3
NO_2

The fact that the nitro group enters the central ring is in keeping with the results of halogenation. Chlorination and bromination always gave the 4'-monohalide.

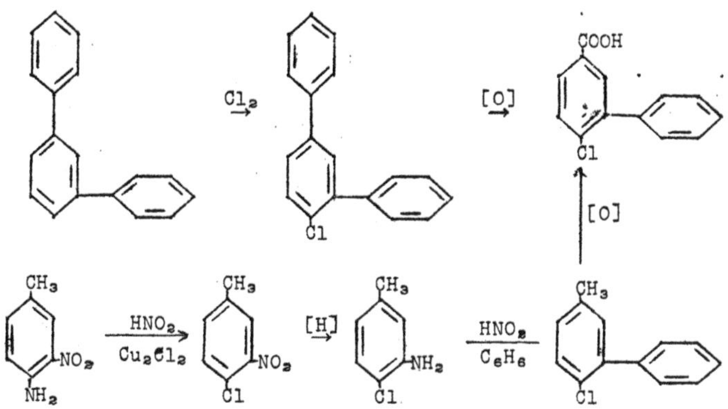

Cl_2
[O]
COOH
Cl
Cl
[O]
CH_3
NO_2
NH_2
HNO_2
Cu_2Cl_2
CH_3
NO_2
Cl
[H]
CH_3
NH_2
Cl
HNO_2
C_6H_6
CH_3
Cl

In contrast to nitration and halogenation, the entering group in a Friedel-Crafts reaction assumes the para position in one of the end rings. Friedel-Crafts reactions were run using acetic anhydride, acetyl chloride, chloroacetyl chloride and benzoyl chloride. In

each case the group entered the 4 position. Proof was furnished by the oxidation products which were isolated.

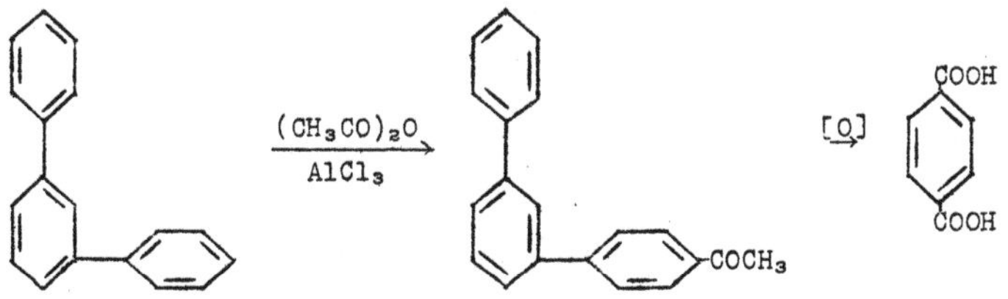
(CH3CO)2O
AlCl3
COCH3
[O]
COOH
COOH

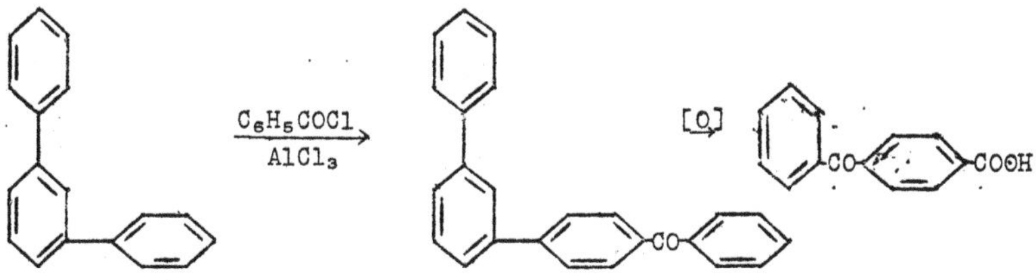
C6H5COCl
AlCl3
CO
[O]
CO
COOH

para-Terphenyl

More derivatives have been prepared of p-terphenyl than of the other two isomers, but most of them have been prepared in an indirect manner, usually by way of the N-nitrosoacetylamine. Bromination of p-terphenyl yielded 4-bromo-p-terphenyl and further bromination led to the 4,4"-derivative. Proof of structure of the monobromo derivative was accomplished as follows:

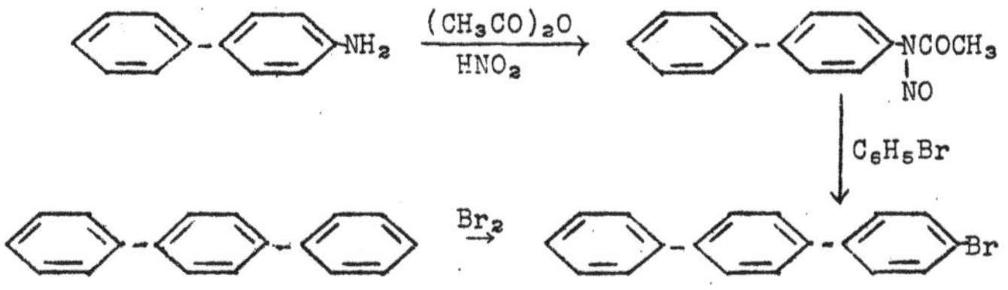
NH2
(CH3CO)2O
HNO2
NCOCH3
NO
C6H5Br
Br2
Br

A mononitro product could not be prepared by direct means, since dinitration to 4,4"-dinitro-p-terphenyl was always observed. An identical dinitro compound was prepared from the reaction of dinitrosodiacetyl-p-phenylenediamine with nitrobenzene, thus proving the structure.

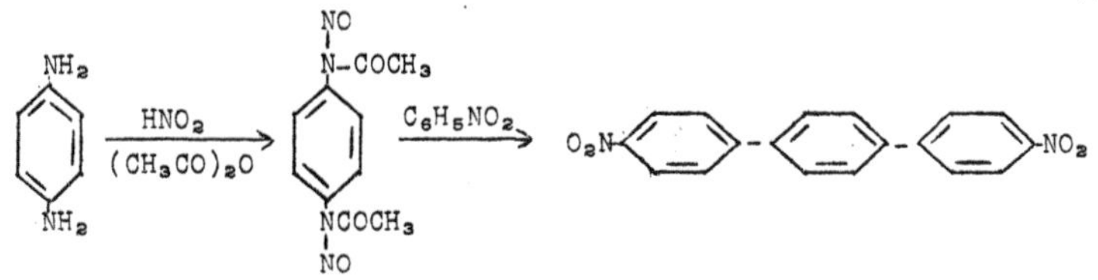
NH2
NH2
HNO2
(CH3CO)2O
NO
N-COCH3
NCOCH3
NO
C6H5NO2
O2N
NO2

Under more drastic nitrating conditions the 4,2',4"-trinitro-p-terphenyl was obtained.

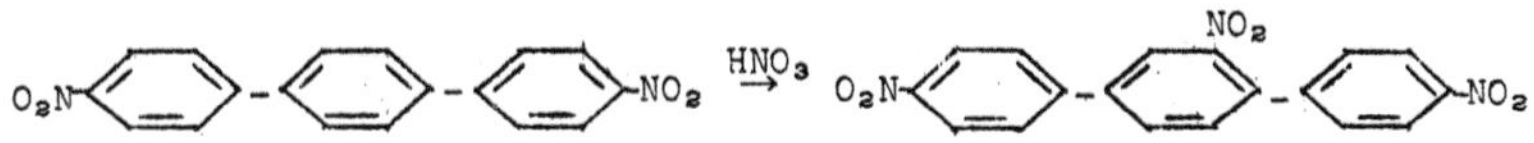
O2N
NO2
HNO3
O2N
NO2
NO2

Bibliography

Allen and Pingert, J. Am. Chem. Soc., 64, 1365 (1942).
Allen and Pingert, ibid., 64, 2639 (1942).
Wardner and Lowy, ibid., 54, 2510 (1932).
Goodman and Lowy, ibid., 60, 2155 (1938).
France, Heilbron and Hey, J. Chem. Soc., 1364 (1938).
France, Heilbron and Hey, ibid., 1288 (1939).
Nenitzescu and Curcaneanu, Ber., 70B, 346 (1937).
Basford, J. Chem. Soc., 1593 (1936).
Cook and Cook, J. Am. Chem. Soc., 55, 1212 (1933).

Reported by W. E. Wallace
December 9, 1942

THE PFITZINGER REACTION

There are many ways to carry out the synthesis of quinoline and its derivatives and much work has been reported in the literature because of the importance of quinoline compounds. Many of the syntheses reported, however, are useful only in a few special cases or involve starting materials that are difficult to obtain. Actually, there are not many good general methods of preparing quinoline derivatives.

The Skraup reaction is by far the most useful reaction for the preparation of quinoline derivatives where the substitutents are located in the benzene nucleus and the pyridine nucleus is unsubstituted. Hundreds of arylamines have been found to undergo the Skraup reaction. Similarly the Döbner-Miller reaction is very general and useful when substituents are desired in the benzene ring and usually in the α-position of the pyridine nucleus. Beyer's modification of the latter synthesis using ketones instead of aldehydes with aryl amines leads generally to α,γ-disubstituted quinolines.

When quinolines substituted in the α- , or β- , or both α- and β- positions are desired the Pfitzinger reaction is often the most useful. In 1882 Friedländer discovered that o-aminobenzaldehyde condenses with acetaldehyde to yield quinoline. In 1886 Wilhelm Pfitzinger substituted isatic acid (o-aminobenzoylformic acid) for o-aminobenzaldehyde and found that this compound undergoes a reaction with acetone to produce α-methylcinchoninic acid.

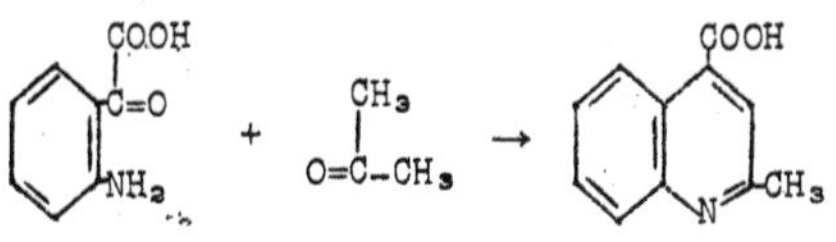

COOH
C=O
NH2
+
CH3
O=C-CH3
→
COOH
N
CH3

At the time of its discovery the method was more interesting than important since isatic acid was as difficult to obtain as o-aminobenzaldehyde. With the production of synthetic indigo a few years later, however, isatin itself became readily available through the oxidation of indigo with a mixture of nitric and chromic acids. (Yield 85%). Isatin is readily hydrolized in the presence of aqueous alkali solution to isatic acid which is used without isolation.

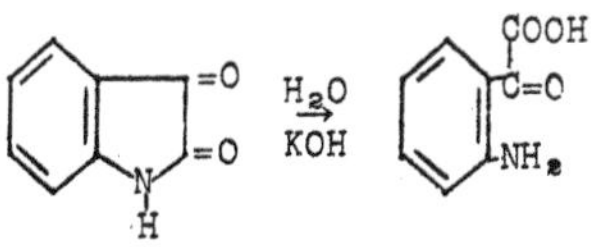

=O
=O
N
H
H2O
KOH
COOH
C=O
NH2

Aldehydes and Ketones.--A great many of these were used by Pfitzinger and other earlier investigators. They lead to β-alkyl or α,β-dialkyl substituted cinchoninic acids. Acetaldehyde does not react due to the strongly alkaline conditions under which the reaction is carried out. By first preparing the oximes of acetaldehyde and the other lower aldehydes, the reaction takes place readily, the resinifying effect of the alkali being greatly reduced, and β-alkyl cinchoninic acids are produced. Oximes of aryl aldehydes also react readily provided, of course, there is an α-CH_2 group present as in phenyl acetaldehyde.

Ketones may be used directly. They give rise to α,β-alkyl or aryl substituted cinchoninic acids except acetone and acetophenone which yield α-methyl and α-phenyl cinchoninic acids, respectively. It will be noted that with methylethyl ketone two products are theoretically possible.

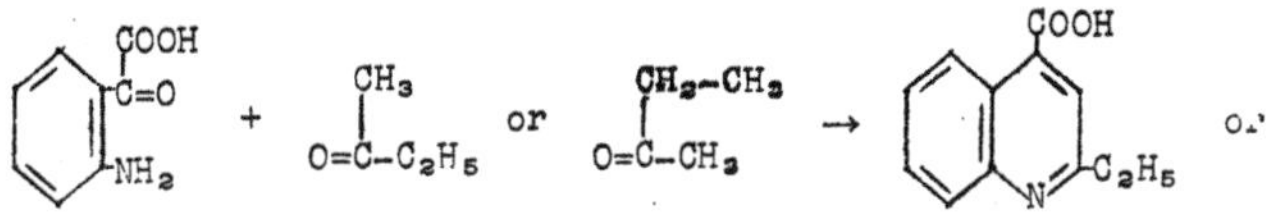

I

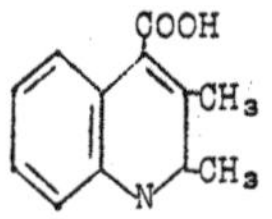

II

Of these, II is obtained in good yields but I is found only in small amounts. This reaction led Pfitzinger to generalize that if two alkyl groups R or R' of the ketone R-CO-R' can react, then the larger group will react with the carbonyl group of isatic acid. Many subsequent experiments have substantiated this generalization.

Isonitroso derivatives of ketones react also, isonitroso acetone, for example, producing the oxime of 2-aldehydocinchoninic acid. This is easily dehydrated with acetyl chloride to produce 2-cyanocinchoninic acid.

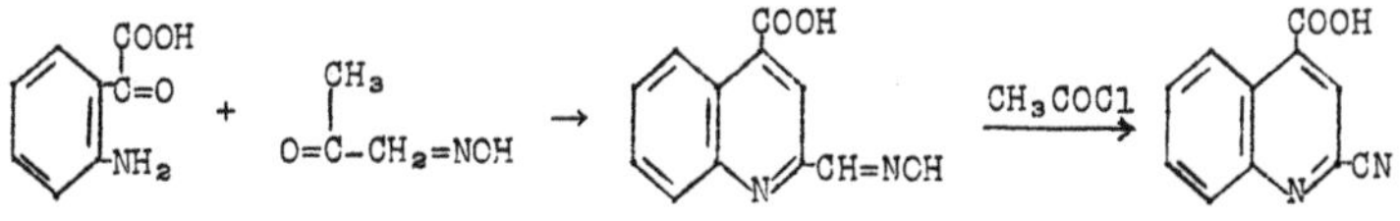

Cyclic ketones react, cyclohexanone, for example, producing tetrahydroacridine.

Ketoacids.--These produce carboxyl derivatives of cinchononic acid. Pyruvic acid yields 2,4-quinolinedicarboxylic acid.

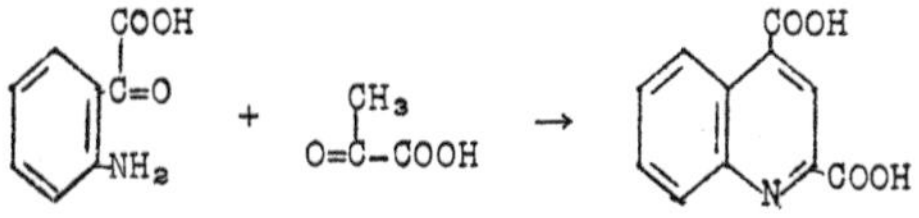
COOH
C=O
NH2
+
CH3
O=C-COOH
COOH
N
COOH

Levulinic acid yields only 2-methyl-4-carboxyl-3-quinolineacetic acid and none of the isomeric 4-carboxyl-2-quinoline-β-propionic acid, which seems to extend Pfitzinger's rule to this class of compounds. Further verification is found in the reaction of acetoacetic ester to produce only the α-methyl-β,γ-quinolinedicarboxylic acid.

Acids.--When active methylene groups are present, acids react to produce hydroxy derivatives of cinchoninic acid,

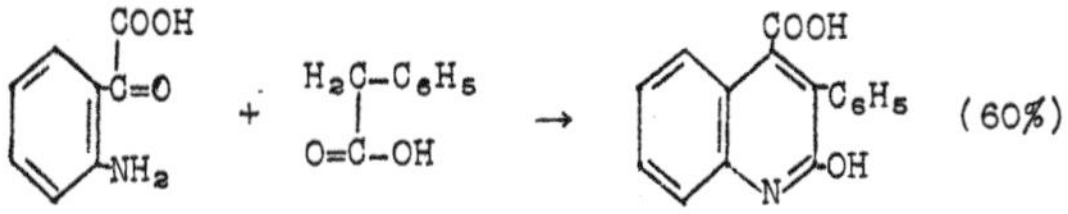
COOH
C=O
NH2
+
H2C-C6H5
O=C-OH
COOH
C6H5
N
OH
(60%)

Malonic acid reacts similarly to produce α-hydroxy-β,γ-quinolinedicarboxylic acid in about 65% yields. The β-carboxyl group is very readily lost in the reaction. The mechanism of this reaction was formerly thought to involve first a condensation with the β-keto group of isatin and subsequent ring opening and closing.

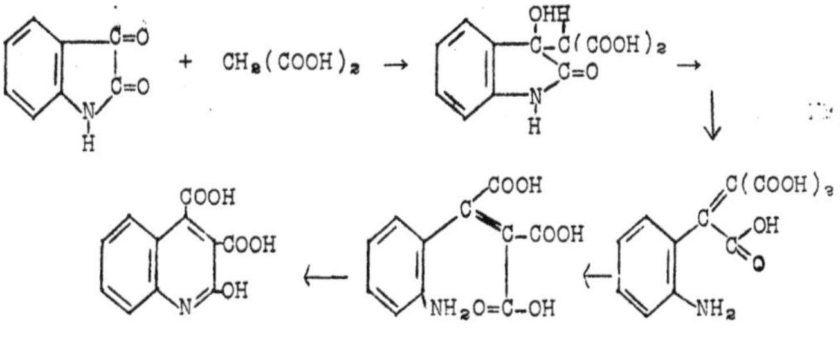
C=O
C=O
N
H
+
CH2(COOH)2
OHH
C-C(COOH)2
C=O
N
H
C(COOH)2
C
OH
C=O
NH2
COOH
C
C-COOH
NH2 O=C-OH
COOH
COOH
N
OH

More recent views indicate **first** a ring opening and condensation with the amino group.

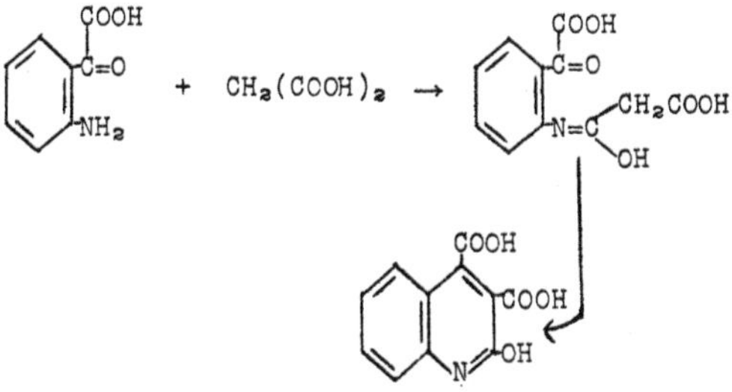

Ketoethers.--Much of the recent work involving the Pfitzinger reaction has been concerned with the aryloxy and alkoxy ketones, i.e. the ketoethers. These react quite readily to produce generally β-aryloxy or -alkoxy derivatives of cinchoninic acid in good yields.

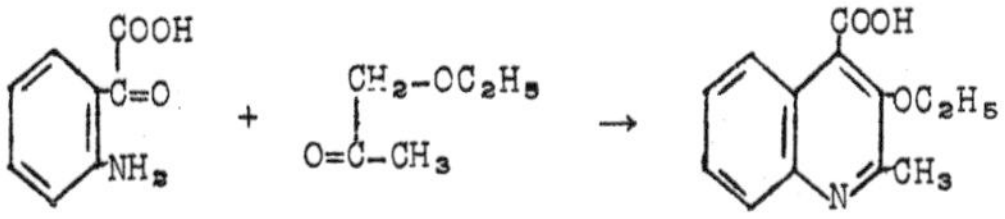

All of the cinchoninic acids are readily decarboxylated **to yield** the corresponding quinoline derivative. The ether linkage is cleaved with concentrated hydrochloric acid to yield the corresponding hydroxy quinoline.

Even though Pfitzinger's rule is followed with the ketoethers, an interesting reaction occurs between 1-methoxydiethylketone and isatin. Here the methylene group that would normally react is substituted and ethoxy-ethyl group is forced into the α-position of the resulting quinoline.

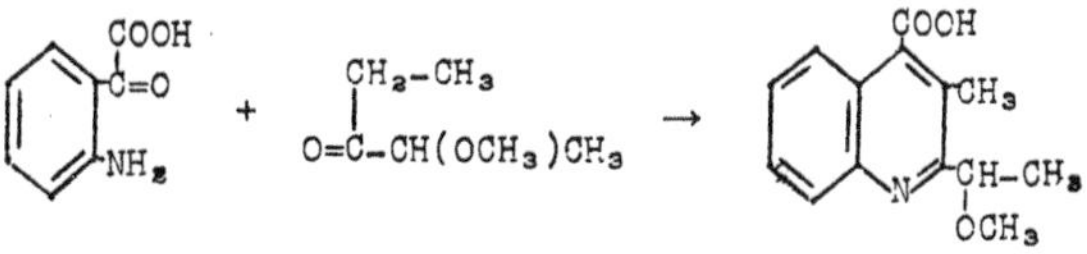

This compound undergoes some unusual reactions.

Some Typical Examples of the Pfitzinger Reaction

Product

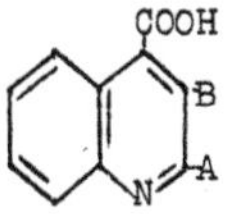
COOH
B
A
N

Isatin and Carbonyl compound	A	B	Yield	Ref.
CH_3CHO			0%	1
$CH_3CH{=}NOH$	H	H		1
$CH_3-CH_2-CH_2-CH{=}NOH$	H	C_2H_5		11
$CH_3-CO-CH_3$	CH_3	H	80%	1
$CH_3-CO-CH-NOH$	$-CH{=}NOH$	H	>80%	1
$CH_3-CO-C_2H_5$	(CH_3	(CH_3		1
	(C_2H_5	(H		8
$C_6H_5-CO-CH_3$	C_6H_5	H	>90%	1,2
$C_6H_5-CH_2-CH{=}NOH$	H	C_6H_5	65%	4
$C_6H_5-CO-CH_2-C_6H_5$	C_6H_5	C_6H_5		1
$C_6H_5-CH_2-CO-CH_2-C_6H_5$	$C_6H_5CH_2$	C_6H_5		2
Acetothienone	α-Thienyl	H		5
$C_6H_5-CH_2-COOH$	OH	C_6H_5	60%	4
$CH_3-CO-COOH$	COOH	H		1
$CH_2(COOH)_2$	OH	COOH	65%	3
$CH_3-CO-CH_2-COOH$	CH_3	COOH		1,2
$CH_3-CO-CH_2-CH_2-COOH$	CH_3	CH_2COOH		2
$HOOC-CO-CH_2-COOH$			0%	2
$HOOC-CH_2-CO-CH_2-COOH$	$-CH_2COOH$	$H(-CO_2)$		2
Methyl β-tetralyl ketone	β-tetralyl	H	80%	6
$C_6H_5-CO-CH_2-COOH$	C_6H_5	COOH		2
Cyclohexanone			90%	7

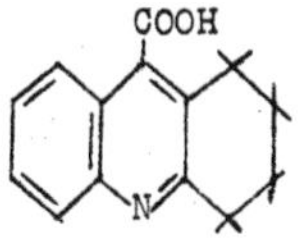
COOH
N
(end product)

Isatin and Carbonyl compound	A	B	Yield	Ref.
$C_6H_5O-CH_2-CO-CH_3$	CH_3	C_6H_5O-	81%	14
$C_{10}H_7O-CH_2-CO-CH_3$	CH_3	α- or β- $C_{10}H_7O-$	68%	14
Thymoxyacetone	CH_3	Thymoxy	33%	14
Ethoxyacetone	CH_3	C_2H_5O-	44%	15
Ethoxymethylethyl ketone	C_2H_5	C_2H_5O-	92%	15
1-Methoxydiethyl ketone	$-CH(OCH_3)CH_3$	CH_3	74%	16
$C_6H_5CH_2-CO-CH(CH_3)_2$	$-CH(CH_3)_2$	C_6H_5	33%	12
$C_6H_5CH_2-CO-OOOH$	$-COOH$	C_6H_5	77%	13

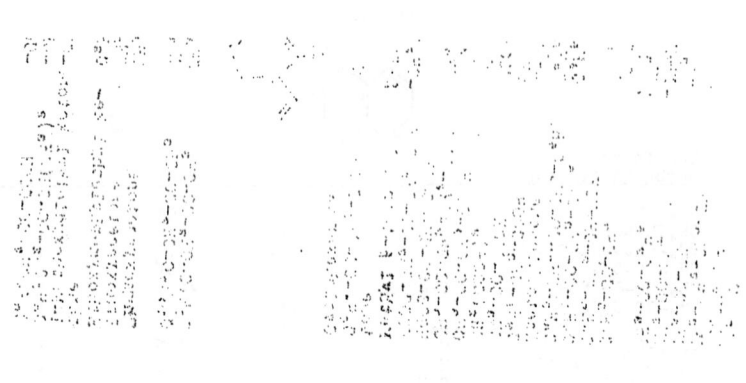

Bibliography

1. W. Pfitzinger, J. prakt. Chem., 33, 100, (1886); 38, 582, (1888); 56, 283, (1897); 66, 263, (1902).
2. Engelhard, ibid., 57, 467, (1898).
3. Borsche and Jacobs, Ber., 47, 354, (1914).
4. Hubner, ibid., 41, 482-7, (1908); 39, 983, (1906).
5. Hartmann and Wybert, Helv. Chim. Acta., 2, 60, (1919).
6. V. Braun, Hahn, and Seemann, Ber., 55, 1687, (1922).
7. W. Borsche, ibid., 41, 2207, (1908).
8. V. Braun, Gmelin, and Schultheiss, ibid., 56, 1338, (1923).
9. J. Halberkann, ibid., 54B., 3090, (1921).
10. V. Braun et al, ibid., 56, 2345, (1923).
11. Mulert, ibid., 39, 1904, (1906).
12. W. Borsche and Sinn, Ann., 532, 146, (1937).
13. Borsche and Noll, ibid., p. 127.
14. Calaway and Henze, J.A.C.S., 61, 1355, (1939).
15. Cross and Henze, ibid., 61, 2731, (1939).
16. Lesesne and Henze, ibid., 64, 1897, (1942).

Reported by R. D. Lipscomb
December 9, 1942

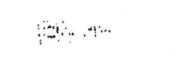

USES OF PHOSPHORIC ACID IN ORGANIC CHEMISTRY

With the new methods which have brought a better quality of phosphoric acid into production at a lower price has come an increased interest in the use of this acid as applied to organic chemistry.

I. Dehydration.--In 1901 G. S. Newth announced the use of phosphoric acid as a media for the continuous production of ethylene from alcohol in the laboratory. The sulfuric acid method had not been satisfactory due to charring, frothing, side reactions and the simultaneous production of sulfur dioxide and carbon dioxide XXXXXXXXXXXXXXXXXX and was not continuous. His reaction was further developed until a 90% yield of 99.5% ethylene was obtained at 250° C.

Walton has made a study of the rate of dehydration of formic acid by various phosphoric acids and the data are summarized below:

$$HCOOH \xrightarrow[H_3PO_4]{120^\circ\ C.} CO + H_2O \quad (100\%\ yield)$$

	$K \times 10_3$		$K \times 10^3$
85% H_3PO_4	2	Acid A: 90% H_3PO_4, 5% $H_4P_2O_7$ + HPO_3, 5% H_2O	274
95% H_3PO_4	21	Acid A + 25% meta	140
100% H_3PO_4	72	Acid A + 25% P_2O_5	413
100% $H_4P_2O_7$	121	Acid A + 1% H_2SO_4	694

Acid A was later found by Walton to be the optimum catalyst for the dehydration of alcohol at 250° C.

Phosphoric acid has been used in the dehydration of the higher homologues of ethanol by Newth and others, but investigators, such as Adams and Marvel, have found modifications of the sulfuric acid method to be on the whole more satisfactory. However, Dehn and Jackson claim phosphoric acid is a superior media for the dehydration of various terpene alcohols.

$$\text{Menthol} \xrightarrow[85\%\ H_3PO_4]{160^\circ\ C} \text{Menthene} \quad (94\%\ yield)$$

$$\text{Cyclohexanol} \rightarrow \text{Cyclohexene} \quad (96\%\ yield)$$

II. Condensations.--Dreyfus claims that phosphoric acid may be used advantageously in place of sulfuric acid in a number of chemical reactions involving the splitting out of water.

1. Ether Formation.--$2C_2H_5OH \xrightarrow[\text{Acid A}]{160^\circ\ C} H_2O + (C_2H_5)_2O$ (65% yield)

2. Esterification.—$C_6H_5COOH + C_2H_5OH \xrightarrow[\text{Acid A}]{<100^\circ\ C} C_6H_5COOC_2H_5$ (100% yield)

3. Nitration.--Kranz and Blechta compare the use of phosphoric - nitric acid and sulfuric - nitric acid mixtures for the nitration of cellulose. They conclude that phosphoric - nitric acid is to be recommended for laboratory preparations of this sort, because with nitric - sulfuric acid mixtures sulfates are also formed and the products tend to explode spontaneously. Dreyfus using the Acid A + 1% sulfuric acid catalyst reports the successful trinitration of toluene, glycerol and phenol.

4. Acetylation.—Dreyfus reports the O-acetylation of phenols:

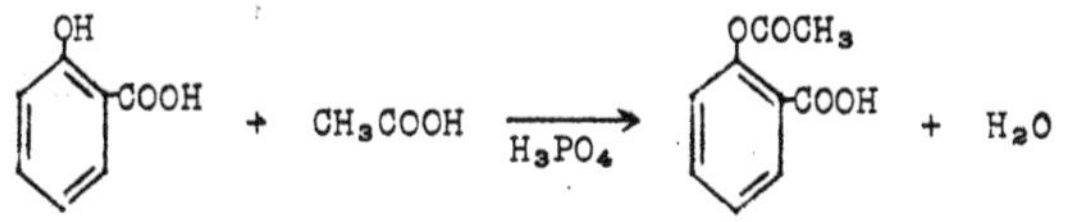
OH
COOH
+ CH₃COOH
H₃PO₄
OCOCH₃
COOH
+ H₂O

Vyazkova finds that the presence of phosphoric acid greatly speeds the acetylation of phenol by acetic anhydride in the usual manner. Acetylation is very rapid at low temperatures.

5. Alkylation (Friedel - Crafts).—A. E. Tchitchibabine made an exhaustive study of the alkylation of phenols with alcohols in the presence of phosphoric acid. He came to the conclusions that Acid A is an excellent agent for condensing phenols and their simple ethers with (a) all secondary and tertiary alcohols, and primary alcohols of the benzyl type (b) olefins (especially branched ones of the type $R-\underset{R}{C}=CH-R$), and that the ortho substitution product was to be expected.

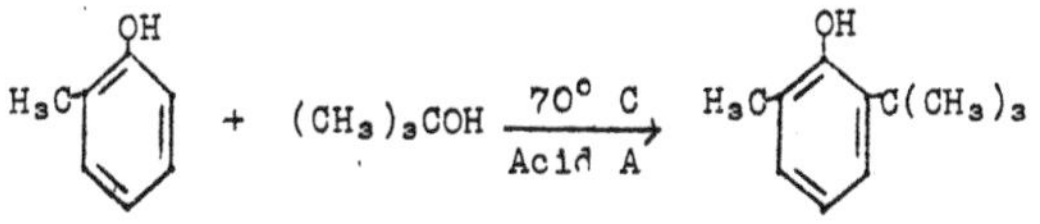
OH
H₃C
+ (CH₃)₃COH
70° C
Acid A
OH
H₃C
C(CH₃)₃

(78% yield)

Howard and Sons have patented this method for the production of thymol:

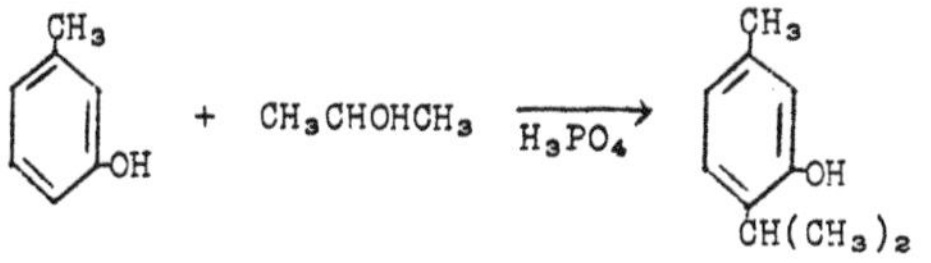
CH₃
OH
+ CH₃CHOHCH₃
H₃PO₄
CH₃
OH
CH(CH₃)₂

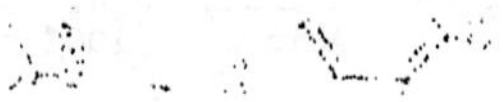

Recently, Burwell has used this method for the alkylation of benzene:

$$C_6H_6 + CH_3CHOHCH_2CH_3 \xrightarrow[H_3PO_4\ (100\%)]{70^\circ\ C} C_6H_5CH(CH_3)CH_2CH_3 \quad (12\%\ \text{yield})$$

Ipatieff has used 87% phosphoric acid at high temperatures and pressures for what might be called an exhaustive alkylation of benzene, tetralin, florene, phenols, etc.

$$C_6H_6 + CH_2{=}CH_2 \xrightarrow[\text{pressure},\ H_3PO_4]{300^\circ\ C} C_6H_5C_2H_5\ (18\%\ \text{yield}) + \text{polysubstituted benzenes}$$

Berman and Lowy obtain ethylbenzene in 62% yield by the following reaction:

$$3C_6H_6 + (C_2H_5O)_3PO \xrightarrow{AlCl_3} 3C_6H_5C_2H_5 + AlPO_4 + 3HCl$$

6. The synthesis of Triphenylmethane Derivatives.--Tanasescu and Simonescu report that the yields and purity of oxytriphenylmethane compounds obtained by the use of phosphoric acid are particularly good. The type of reaction is represented by the following reactions:

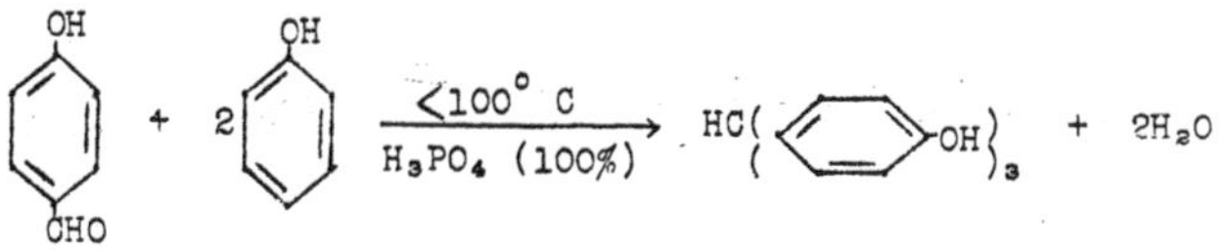

OH
OH
+ 2
<100° C
H_3PO_4 (100%)
HC(
OH)
3
+ $2H_2O$
CHO

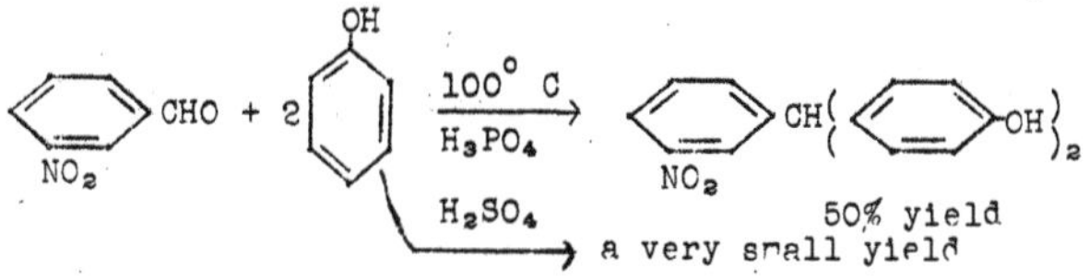

CHO + 2
OH
100° C
H_3PO_4
CH(
OH)
2
NO_2
NO_2
H_2SO_4
50% yield
a very small yield

III. <u>Hydrolysis</u>

1. Ketonic Hydrolysis of Alkyl Acetoacetates.--Dehn and Jackson state that 85% phosphoric acid is the ideal catalyst for the ketonic hydrolysis of alkylacetoacetates:

$$CH_3\overset{O}{\overset{\|}{C}}\text{-}\underset{nC_4H_9}{CH}\text{-}COOC_2H_5 \xrightarrow[160^\circ\ C]{85\%\ H_3PO_4} CH_3\overset{O}{\overset{\|}{C}}\text{-}C_5H_9 + C_2H_5OH + CO_2 \quad (95\%\ \text{yield})$$

2. Nitriles.--Berger and Olivier in 1927, having tried all known methods for the saponification of nitriles on the nitrile of 2,6-dimethylbenzoic acid without success, used Acid A at 150° C, and obtained a 70% yield of the acid in one-half hour. This was such a striking success that they went on and developed it as a general method.

IV. Isomerization

1. Terpenes.--According to Dehn and Jackson, phosphoric acid has found extensive use in the isomerization of terpenes:

$$\text{Turpentine}\begin{cases}\alpha\text{-pinene}\\ \beta\text{-pinene}\end{cases}\xrightarrow[H_3PO_4]{200^\circ\ C.}\ \underset{(72\%\ \text{yield})}{\text{dipentene}}\ \xrightarrow[\text{tube}]{\text{hot } 700^\circ\ C}\ \underset{(67\%\ \text{yield})}{\text{isoprene}}$$

2. Benzoin Rearrangement.--Jones and Lyons found that, by adding alumina to the phosphoric acid customarily used, a 50% yield was obtained:

$$C_6H_5CHOHCOC_6H_5 \xrightarrow[H_3PO_4]{260^\circ\ C} (C_6H_5)_2CHCOOH$$

V. Purification of Optically Active Alcohols.--Blagden and Huggett state that the reaction of menthol with phosphoric acid is used industrially for the purification of synthetic dl-menthol:

$$3(\underline{dl}\text{-menthol}) + 85\%\ H_3PO_4 \xrightarrow{25^\circ\ C} 3(\underline{dl}\text{-menthol}\cdot H_3PO_4 \quad \text{m.p. } 74^\circ\ C.$$

Such salts are of similar use in the cases of most terpene and hydroaromatic alcohols and some simple aliphatic alcohols.

VI. The Solid Phosphoric Acid Catalyst.--Dunsten discusses the use of the "solid phosphoric acid" catalyst in the polymerization of the low-molecular-weight hydrocarbons obtained from the cracking of petroleum. This catalyst was developed by Ipatieff and Egloff about 1934. It is prepared by heating diatomaceous earth and phosphoric acid to 250° C. for several hours and has the approximate composition: 60% total phosphorus as phosphorus pentoxide and 25% free phosphoric acid. Its use has become widespread, a typical "poly" plant condensing some 3,000,000 cu. ft. of gases daily to 15000 gallons of 82 octane gasoline. In the production of high octane gasoline, use is made of the C_3 and C_4 fractions. Thus recently, according to Ipatieff:

$$\text{Propene + isobutene}\ \xrightarrow[130^\circ\ C.]{\substack{40\ \text{atm.}\\ H_3PO_4}}\ \text{isoheptene in good yield}$$
$$\xrightarrow[300^\circ\ C.]{}\ \text{aromatic compounds}$$

Holm and Shiffler report the development of an isomerization-polymerization catalyst employing a film of phosphoric acid on a non-porous metal.

In 1938, Ipatieff reported the use of "solid phosphoric acid" in the distillation of West Texas Crude at 200° C., and an unusually high grade of distillate is produced. At higher temperatures and pressures, cracking also occurs, which further improves the distillate.

Recently, McAllister reports a general cleavage for α,β-unsaturated ketones, using the "solid phosphoric acid" catalyst.

1. $CH_3C(CH_3)(OH)CH_2COCH_3 \xrightarrow[H_3PO_4]{265^\circ\ C} \underset{CH_3}{CH_3C}{=}CH_2 + CH_3COOH$

(85% yield) (85% yield)

2. $\underset{CH_3}{CH_3C}{=}CHCOCH \xrightarrow[H_3PO_4]{H_2O} \underset{CH_3}{CH_3C}{=}CH_2 + CH_3COOH$

(85% yield) (85% yield)

Hardy has carried out the following reactions:

1. $CH_3COCH_3 + CO + H_2O \xrightarrow[87\%\ H_3PO_4]{200\ atm.-200^\circ\ C} (CH_3)_3CCOOH + CH_3COOH$

200 g. 250 g.

2. $\underset{CH_3}{CH_3C}{=}CH + CO + H_2O \rightarrow (CH_3)_3CCOOH$

BIBLIOGRAPHY

Carothers, Chem. Ind., 42, 523, (1938).
Newth, J. Chem. Soc., 79, 916, (1901).
Walton and Stark, J. Phys. Chem., 34, 359, (1930).
Walton and Weber, ibid., 34, 2693, (1930).
Adams, Kamm and Marvel, J. Am. Chem. Soc., 40, 1951, (1918).
Dehn and Jackson, ibid., 55, 4284, (1933).
Dreyfus, U. S. 1,872,700 (1932).
Hofmann and Josephy, D.R.P. 292,543 or C.A. 11, 1276, (1917).
Kranz and Blechta, Chem. News, 134, 1 and 17, (1927).
Vyazkova, J. Applied Chem. U.S.S.R., 8, 471, (1935); C.A. 30, 3492 (1936)
Tchitchibabine, Bull. Soc. Chim., 5, (2), 497, (1935).
Howard Sons Ltd. Bigden, Brit.pat. 200,151 (1923); C.A. 18, 274, (1924)
Burwell and Archer, J. Am. Chem. Soc., 64, 1032, (1942).
Ipatieff, Pines and Komarewsky, Ind. Eng. Chem., 28, 222, (1936).
Ipatieff, Pines and Schmerling, J. Am. Chem. Soc., 60, 1161, (1938).
Berman and Lowy, ibid., 60, 2596, (1938).
Tanasescu and Simonescu, J. Prakt. Chem., 141, 311, (1934).
Berger and Olivier, Rec. Trav. Chim., 46, 600, (1927).
Olivier, ibid., 48, 568, (1929).
Jones and Lyons, J. Org. Chem., 3, 273, (1938).
Blagden and Huggett, J. Chem. Soc., 1934, 317.
Dunstan, Trans. Faraday Soc., 32, 227, (1936).
Ipatieff and Corson, Ind. Eng. Chem., 30, 1316, (1938).
Ipatieff and Schaad, U. S. 2,270,302 (1942); C.A. 36, 3189, (1942).
Holm and Shiffler, U. S. 2,186,021 (1940); C.A. 34, 2862, (1940).
McAllister, Bailey and Bouton, J. Am. Chem. Soc., 62, 3210, (1940).
Hardy, J. Chem. Soc., 1938, 464.
Schotz, Synthetic Rubber, (London: Ernest Benn, Ltd., 1926) pp. [illegible]

Reported by W. J. Shenk, Jr.
December 9, 1942

STEREOCHEMISTRY OF CATALYTIC HYDROGENATION

THE HYDROGENATION OF 9-PHENANTHROL AND PHENANTHRAQUINONE

In a series of seven papers Linstead and his coworkers have recorded the results of catalytic hydrogenation of several aromatic compounds related to phenanthrene. In an earlier seminar by R. L. Frank papers II - V dealing with the stereochemistry of perhydrodiphenic acids were reported. This seminar reports in particular papers VI, the catalytic hydrogenation of 9-phenanthrol and related substances, VII, the hydrogenation of 9,10-phenanthraquinone, and I, a summary and suggested mechanism for catalytic hydrogenation of aromatic rings.

The nomenclature is that reported in the earlier seminar. In a 9 substituted phenanthrene the configuration on the side of the constituent is named first. In the formulas, dots indicate hydrogen on the top side of the molecule. The system of nomenclature is illustrated in compounds I and II.

I

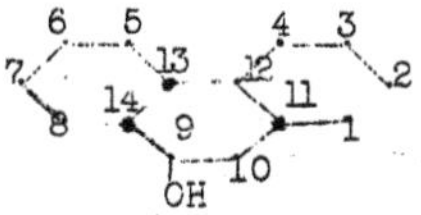
6
5
13
4
3
7
12
2
14
11
8
9
1
10
OH

cis-*anti*-*trans*-perhydro-9-phenanthrol

II

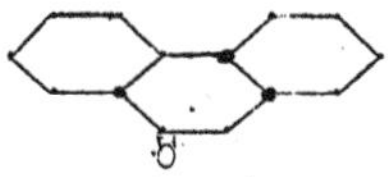
O

trans-*anti*-*cis*-9-keto-perhydrophenanthrene

The configuration of the perhydrophenanthrene skeleton of the compounds obtained in the present study were conclusively proved by oxidation of the perhydrophenanthrene derivative to the previously determined perhydrodiphenic acids.

The Hydrogenation of 9-Phenanthrol.--Perhydrophenanthrene can theoretically exist in six diastereoisomeric modifications. These correspond to the perhydrodiphenic acids previously reported. Corresponding to each of four of the perhydrophenanthrenes (*cis*-*syn*-*cis*, *trans*-*syn*-*trans*, *cis*-*anti*-*cis*, and *trans*-*anti*-*trans*) there will be one ketone with the carbonyl at C_9. Corresponding to each of the other two modifications (*cis*-*syn*-*trans* and *cis*-*anti*-*trans*) there will be two ketones with the carbonyl at C_9, since C_9 may lie adjacent to either a *cis* or *trans* junction. Thus eight diastereoisomeric ketones are possible. Each ketone can give rise to two epimeric alcohols so that sixteen racemic alcohols are possible.

Three of the eight possible 9-keto-perhydrophenanthrenes have been prepared and their configuration determined by oxidation to the corresponding perhydrodiphenic acid. Those prepared are the <u>trans</u>-<u>anti</u>-<u>trans</u>, <u>cis</u>-<u>syn</u>-<u>cis</u>, and the <u>trans</u>-<u>syn</u>-<u>cis</u> isomers.

The <u>trans</u>-<u>anti</u>-<u>trans</u>-9-keto-perhydrophenanthrene, m. p. 49°, and the related secondary alcohol, m. p. 119°, have been prepared previously by Linstead and Walpole. The configuration of the ketone was definitely proved by oxidation to <u>trans</u>-<u>anti</u>-<u>trans</u>-perhydrodiphenic acid, m. p. 244°. The synthesis of this ketone involved the following preparation of Rapson and Robinson.

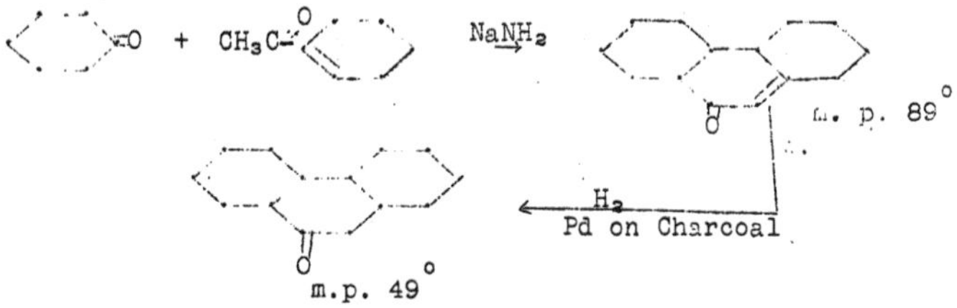
+ CH_3C
$NaNH_2$
m. p. 89°
H_2
Pd on Charcoal
m.p. 49°

The other two ketones have been obtained through a study of the catalytic hydrogenation over platinum of 9-phenanthrol. Hydrogenation of 9-phenanthrol over platinum in acetic acid yields three fractions, (1) sym-octahydro-9-phenanthrol, m. p. 135°, (2) a <u>cis</u>-<u>syn</u>-<u>cis</u>-perhydro-9-phenanthrol, m. p. 111°, (3) and a small amount of a hydrocarbon fraction.

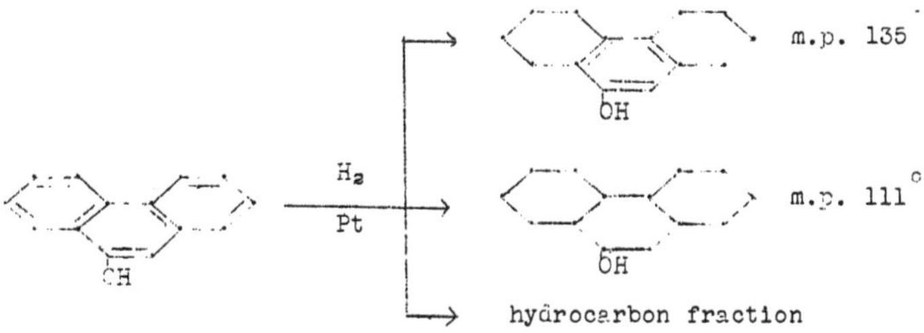
m.p. 135
OH
H_2
Pt
m.p. 111°
OH
OH
hydrocarbon fraction

The configuration of the perhydrophenanthrol (apart from the orientation of the hydroxyl group) was proved by nitric acid oxidation to <u>cis</u>-<u>syn</u>-<u>cis</u>-perhydrodiphenic acid, m. p. 289°.

Oxidation of *cis-syn-cis*-perhydro-9-phenanthrol, m. p. 111°, with chromic and acetic acids at room temperature yielded a 9-keto-perhydrophenanthrene, m. p. 44°. On further oxidation with nitric acid this yielded *cis-syn-cis*-perhydrodiphenic acid, m. p. 289°. The ketone yielded the parent alcohol upon catalytic reduction in ethyl alcohol over platinum.

Oxidation of *cis-syn-cis*-perhydrophenanthrol, m. p. 111°, with a mixture of chromic and acetic acids on a steam cone yielded a stereoisomeric ketone, m. p. 57°. Catalytic hydrogenation of this ketone yielded a new secondary alcohol, m. p. 88-89°, and oxidation of the ketone with nitric acid yielded *cis-syn-trans*-perhydrodiphenic acid, m. p. 199°. The ketone must have been formed by inversion of the configuration of one asymetric carbon atom, and this carbon must be the one adjacent to the carbonyl group where inversion can take place through the process of enolization. The ketone and the derived alcohol can therefore be assigned the *trans-syn-cis* configuration.

The present state of the 9-substituted perhydrophenanthrenes of known structure is summarized in the following table.

9-Keto-perhydrophenanthrene		Perhydro-9-phenanthrol
cis-syn-cis, m.p. 44° (L.W.L.)	→	m.p. 111° (L.W.L.)
	→	---------
trans-syn-cis, m.p. 57° (M.W., L.W.L.)	→	m.p. 89° (L.W.L.)
	→	? m.p. 67° (M.W.)
trans-anti-trans, m.p. 49° (L.W.)	→	m.p. 119° (L.W.)
	→	---------

References:
L.W.L. - Linstead, Whetstone and Levine, J.A.C.S., 64, 2018(1942)
M.W. - Marvel and White, J.A.C.S., 62, 2739(1940)
L.W. - Linstead and Walpole, J.Chem.Soc., 842(1939)

The Hydrogenation of Phenanthraquinone.--Corresponding to the six inactive stereoisomeric forms of the fundamental perhydrophenanthrene skeleton there are twenty inactive 9,10-glycols. Each of the *cis-cis* and *trans-trans* forms (i. e., the *cis-syn-cis*, *trans-syn-trans*, *cis-anti-cis*, and *trans-anti-trans* configurations) can give rise to three glycols (see III below) of which two are meso and one is a racemic pair. The remaining two perhydrophenanthrene structures (*cis-syn-trans* and *cis-anti-trans*) can give rise of four glycols each, all being racemic pairs (see IV below). These possibilities are outlined for the cases of the *cis-syn-cis* and the *cis-syn-trans* series. Only the central ring is shown.

Oxidation of cis-syn-cis perhydro-[illegible] m.p. 111° with [illegible] and acetic acid [illegible] yielded a [illegible] m.p. [illegible]. On further oxidation with nitric acid it yielded cis-syn-cis-[illegible] acid, m.p. [illegible]. The ketone yielded the parent alcohol upon [illegible] reduction in [illegible] acetic over platinum.

Oxidation of cis-anti-cis-perhydro[illegible] m.p. 111°, with a solution of chromic and acetic acids [illegible] yielded a [illegible] ketone, m.p. [illegible]. [illegible] this ketone yielded a new [illegible] alcohol [illegible] and oxidation of the ketone with nitric acid [illegible] perhydrodiphenic acid, m.p. [illegible]. The ketone [illegible] formed of the [illegible] configuration [illegible] carbon atom, and this carbon must be the [illegible] the [illegible] probably [illegible] takes place through the [illegible]. The ketone and the derived alcohol can [illegible] the [illegible] configuration.

[illegible] of the [illegible] perhydro[illegible] [illegible] will be [illegible].

[illegible] [illegible]

[illegible]

[illegible]

[illegible]

References

[illegible]

[illegible]

III *cis-syn-cis* Series

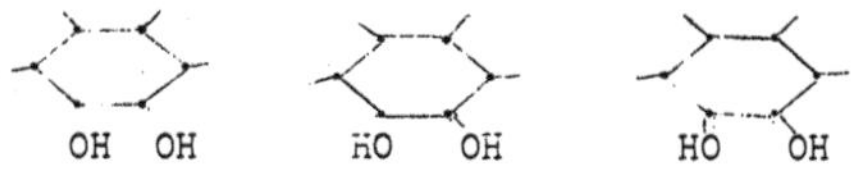
OH OH
HO OH
HO OH

meso meso racemic

IV *cis-syn-trans* Series

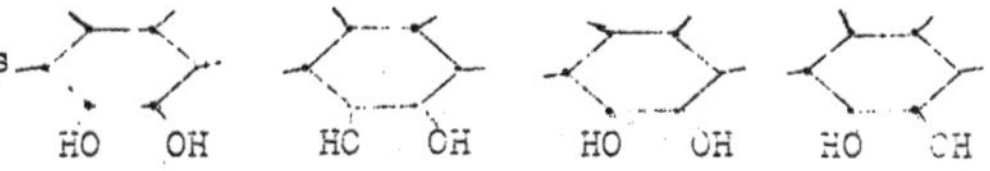
HO OH
HO OH
HO OH
HO OH

racemic racemic racemic racemic

Linstead and Levine have obtained four of these glycols by hydrogenation of phenanthraquinone. Over Adams's platinum oxide catalyst in acetic acid at room temperature a glycol, m. p. 174°, was obtained. Over Raney nickel in ethanol at 160° and 170 atmospheres pressure the reaction yielded principally two glycols, m. p. 174° and m. p. 155°, together with a small amount of a third glycol, m. p. 184°. The Pt 174° glycol and the Ni 174° glycol give a strongly depressed mixed melting point indicating that they do not have the same structure.

The configuration of the perhydrophenanthrene skeleton of the perhydro glycols was determined by oxidation to the corresponding perhydrodiphenic acids. The same perhydrodiphenic acid was obtained from the oxidation of the easily formed glycols, m. p.'s. Pt 174, Ni 174, and Ni 155°. This was *cis-syn-cis* perhydrodiphenic acid, m. p. 289°. The three glycols thus have the same skeletal configuration and differ only in the configuration of the hydroxyl groups. All the possible glycols of the *cis-syn-cis* series have therefore been prepared.

The glycol, m. p. 184, isolated in small amount from the hydrogenation over nickel was oxidized to *cis-syn-trans* perhydrodiphenic acid, m. p. 198-200°.

From the evidence of the products isolated phenanthraquinone can be said to hydrogenate *cis-syn-cis* over platinum and almost completely *cis-syn-cis* over nickel along with a small amount of *cis-syn-trans* hydrogenation over the latter catalyst.

Interpretation of Results.--All the hydrogenations studied by Linstead and his coworkers gave largely *cis* and *syn* material. Other products with other orientations were isolated, but in all cases the main product of the hydrogenation was *cis* and *syn*. The regularity of the results has lead to the postulation of three hypotheses to explain the phenomena observed.

1. When one or more aromatic rings are hydrogenated during a single period of adsorption of the molecule on the catalyst, the hydrogen atoms add to one side of the molecule. The hydrogenation process may be divided into two steps: (1) the adsorption of the aromatic molecule on a suitable part of the catalyst; and (2) the addition of hydrogen to the underside of the molecule so that all the hydrogen atoms appear on the same side of the molecule. In polynuclear compounds this picture may be more complicated due to the formation of comparatively stable compounds which are not completely hydrogenated. Thus perhydrogenation of aromatic compounds may take place during two or more periods of adsorption leading to products in which all the hydrogens on the asymmetric carbon atoms are not on the same side of the molecule.

2. A second hypothesis to explain the preferred cis and syn orientation of the hydrogens has been given the name "catalyst hindrance", i. e., a steric hindrance between the catalyst and the molecule. In order that an aromatic compound can be hydrogenated the aromatic ring must be closely adsorbed on the catalyst. This would require that the saturated ring, or rings, of a partially saturated polynuclear compound be turned away from the catalyst to prevent interference with the adsorption of the aromatic portion. When this happens the hydrogen at the saturated bridge carbon will be turned down towards the catalyst, and since hydrogen adds to the underside of the molecule, it follows that the hydrogens will all be on the same side of the compound. This would lead to predominantly cis-syn-cis orientation even when hydrogenation takes place in several steps.

3. In the perhydrogenation of diphenic acid and its derivatives the cis-syn-cis configuration was preferred. This has lead to the postulate that diphenic acid and its derivatives are adsorbed on the catalyst in the coiled or psuedo-tricylclic phase (i. e., the carboxyl groups are on the same side of the biphenyl nucleus rather than on opposite sides.). No explanation of this behavior has been given.

Bibliography

Linstead and Walpole, J.Chem.Soc.,1939,842
Rapson and Robinson,ibid,1935,1285
Marvel and White,J.Am.Chem.Soc.,62,2739(1940)
Linstead,et al,ibid.,64,1985,1991,2003,2006,2009,2014,2022(1942)

Reported by J. R. Elliott
January 6, 1943

ALKYLATIONS BY MEANS OF LEAD TETRAESTERS AND ACYL PEROXIDES

The use of lead tetraesters as alkylating agents was discovered by chance in attempting to improve on a known procedure for converting butadiene-toluquinone (I) into 2-Methyl-1,4-naphthoquinone (IV):

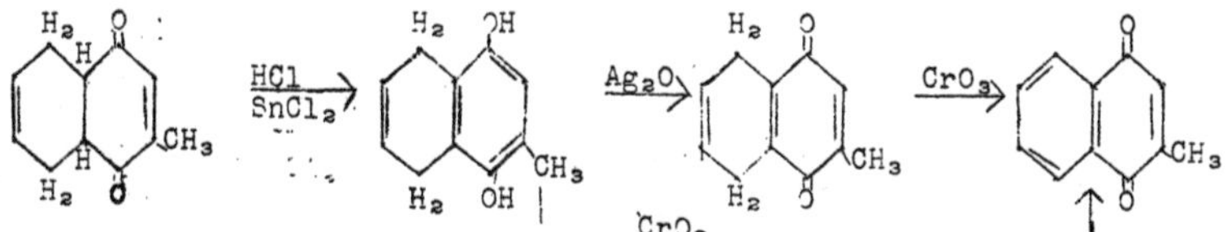
HCl
$SnCl_2$
Ag_2O
CrO_3
CrO_3
CH_3

I II III IV

The use of lead tetraacetate as an oxidizing agent might improve the reaction since the intermediate quinone (III) might undergo acetoxylation on one of the active methylene groups or might undergo addition to the double bond, in either case acetic acid would split out readily to give the desired 2-methyl-1,4-naphthoquinone (IV). When the hydroquinone (II) was warmed in acetic acid with two moles of lead tetraacetate a compound melting slightly to high for 2-methyl-1,4-naphthoquinone (M.P. 107°) was obtained. Since this resembled a partial oxidation product previously encountered, the reaction was repeated using three moles of lead tetraacetate, the reaction product melted at 127° and was shown to be 2,3-dimethyl-1,4-naphthoquinone. The reaction proceeded in the same manner with the 5,8 dihydroquinone (III). In both cases large amounts of carbon dioxide were evolved. However when, 2-methyl-1,4-naphthoquinone was treated in the same way there was no evolution of gas, and the starting material was recovered unchanged. On further investigation it was found that the isomerized addition product of 2,3 dimethylbutadiene and toluquinone gave on treatment with lead terascetate, 2,3,6,7-tetramethyl-1,4-naphthoquinone rather than 2,6,7-trimethyl-1,4-naphthoquinone.

The source of the methyl radical was then investigated; the fact that it might come from the lead tetraacetate was at first discounted since this reagent was without action on 2-methyl-1,4-naphthoquinone. That the 5,8-dihydride III will react, suggested that it may serve the dual role of providing the carbon substituent, through an allylic intermediate, as well as acting as an acceptor of the methyl group. In an attempt to separate the two functions, malonic acid was used as the active methylene component and 2-methyl-1,4-naphthoquinone as the acceptor. In the presence of the malonic acid the quinone was methylated with lead tetraacetate to give 45-50% yields of the dimethylquinone.

Other active-hydrogen compounds were tried with varying results, for example ethyl acetoacetate was found to function as well as malonic acid, while diethyl malonate was not satisfactory. A sufficiently high temperature would also start the reaction. Methylmalonic acid also promoted the reaction. If the alkyl group is derived from the malonic acid, methylmalonic acid should give 2-methyl-3-ethyl-1,4-naphthoquinone, however the product obtained was the 2,3-dimethyl compound. Similarly ethyl ethylacetoacetate when used

as a promotor also gave the dimethyl compound. Therefore, the active-hydrogen component, although it promotes the reaction is not the alkylating agent.

A method of testing the possibility that lead tetraacetate is the methylating agent was to attempt to introduce higher alkyl groups by means of appropriate esters of tetravalent lead. Thus lead tetrapropionate should give an ethyl substituent. To simplify the procedure, lead tetrapropionate was formed in the reaction mixture by the action of red lead on propionic acid. 2-Methyl-1,4-naphthoquinone was dissolved in propionic acid, an equivalent amount of ethyl acetoacetate added and red lead added in small portions to the heated, stirred solution. A 50% yield of 2-methyl-3-ethyl-1,4-naphthoquinone was obtained. This same compound was made by methylating 2-ethyl-1, 4-naphthoquinone with lead tetraacetate.

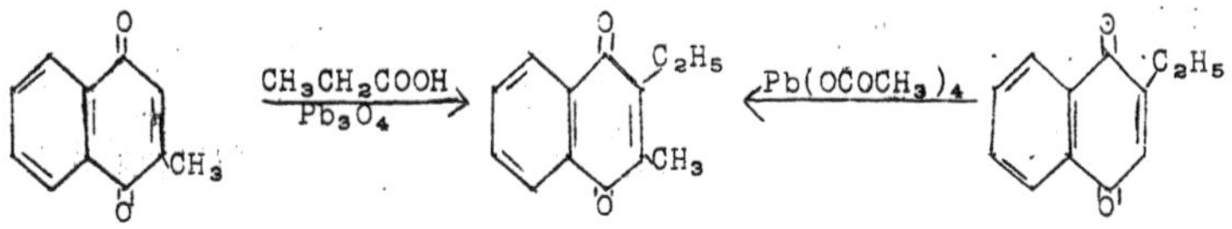
O
CH₃
O
CH₃CH₂COOH
Pb₃O₄
O
C₂H₅
CH₃
O
Pb(OCOCH₃)₄
O
C₂H₅
O

Thus the organo-metallic ester is identified as the alkylating agent.

With n-butyric acid, 2-n-propyl-3-methyl-1,4-naphthoquinone was obtained, while with iso-butyric acid the 2-isopropyl-3-methyl compound was obtained, thus showing that the substituent does not undergo isomerization in the course of the reaction. As further proof the same compounds were made by the methylation of the corresponding propyl quinones. In the preparation of these compounds it was observed that alkylation proceeds less readily when the quinoid ring carries a substituent larger than methyl, than when a 2-methyl group is initially present. A higher temperature is required and the yields are lower. However, there appeared to be no limitation to the introduction of higher alkyl groups into 2-methyl-1,4-naphthoquinone. Alkylation with n-caprylic acid and red lead gave a 34% yield of 2-methyl-2-n-heptyl-1,4-naphthoquinone. Quinones of this type have heretofore only been obtainable by a rather elaborate sequence of synthetic reactions. Similarly phenylacetic acid gave a 65% yield of the 2-methyl-3-benzyl compound.

Since the reaction involves the evolution of carbon dioxide and possible the formation of free hydrocarbon radicals it appears to be analogous to the Kolbe electrolysis of salts such as sodium acetate. The current view of the mechanism of the Kolbe electrolysis is that it proceeds through the intermediary formation of a diacyl peroxide. Then it might be supposed that lead tetraacetate undergoes decomposition to the same intermediate:

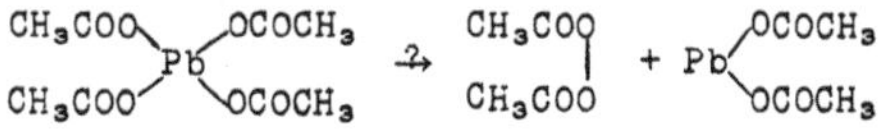
CH₃COO
OCOCH₃
Pb
CH₃COO
OCOCH₃
CH₃COO
CH₃COO
+ Pb
OCOCH₃
OCOCH₃

Indeed diacetyl peroxide does methylate 2-methyl-1,4-naphthoquinone

to the dimethyl compound in 55% yield. No promoter is required in the reaction and the yields are better with one equivalent than with a three to four fold excess which was best with the lead tetraesters.

Many examples of the thermal decomposition of diacyl peroxides in the presence of solvents which serve as acceptors of the hydrocarbon residue have been reported. For example Gelissen and Hermans have shown that dibenoyl peroxide reacts with boiling benzene to give biphenyl, carbon dioxide, benzoic acid, as well as some phenyl benzoate, terphenyl and quarterphenyl:

$$C_6H_3COO\text{-}OCOC_6H_5 + C_6H_6 \rightarrow C_6H_5C_6H_3 + CO_2 + C_6G_5COOH$$

Fieser's reactions are novel in that the reaction is applied to a type of acceptor so favorable for the reaction that the alkylation can be conducted with equivalent amounts of reactants in a solvent essentially inert to the peroxide.

The yields are frequently excellent: thus distearoyl peroxide reacts with 2-methyl-1,4-naphthoquinone to give a 60% yield of 2-methyl-3-heptadecyl-1,4-naphthoquinone. For the introduction of higher alkyl groups the use of diacyl peroxides is more convenient than the use of a mixture of red lead and the fatty acid. Also, the method can be applied to the introduction of at least certain unsaturated hydrocarbon residues. The peroxides from erucic, chaulmoogric, and undecenoic acids were used for the synthesis of the corresponding 3-substituted-2methyl-1,4-naphthoquinones. Even the $\underline{\alpha},\underline{\beta}$-unsaturated acid, 2-heptadecenoic acid, yielded a peroxide which on reaction with 2-methyl-1,4-naphthoquinone gave 2-methyl-3-hexadecenyl (1')-1,4-naphthoquinone in 25% yield. The diacyl peroxides are easily obtained by the action of sodium peroxide on a petroleum ether solution of the acid chloride.

As with the lead tetraesters, it was found that an alkyl group larger than methyl at the 2-position of a 1,4-naphthoquinone impedes the introduction of a second alkyl group. Thus the reaction of 1,4-naphthoquinone with dipalmitoyl peroxide gave 2-pentadecyl-1,4-naphthoquinone. The reaction was also carried out with certain benzoquinones. Cumoquinone was converted into duroquinone and 2,3,5 trimethyl-6-pentadecylbenzoquinone.

It was found that 2-methoxy-1,4-naphthoquinone did not react with diacetyl peroxide, while 2-hydroxy-1,4-naphthoquinone gave phthiocol in good yield. Similarly while 2,5-dimethoxybenzoquinone did not react with dipalmitoyl peroxide: the 2,5 dihydroxy compound gave 2,5-dihydroxy-3-pentadecylbenzoquinone. The reaction may be applicable to the synthesis of the naturally occurring anthelmintic pigment, embelin, 2,5-dihydroxy-3-undecylbenzoquinone.

Tribromoquinone was converted to tribromotoluquinone in acetic acid solution in 68% yield. To test the possibility of using bromine to block nuclear positions temporarily and thus expand the scope of the method, Fieser hydrogenated tribromotoluquinone in acetic acid in the presence of palladium-barium sulfate and sodium acetate and was able to isolate toluhydroquinone in about 70% yield.

Arylation with dibenzoyl peroxide gave unpromising results but the reaction was not completely investigated and the poor results may be due in part to choice of unsuitable experimental conditions.

The reaction was also applied to trinitrotoluene. It was found that lead tetraacetate gave trinitro-m-xylene in 28% yield. Trinitrobenzene reacted with lead tetraacetate to give a mixture of a mixture of trinitro-m-xylene and trinitrotoluene. Nitrobenzene reacts with lead tetraacetate somewhat less readily to give a mixture of products. Oxidation of the mixture with potassium permanganate gave a mixture of o- and p-nitrobenzoic acids. Benzene also reacted to give benzyl acetate, evidently the reaction proceeding in two steps first methylation and then acetoxylation. It had been previously shown by Dimroth and Schweizer that toluene can be acetoxylated by lead tetraacetate. Chlorobenzene gave p-chlorobenzyl acetate. However, naphthalene gave only 1-acetoxynaphthalene.

In general the substances capable of being methylated by lead tetraacetate are unsaturated cyclic compounds which are rather resistant to ordinary aromatic substitution and which do not appear to be susceptible to the acetoxylating action of the reagent. Further evidence of the free radical mechanism of the reaction is the methylation of nitrobenzene in the ortho and para positions rather than the expected meta position.

Acyl peroxides have also been used to alkylate trinitrotoluene giving trinitro-m-xylene in about 11% yield. As further proof for the analogy with the Kolbe synthesis, it was found that when a solution of trinitrotoluene in acetic acid saturated with sodium acetate was electrolyzed, a 9% yield of trinitro-m-xylene was obtained.

The synthesis provides a practical method for the synthesis of 2,3-disubstituted-1,4-naphthoquinones analogous to Vitamin K and other natural products, with the probable extension to other compounds such as monoalkylated dihydroxy benzoquinones, polyalkyl, polynitro benzenes and alkyl hydroquinones.

BIBLIOGRAPHY

Fieser and Chang, J. Am. Chem. Soc., 64, 2043 (1942)
Fieser, Clapp, and Daudt, ibid., 2052
Fieser and Oxford, ibid., 2060
Hey and Waters, Chem. Rev., 21, 186 (1937)
Glasstone and Hickling, Trans. Electrochem. Soc., 75, 333 (1939)

Reported by Zeno W. Wicks, Jr.
January 6, 1943

FISCHER INDOLE SYNTHESIS

Emil Fischer, in 1886, converted the phenylhydrazones of the more common ketones and aldehydes to their corresponding indole derivatives by fusion with zinc chloride. Since his time a great number of these compounds have been prepared using varied procedures which involved condensing agents other than zinc chloride. It is the purpose of this seminar to try to present some idea as to the variety of indole derivative which may be prepared, the methods which have been used, the mechanism involved, and a limited guide to the literature. No attempt has been made to discuss other methods of syntheses for indole and its derivatives.

An examination of the types of compounds which can be used and the indole derivatives they form indicates the variety of products formed.

1. 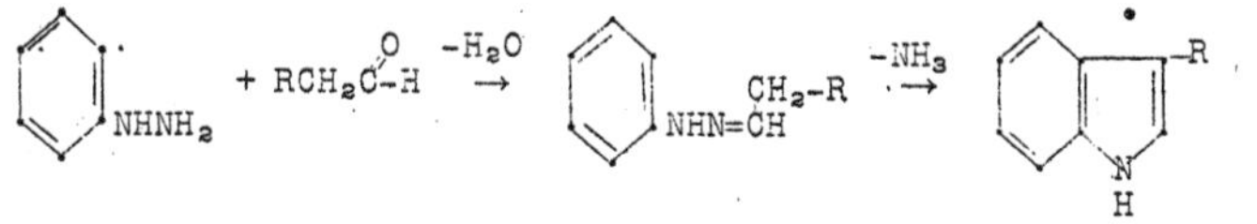
NHNH₂
+ RCH₂C-H
O
-H₂O
CH₂-R
NHN=CH
-NH₃
R
N
H

2. 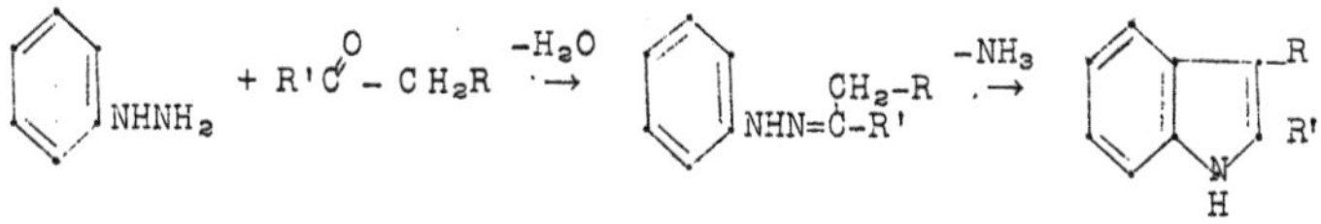
NHNH₂
+ R'C - CH₂R
O
-H₂O
CH₂-R
NHN=C-R'
-NH₃
R
R'
N
H

3. 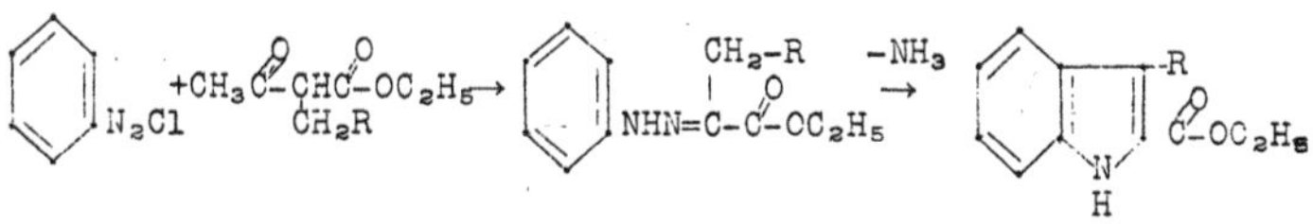
N₂Cl
+CH₃C-CHC-OC₂H₅
CH₂R
CH₂-R
NHN=C-C-OC₂H₅
-NH₃
R
C-OC₂H₅
N
H

4. 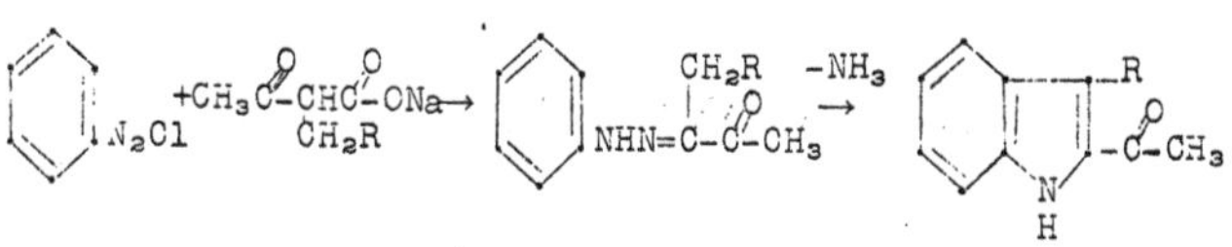
N₂Cl
+CH₃C-CHC-ONa
CH₂R
CH₂R
NHN=C-C-CH₃
-NH₃
R
-C-CH₃
N
H

Such substituents as methyl, ethyl, methoxyl, nitro, chloro, bromo, and iodo groups may be placed anywhere on the benzene ring by starting with the correspondingly substituted phenylhydrazine or benzene diazonium chloride. Naphthylhydrazine has also been used. The phenylhydrazine may have alkyl groups attached to the nitrogen atom adjacent to the ring in which case N-alkylindole derivatives result. Among others, such groups as methyl, ethyl,

phenyl, and acetic acid have been used. These N-alkylphenylhydrazines are usually synthesized by reduction of the N-nitrosoalkylanilines. As for the ketones, R' may be methyl, ethyl, propyl, isopropyl, or larger alkyl group, phenyl, benzyl, carbethoxy, or carboxyl. The R group of the aldehyde, ketone, or ester may be hydrogen, methyl, ethyl, or larger alkyl group, phenyl, methylene carboxylic acid, polymethylene carboxylic acid, or phenoxymethylene. In order to use 4-aminobutanal, the diethyl acetal was intimately mixed with phenylhydrazine and zinc chloride and the mixture fused. Many attempts have been made to convert the phenylhydrazone of acetaldehyde to indole by this method but they have all failed.

The reaction of aromatic diazonium chlorides with substituted acetoacetic esters or the salts of the hydrolyzed esters is known as the Japp-Klingemann reaction and leads to the production of phenylhydrazones of α-keto esters or α-ketophenylhydrazones. In an analogous manner, benzenediazonium chloride reacts with α-carbethoxycyclopentanone and α-carbethoxycyclohexanone to produce the phenylhydrazone of the monoethyl ester of α-keto adipic and α-keto pimelic acids, respectively.

The α-carbethoxyindole derivatives may be hydrolyzed easily to the acids, which may be converted to the chloride by the action of phosphorus pentachloride in acetyl chloride, or which may be decarboxylated, by heating at a temperature near the melting point of the acid. Ethylindole-3-acetate may be reduced to the corresponding alcohol by sodium in ethanol. The carboxyl group of indole-3-acetic acid is also easily removed by heating at the melting point but the higher homologues are not so easily decarboxylated.

Mechanisms for the Fischer indole synthesis have been suggested by Cohn, Reddelien, Bamberger and Landau, and Hollins. Those of Cohn, and Bamberger and Landau may be easily dismissed as incomplete. Robinson and Robinson have presented what they consider to be evidence against the theory of Reddelien, as modified by Hollins.

The mechanism most recently suggested and best supported by experimental results is that of Robinson and Robinson.

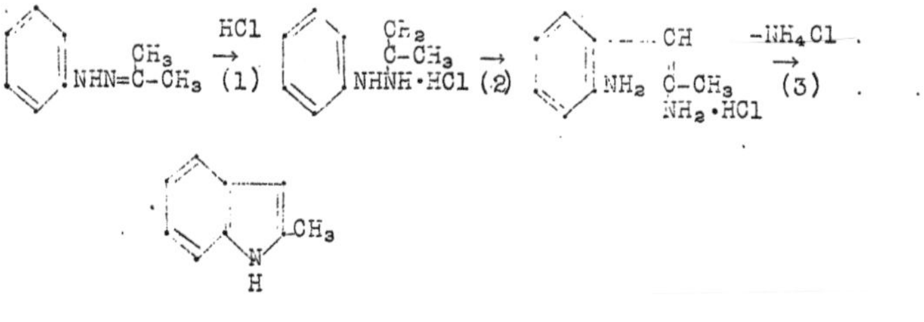
HCl
(1)
NHN=C-CH3
CH3
CH2
C-CH3
NHNH·HCl
(2)
CH
NH2
C-CH3
NH2·HCl
-NH4Cl
(3)
CH3
N
H

According to this mechanism the steps involved in the production of a derivative of indole from an arylhydrazone, in succession, (1) rearrangement to an unsaturated hydrazine, which is the isomeric change of an enimic into an enamic modification; (This is assumed to occur by the addition of the acid reagent and decomposition of the additive product.) (2) an isomerization of the ortho-benzidine type such as may be realized in the naphthalene series, and finally, (3) the elimination of ammonia from the resulting diamine. It will at once be recognized that the conditions necessary for the reaction, namely, an acidic reagent and an elevated temperature, are those which would be expected to favor each of the above steps if considered separately. In accordance with the experience of molecular transformation due to the intervention of an acidic reagent it is noted that each stage is more basic than the last until, finally, the basic character is neutralized by the accident of the formation of a ring of benzenoid character.

Experimental analogies for the various steps in this mechanism can be readily drawn.

1. The phenylhydrazones of ketones and aldehydes which would enolize readily are more easily converted to indole derivatives that those which have little tendency to enolize.

2. The ortho-benzidine type of rearrangement has been realized in the case of certain maphthalene derivatives and the following changes are well established.

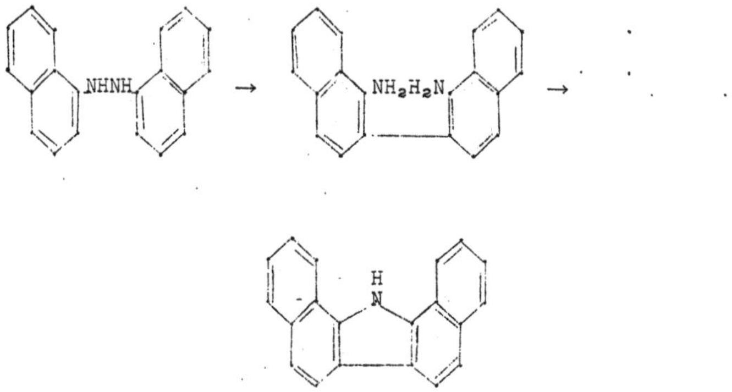
NHNH
NH₂H₂N
H
N

3. The formation of piperidine from pentamethylene diamine monohydrochloride and the formation of secondary aryl amine and its hydrochloride have been accomplished.

Two interesting analogous reactions have been carried out by Robinson and Robinson.

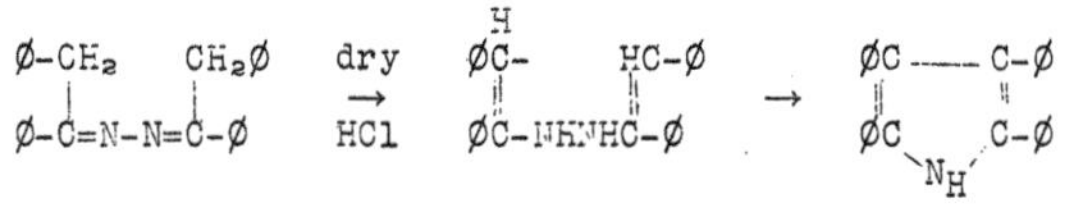

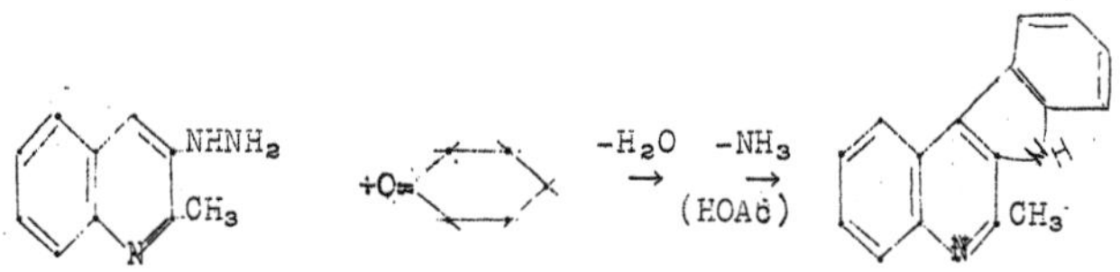

The following tables contain examples of the various procedures which may be used in the Fischer indole synthesis. In these tables P. H. represents phenylhydrazine and B. D. C. represents benzenediazonium chloride.

I. Ethanolic HCl

Dry HCl is rapidly passed into a solution of the phenylhydrazone which heats up and precipitates ammonium chloride.

Starting Materials	Product	% Yield	Reference
α-ketoglutaric acid; m-methoxy P.H.	ethyl 6-methoxy-2-carbethoxy-indole-3-acetate	Poor	27
pyruvic acid, p-methoxy-N-methyl P.H.	1-methyl-2-carbethoxy-5-methoxyindole		28
α-ketoglutaric acid; N-acetic acid P.H.	diethyl 2-carbethoxy-indole-1,3-diacetate		28
ethyl α-acetylglutarate, B.D.C.	ethyl 2-carbethoxy-indole-3-acetate	45	21
benzylacetoacetic ester; o-nitro B.D.C.	2-carbethoxy-3-phenyl-7-nitroindole	60	23
methylacetoacetic ester; 3,4-dimethoxy B.D.C.	2-carbethoxy-5,6-dimethoxy indole	65	24
sodium salt of γ-phthalimidopropyl acetoacetic acid; B.D.C.	2-acetyl-3-(β phthalimidoethyl)-indole		30
α-carbethoxycyclopentanone; o-methoxy B.D.C.	ethyl 2-carbethoxy-6-methoxyindole-3-propionate	19.2	7

II. Aqueous HCl

Starting Materials	Conc.	Product	% Yield	Reference
cyclohexanone; P.H.	dil.	tetrahydrocarbazole		4
cyclohexanone-β-carboxylic acid; P.H.		tetrahydrocarbazole-1-carboxylic acid		5
pyruvic acid; N-methyl P.H.	10%	1-methylindole-2-carboxylic acid		16
pyruvic acid; N-ethyl P.H.	20%	1-ethylindole-2-carboxylic acid		16
diethylketone; o-nitro P.H.		2-ethyl-3-methyl-7-nitroindole		23
sodium salt of benzylacetoacetic acid, BDC	Conc.	2-acetyl-3-phenylindole	98	30

III. Ethanolic H_2SO_4

Starting Materials	% Conc.	Product	% Yield	Reference
β-aldehydopropionic acid; P.H.	10	ethyl indole-3-acetate	47	13
2-carbethoxycyclohexanone; B.D.C.	10	ethyl 2-carbethoxyindole-3-butyrate	38	25
2-carbethoxycyclopentanone; B.D.C.	20	ethyl 2-carbethoxyindole-3-propionate	63.4	18
2-carbethoxycyclohexanone; p-nitro B.D.C.	50	ethyl 2-carbethoxy-5-nitroindole-3-propionate	76	19
2-carbethoxycyclopentanone; 3,4,5-triiodo B.D.C.	30	ethyl 2-carbethoxy-4,5,6-triiodoindole-3-propionate	65.5	19
ethyl α-acetylphenoxyvalerate	40	2-carbethoxy-3-β-phenoxyethylindole	32	29
α-ketobutyric; p-methoxy P.H.		2-carbethoxy-3-methyl-5-methoxyindole	37.5	8

IV. Aqueous H_2SO_4

Starting Materials	% Conc.	Product	% Yield	Reference
3-methylcyclohexanone; p-tolylhydrazine		dimethyl 4,9-1,2,3,4-tetrahydrocarbazole		9
cyclohexanone; α-naphthylhydrazine		tetrahydrobenzo-α-naphthindole		9
cyclohexanone; p-ethoxy P.H.		9-ethoxytetrahydrocarbazole		9

V. Fusion with $ZnCl_2$

Starting Materials	Product	% Yield	Reference
acetophenone; P.H.	2-phenylindole	72-80	31
γ-aminobutyraldediethyl-acetal; P.H.	3-(β-aminoethyl)-indole	45	14
γ-aminobutyraldehyde-diethlacetal; m-methoxy P.H.	3-(β-aminoethyl)-6-methoxyindole	38.2	36
acetone; P.H.	2-methylindole	18	15,34
acetone, N-methyl P.H.	1,2-dimethylindole		11
propionaldehyde; P.H.	3-methylindole (skatole)		12
methyl ethyl ketone; P.H.	2,3-dimethylindole		15
propionaldehyde; N-methyl P.H.	1,3-dimethylindole		11

VI. $ZnCl_2$ in Ethanol

α-ketopimelic acid; o-nitro P.H.	ethyl 2-carbethoxy-indole-3-butyrate	5-10	23

VII. $ZnCl_2$ in Cumene

propylacetoacetic ester; o-nitro B.D.C.	2-carbethoxy-3-ethyl-7-nitroindole		23

VIII. Concentrated H_2SO_4

The hydrazone is dissolved in 10 times its weight of cold concentrated sulfuric acid, allowed to stand 24 hours and poured onto crushed ice.

butylacetoacetic ester; o-nitro B.D.C.	3-propyl-7-nitro-indole-2-carboxylic acid		23

IX. Acetic Acid

Starting Materials	Combined With	Product	% Yield	Reference
diethyl α-keto-glutarate; N-methyl P.H.		1-methylindole-2-carboxylic-3-acetic acid	72.6	21
α-acetylbutyrolactone; B.D.C.	HCl	lactone of 3-methylolindole-2-carboxylic acid	33	22
benzylacetoacetic ester; o-nitro B.D.C.	HBr	3-phenyl-7-nitroindole-2-carboxylic acid	45	23
1,2-cyclohexanedione P.H.	HCl	1-keto-1,2.3,4-tetrahydro-carbazole	75	20

X. Thermal Decomposition Using Catalytic Amounts of Salts

Starting Materials	Catalyst	Product	% Yield	Reference
methylpropylketone; P.H.	CuCl	2-methyl-3-ethyl-indole		1
di-n-propyl ketone; P.H.	CuCl	2-n-propyl-3-ethylindole		1
diethyl ketone;P.H.	$ZnCl_2$	2-ethyl-3-methylindole	69	2
methylethylketone; P.H.	CuCl	2,3-dimethyl-indole		3
propionaldehyde; P.H.	CuCl	3-methylindole	60	3
propionaldehyde; P.H.	$PtCl_2$	3-methylindole	61	3
propionaldehyde; P.H.	$ZnCl_2$	3-methylindole	74	3

Note: In cases where no yield is cited, the compound was obtained but the authors did not give their yield.

Bibliography

1. Arbusow and Frühauf, J. Russ. Phys. Chem. Soc., 45, 694 (1913)
2. Arbusow and Rotermel, J. Gen. Chem., (U.S.S.R.) 2, 397 (1932)
3. Arbusow and Tichwinsky, Ber., 43, 2301 (1910)
4. Baeyer, Ann. 278, 105 (1894)
5. Baeyer and Noyes, Ber., 22, 2184 (1889)
6. Bamberger and Landau, Ber., 52, 1097 (1919)
7. Barrett, Perkin and Robinson, J. Chem. Soc., 2942 (1929)
8. Blaike, Perkin, J. Chem. Soc., 125, 296 (1924)
9. Borsche, Ann., 359, 49 (1908)
10. Cohn, "Die Carbazolgruppe", Georg Thieme, Leipzig, 12 (1919)
11. Degen, Ann., 236, 158 (1886)
12. Degen, Ber., 19, 1566 (1886)
13. Ellinger, Ber., 37, 1806 (1904)
14. Ewins, J. Chem. Soc., 99, 270 (1911)
15. Fischer, Ann., 236, 116 (1886)
16. Fischer and Hess, Ber., 17, 559 (1884)
17. Hollins, J. Am. Chem. Soc., 44, 1598 (1922)
18. Kalb, Schweizer and Schimpf, Ber., 59, 1858 (1926)
19. Kalb, Schweizer, Zellner and Berthold, Ber., 59, 1860 (1926)
20. Kent, J. Chem. Soc., 976 (1935)
21. King and L'Ecuyer, J. Chem. Soc., 1901 (1934)
22. Lions and Harradence, J. Proc. Roy.Soc.,New South Wales,72,221 (1938)
23. Lions, Hughes and Ritchie, J. Proc. Roy. Soc. New South Wales, 72, 209 (1938)
24. Lions and Spurson, J. Proc. Roy. Soc., New South Wales, 66, 171 (1932)
25. Jackson and Manske, J. Am. Chem. Soc., 52, 5029 (1930)
26. Japp and Klingemann, Ber., 21, 49 (1888); Ann. 247, 190 (1888)
27. Kermack, Perkin and Robinson, J. Chem. Soc., 119, 1619 (1921)
28. Kermack, Perkin and Robinson, J. Chem. Soc., 121, 1872 (1922)
29. Manske, Can. J. Research, 4, 591 (1931)
30. Manske, Perkin, Robinson, J. Chem. Soc., 2 (1927)
31. "Organic Syntheses" Vol. 22, p. 98
32. Reddelien, Ann., 388, 179 (1912)
33. Robinson and Robinson, J. Chem. Soc., 113, 639 (1918)
34. Robinson and Robinson, J. Chem. Soc., 125, 827 (1924)
35. Späth and Lederer, Ber., 63, 120 (1930)
36. Späth and Lederer, Ber., 63, 2102 (1930)

Submitted by Curtis W. Smith
January 6, 1943

The Mechanism of the Sandmeyer
and Gattermann Reactions

In 1884, Sandmeyer, in trying to prepare phenylacetylene from diazobenzene and copper acetylide, obtained chlorobenzene in good yield. Investigating further, he found that the active agent in promoting the change was the cuprous chloride. Previous to this, the best method for the preparation of the chloro and the bromo compounds from the diazo group was by the thermal decomposition of the double salts of the diazonium halide and some inorganic halide. Schwechten's method gave yields of 80% in the preparation of 2,2'-dibromobiphenyl. In concentrated halogen acids, yields are poor for chlorine and bromine substitution.

The explanation of the unique catalytic effect of copper in the Sandmeyer and Gattermann reactions has long eluded theoretical chemists. As late as 1940, Hammett in his book states that the characteristic effect of cuprous compounds in the Sandmeyer reaction has no obvious explanation. Sidgwick says in his book on nitrogen that the action of metallic copper is not understood.

Hodgson, Birtwell, and Walker diazotised eight aryl amines in sulfuric acid and decomposed them in two ways:

(a) By cuprous chloride dissolved in concentrated hydrobromic acid; the weights of the chloro and bromo substitution were in the ratio of about 4:96.

(b) By cuprous bromide dissolved in concentrated hydrochloric acid; the ratio then shifted to 60:40.

To explain these results the following type of mechanism was proposed by Hodgson and his coworkers. In (a) the cuprous chloride forms in hydrobromic acid a complex anion of the type.

$$\left[\begin{array}{c} Cl \\ Br-Cu-Br \\ Br \end{array} \right]^{\equiv}$$

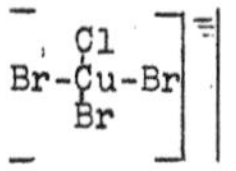

In this complex the halide ions are supposed to be more reactive than in the case of the free halide ions. That is, they give up their electrons more readily. One of these halogen ions, preferably the bromide ion here as bromide ions give up their electrons more readily than chloride ions, loses its electron to the diazonium kation with separation of neutral bromine, evolution of nitrogen and linkage of neutral bromine with the aryl radical. In this case the amount of bromo substitution will largely predominate because there are more bromide ions present and the bromide ions react preferably. In (b) the complex anion

$$\left[\begin{array}{c} Cl \\ Cl-Cu-Cl \\ Br \end{array} \right]^{\equiv}$$

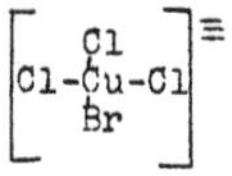

is supposed to be formed. Here, though the chloride ions are in a much larger relative amount over the bromide ions the greater reactivity of the bromide ion makes up for its smaller concentration, so that the products are roughly 50% chloro and 50% bromo substituted.

A similar type of mechanism involving complex anions that activate ions so that they become stronger reducing agents (give up their electrons more readily) is used to explain the dry decomposition of the aryl diazonium borofluorides to give the aryl fluorides, and the action of the hydriodic acid on the diazonium salts to give aryl iodides in the absence of copper.

The trouble with the preceding suggestion of Hodgson for the Sandmeyer reaction mechanism is that he has not been able to suggest why the cuprous double salts decompose in a way which is so very different from the double salts which diazonium halides form with the halides of other metals, such as zinc, cadmium, mercury, arsenic, etc. These double salts decompose in the presence of water to give phenols, and yield appreciable percentages of aryl halides only when they are heated in the absence of water, i.e. in circumstances in which the diazonium halides can themselves give fairly high yields of aryl halides.

Waters presented the following mechanism to account for the unique behaviour of copper. To obtain a neutral aryl radical from a diazonium kation one must supply an additional electron. This can be done by a reducing agent such as a cuprous kation:

$$Cu^{+} + Ar\text{-}N{\equiv}N{:}^{+} \rightarrow Cu^{++} + Ar^{\bullet} + N_2 \qquad (A)$$

This electron transfer can occur within the complex kation of a cuprous diazonium double salt. If a neutral aryl radical is formed in an acid aqueous solution containing halide ions in abundance then an aryl halide molecule can be formed by process (B) which involves an electron release if there is an electron acceptor present.

$$Ar^{\bullet} + {:}\ddot{\underset{..}{Cl}}{:}^{-} \rightarrow Ar{:}\ddot{\underset{..}{Cl}}{:} + e \qquad (B)$$

The electron released in (B) can be accepted by the cupric kation formed in (A).

$$Cu^{++} + e \rightarrow Cu^{+} \qquad (C)$$

This mechanism does explain the experimental fact that cuprous salts are almost unique in promoting this type of reaction.

Waters says that the cuprous ion is the only ion that can cause this series of reactions to occur because it has the ideal oxidation-reduction potential for step (A). Metallic kations of constant valency or kations in their highest state of oxidation cannot act as electron sources. The oxidation potential for the release of an electron from the kations of other transition elements is far too high for the change to occur easily in acid media.

$Mn^{++} = Mn^{++++}$	about 1.5 volts	$Fe^{++} = Fe^{+++}$	.75
$Ni^{++} = Ni^{+++}$	very high	$Cu^{+} = Cu^{++}$	.20
$Co^{++} = Co^{+++}$	1.80		

The stannous kation (about -.4 volts) is too powerful a reducing agent in acid solution and converts diazonium kations into aryl hydrazines, while As^{+++}, Sb^{+++}, and Bi^{+++} are too feeble reducing agents in acid solution.

One of the best proofs that this is the correct picture is that found in the case of the formation of the aromatic nitriles from the double salt of the diazonium cyanide and nickelous cyanide. The oxidation potential of the system $Ni^{++} = Ni^{+++} + e$ is too high to permit reduction of the diazonium kation. But the nickelous ions form complex ions with cyanide ions. This reduces the oxidation potential of the system so that it is now just right for the reduction of the diazonium kation to the aryl free radical. A similar reaction is found in the use of cobalt salts in the preparation of thiocyanates.

The oxidation potential of the iodide ion is not very much higher than that of the cuprous kation, and consequently it is not surprising that the reaction

$$ArN_2^+ + :\ddot{\underset{..}{I}}:^- = Ar\cdot + \cdot\ddot{\underset{..}{I}}: + N_2$$

$$Ar\cdot + \cdot\ddot{\underset{..}{I}}: = Ar:\ddot{\underset{..}{I}}:$$

can be effected without the use of copper salt. Here the iodide ion is the reducing agent.

The fact that symmetrical diaryls are often formed in good yields when cuprous salts are added gradually to aqueous solutions of diazonium salts is in full accord with the mechanism suggested because the frequency of occurence of reaction (C) would be small as the Cu^{++} concentration would be small and so the aryl free radicals would have a chance to dimerize instead of reacting with halogen ions. This reaction leads to the permanent oxidation of Cu^+ to Cu^{++}.

<u>Gattermann Reaction</u>. Hodgson and his co-workers also decomposed diazonium salts with metallic copper in the presence of hydrobromic and hydrochloric acids. For example, p-nitroaniline was diazotised in concentrated sulfuric acid, concentrated aqueous solutions of sodium chloride and sodium bromide were added, and the mixture treated with copper powder. 4-Bromonitrobenzene was formed in 85% yield or better, accompanied by the chloro compound and the phenol. Hodgson made no attempt to explain these results, as his mechanism built around the idea of a complex ion would fall down here. However these results fit in beautifully with Waters' picture.

The Gattermann reaction, in which copper powder replaces the cuprous salt, must be of the same type as the Sandmeyer reaction, since in all details, even in regard to the by-products, it proceeds in a similar manner to the Sandmeyer. The formation of a small quantity of a copper salt is usually noticed when the reaction us carried out, and hence the same scheme as used in the case of the cuprous ion may be used. The free copper acts as the reducing agent in step (A). Since the ionization potential of the

copper is -0.13 volts, this is just about right for the reduction of $Ar{-}N{\equiv}N{:}^{+}$ to $Ar\cdot$.

It appears that Hodgson developed his mechanism in order to explain the change in the ratio of chloro substitution to bromo substitution in the mixed Sandmeyer reactions. But Waters mechanism can explain this more easily, and at the same time explain the large amount of bromo substitution in the Gattermann reaction cited above. An aryl free radical in a mixture of bromide and chloride ions will react preferably with a bromide ion because the bromide ion gives up an electron more readily. $Ar\cdot + :\ddot{X}:^{-} = Ar:\ddot{X}: + e$

Evidence in favour of the transient existence of neutral radicals in reactions of the Gattermann type are that diaryls and polyaryls are frequently by-products and occasionally major products, as in Pschorr's synthesis of phenanthrene derivatives. Though copper is by no means unique in giving up electrons to diazonium chlorides in acetone yet it seems as if copper were almost unique amongst metals for effecting the reaction $ArN_2^{+} + M\cdot \rightarrow Ar\cdot + N_2 + M^{+}$ in aqueous acids. From an inspection of ionization potentials it is evident why many metals, e.g. iron, nickel or platinum cannot replace copper, but in other cases the inertness may be due to surface film formation (e.g. halides of silver and lead). That elementary mercury can play a similar part to copper in initiating the decomposition of diazonium salts, has, however, been demonstrated by McClure and Lowy. Their over all reaction $ArN_2Cl + Hg \rightarrow ArHgCl + N_2$, which is not quantitative, includes combination of the neutral aryl free radical with free mercury.

Hodgson presents some experimental work that is very interesting. He found that good yields of bromo compounds are formed when certain types of amines are diazotised in sulphuric acid and then treated with solutions of sodium bromide (no reaction alone) and cupric sulfate. When 3-bromo-4,6-dinitroaniline was diazotised in sulfuric acid and then was treated with cuprous chloride in hydrochloric acid,

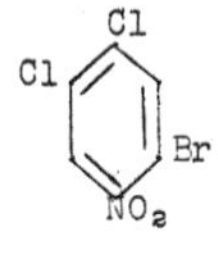
Cl
Cl
Br
NO2

was formed in the cold, and

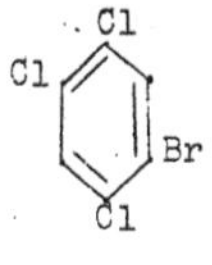
Cl
Cl
Br
Cl

when hot. Copper sulfate treatment gave the normal product

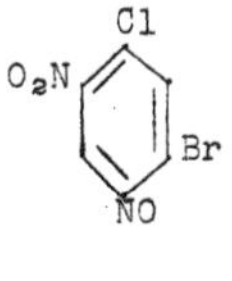
Cl
O2N
Br
NO

. Copper sulfate also is a catalyst for nitro-

naphthyl amines. Waters' mechanism does not explain this catalytic effect of the cupric ion as a cupric ion cannot act as a source of electrons. But this appears to be a highly special case as it occurs only for certain types of amines.

The evidence appears to be in favor of Waters' mechanism, because his explains the unique character of the cuprous ion and free copper in this reaction.

Bibliography

Hodgson, Birtwell, and Walker, J. Chem. Soc., 1941, 770; 1942, 376
Waters, ibid., 1942, 266

Reported by N. K. Sundholm
January 13, 1943.

Julius von Braun's Work on Fluoranthene

TABLE

I.

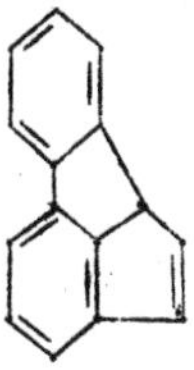

Fluoranthene
(Fittig's structure)

II.

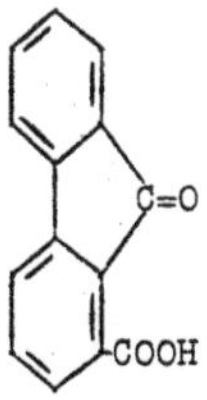

Fluorenone-
1-carboxylic Acid

III.

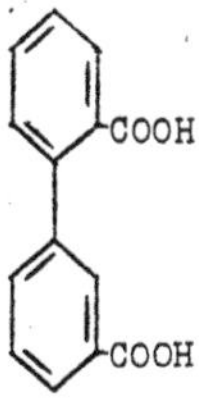

Isodiphenic
Acid

IV.

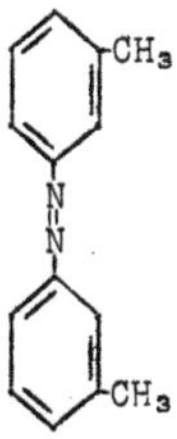

V.

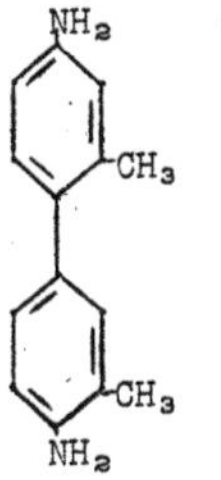

VI.

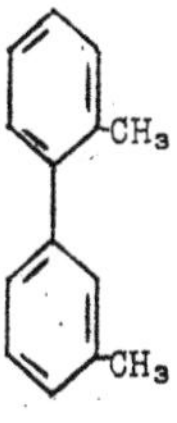

VII.

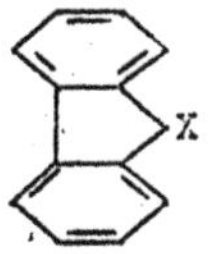

$X=CHCOOC_2H_5$ VII (a)

$X=C(COOC_2H_5)(CH_2CH_2COOC_2H_5)$ VII (b)

$X=CHCH_2CH_2COOH$ VII (c)

VIII.

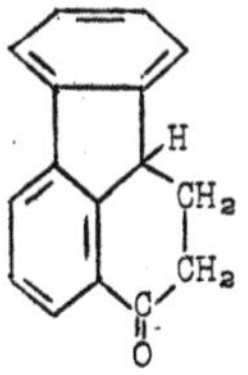

4-keto-1,2,3,4-Tetrahydro-
fluoranthene

IX.

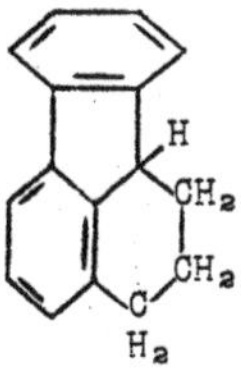

1,2,3,4-Tetrahydro-
fluoranthene

X.

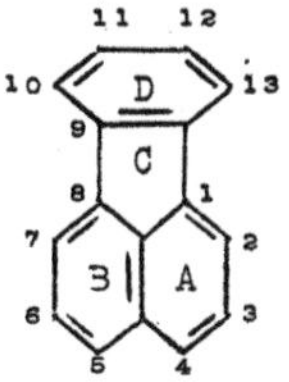

Fluoranthene
(von Braun's structure)

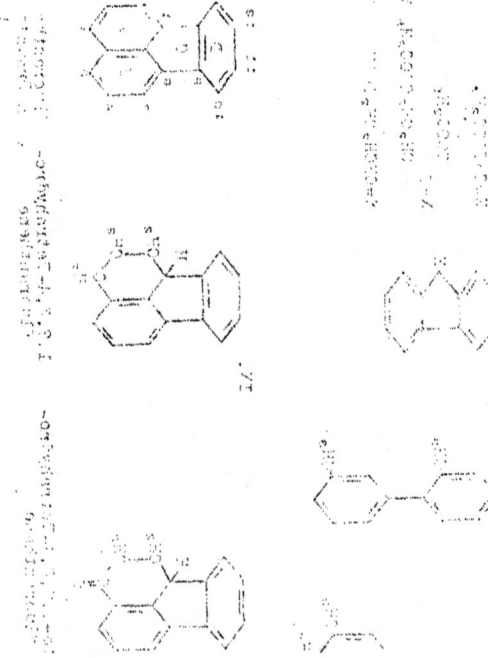

XI.

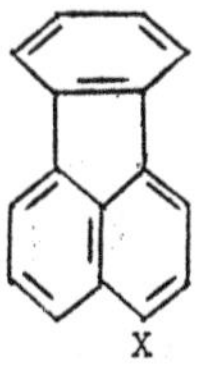

X=Br	XI (a)
X=OH	XI (b)
$X=NH_2$	XI (c)
$X=NHCOCH_3$	XI (d)

XII.

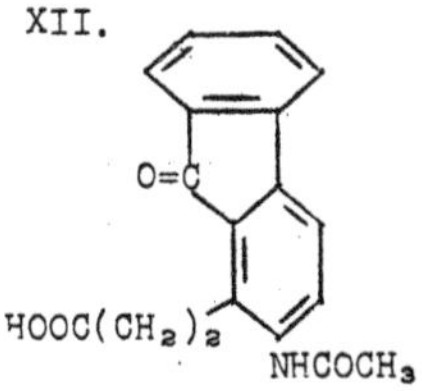
O=C
$HOOC(CH_2)_2$
$NHCOCH_3$

XIII.

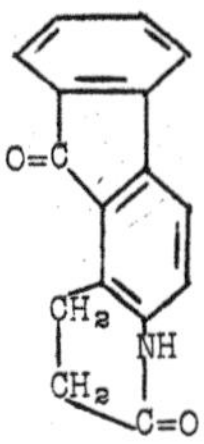
O=C
CH_2
NH
CH_2
C=O

XIV.

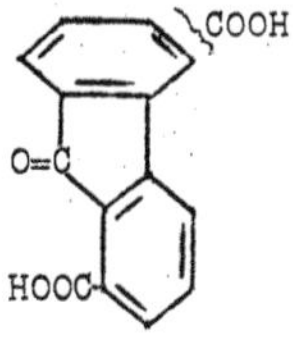
COOH
O=C
HOOC

XV.

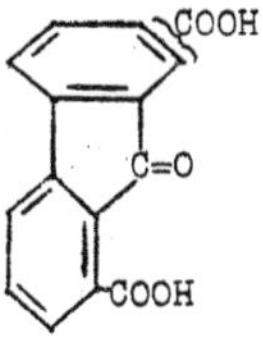
COOH
C=O
COOH

XVI.

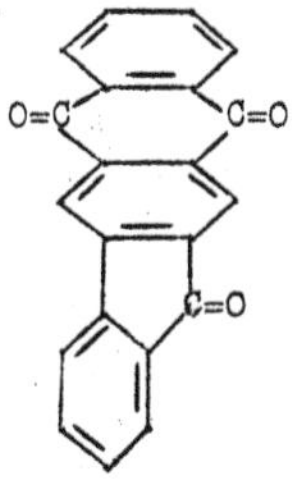
O=C
C=O
C=O

XVII.

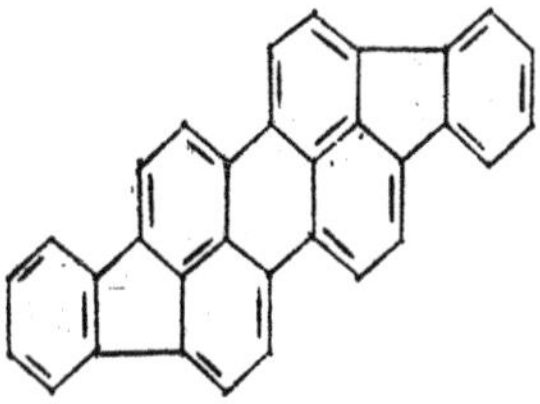

Di-peri-benzoperlyene

I Introduction

In 1873 fluoranthene was first isolated by Fittig and Gebhard as one of the by-products of coal tar distillation. They assigned the structure (I) based on the following observations.

$$\text{I} \xrightarrow[K_2Cr_2O_7]{(O)} \text{II} \xrightarrow{KOH} \text{III}$$

Mayer in 1921 proved conclusively that Fittig's observations had been correct by the reactions below.

$$\text{IV} \xrightarrow[\text{rearrangement}]{[H]} \text{V} \xrightarrow{-NH_2} \text{VI} \xrightarrow{(O)} \text{III} \xrightarrow{H_2SO_4} \text{II}$$

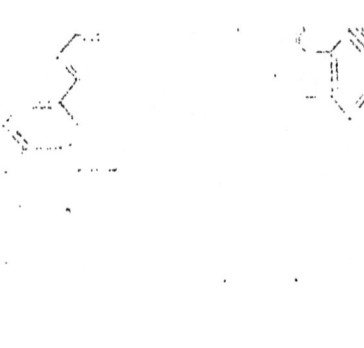

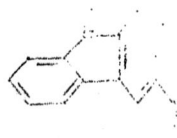

II Structure

In 1929 von Braun, while working with various hydrocarbons, first became attracted to fluoranthene. It was his work which brought about a change in the accepted structure of fluoranthene. He noticed that if the formula of (I) was changed by one carbon, the change in percentage values was still within the limits of experimental error.

$C_{15}H_{10}$	C 94.74%	H 5.26%
$C_{16}H_{10}$	C 95.05%	H 4.95%

This suggested to him the possibility that fluoranthene might not be an indene derivative, but rather a naphthalene derivative. In order to prove this he performed the following reactions.

$$\text{VII (a)} \xrightarrow[\text{ClCH}_2\text{CH}_2\text{COOC}_2\text{H}_5]{\text{Na}} \text{VII (b)} \xrightarrow[-\text{CO}_2]{\text{H}_2\text{O}} \text{VII (c)} \xrightarrow[\text{AlCl}_3]{\text{cyclyzing}} \text{VIII}$$

$$\text{VIII} \xrightarrow{[\text{H}]} \text{IX} \xrightarrow[\text{CO}_2 \text{ over hot PbO}]{\text{distill in}} \text{X}$$

This compound was identical with the product from coal tar, proving fluoranthene to be a naphthalene derivative.

III Hydrogenation

In his next work he studied the hydrogenation products of fluoranthene (X) the order of hydrogenation being in rings A, D, and B consecutively. By controlling the conditions he showed that H_2 adds to one of the naphthalene rings first. With compounds resembling fluoranthene the order was different as in decacyclene and rubicene. Here, apparently, steric hindrance plays an important role.

IV Bromination, Sulfonation and Nitration

These three reactions lead principally to the 4-substituted compounds and small amounts of other mono-substituted compounds. To prove that all three reactions attacked the 4-position he converted the bromo compound to the nitrile to the acid; the sulfonic acid to the sulfonamide, to the nitrile to the acid. He also converted the sulfonic acid to the amine and the nitro compound to the amine and thus showed all entered the same position. To prove that it was the 4-position, he carried out the following degradations.

(1) $\text{XI (a)} \xrightarrow[\text{amalgam}]{\text{Na}} \text{IX}$

(2) VIII was reduced to the alcohol and XI (b) was hydrogenated to the alcohol and both gave the same derivative with phenylisocyanate.

(3) XI (c) $\xrightarrow{CH_3COCl}$ XI (d) $\xrightarrow{(O)}$ XII $\xrightarrow[-H_2O]{heat}$ XIII

The fact that the bromine was eliminated in reaction (1) showed that it must be on the napthalene ring. Step (2) was definite proof and the lactam formation was strong supporting evidence.

V Friedel - Craft Reactions

The reactions of oxalyl chloride, benzoyl chloride and phthalic anhydride on (X) in the presence of aluminum chloride each yielded two compounds (α and β). The β-compound was shown to be the 4-isomer by converting to the corresponding amine. The α-compound was shown to be the 12-isomer by the following reactions.

(1) When the acyl group was transformed to an amino group and the resulting amine degraded by acylation and oxidation according to scheme 3 in section IV above, the resulting amino acid did not cyclize to the lactam.

(2) By hydrogenation he obtained a 1,2,3,4-tetrahydro aryl amino compound.

(3) The α-acid was carefully oxidized to fluorenone dicarboxylic acid and gave a mixture of isomers (XIV and XV).

(4) The phthalic anhydride condensation product produced two isomers on cyclization. By further oxidizing these cyclized products and subsequent decarboxylation he was able to isolate phthaloyl fluorenone (XVI).

It is interesting to note that bromination etc., goes mainly to the 4-position while aluminum chloride catalyzed reactions go mostly to the 12-position. When brominating α-phenylnapthalene the 4-position is most reactive. This is characteristic of the benzene ring, i.e., if it is tied by a double biphenyl ring it is more reactive than in a single biphenyl ring.

VI Reaction with Sodamide

In attempting to make XI (c) directly, rather than thru the intermediate nitro compound, using xylene as a solvent he recovered practically all of the starting material save 0.5% of a red neutral substance. By working in an autoclave he obtained a 50% yield of this material which corresponded to the formula $C_{32}H_{16}(2C_{16}H_{10}-4H)$. In view of the reactivity of the α-hydrogen atoms of napthalene, the most probable and only possible structure is (XVII). This was confirmed by two observations.

(1) (XVII) is formed easily by dehydrogenation with sodamide the 4,4'-bifluoranthyl obtained from 4-bromo fluoranthene with sodium.

(2) 4-Methyl- and 4-phenylfluoranthene prepared from (VIII) by the Grignard method would not give (XVII). The structure is further supported by its behavior on hydrogenation and especially its color. Condensation of two moles of fluoranthene in any other

manner could hardly affect the color. Up to now the only method for making (XVII) is with sodamide. Reaction of (XVII) with nitric acid gave an amorphous nitro compound, with sulfuric acid a sulfonic acid soluble in water, which is used as a dye.

To date, the most important use of fluoranthene is in the manufacture of dyes and dye intermediates.

Bibliography

Fittig and Gebhard, Ann., 193, 142 (1873).
Fittig and Liepmann, ibid., 200, 1 (1880).
Mayer and Frietag, Ber., 54, 347 (1921).
von Braun and Anton, ibid., 62, 145 (1929).
von Braun and Monz, ibid., 63, 2608 (1930).
von Braun and Monz, Ann., 488, 111 (1931).
von Braun and Monz, ibid., 496, 170 (1932).
von Braun and Anton, Ber., 67, 1051 (1934).
von Braun and Monz, ibid., 70, 1603 (1937).

Reported by B. H. Velzen
January 13, 1943

APPLICATION OF THE DIENYNE DOUBLE ADDITION AND RELATED REACTIONS TO THE SYNTHESIS OF CONDENSED RING COMPOUNDS

The Diels-Alder reaction has the advantage of always proceeding stereoselectively to give polycyclic compounds with the _cis_ configuration at the ring junctions. Two successive additions lead exclusively to one of the possible _cis_, _cis_ isomers. Dane and coworkers, Cook and Lawrence, Meggy and Robinson, and Goldberg and Müller have reported applications of the reaction which might be extended to the synthesis of steroids.

A scheme for the synthesis of a steroid from 1,3,5-hexatriene is as follows:

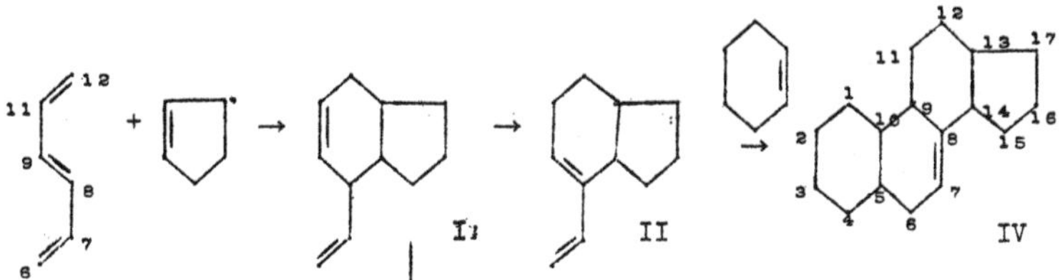
11
12
9
8
7
6
I
II
IV
1
2
3
4
5
6
7
8
9
10
11
12
13
14
15
16
17

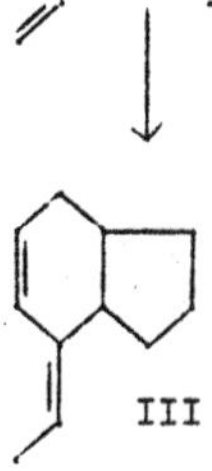
III

The hexatriene thus forms the backbone of the steroid skeleton; Carbon atoms 6,7,8,9,11, 12.

Conditions for success are:

(a) The additions must take place without side reactions, such as polymerization.

(b) The 3-vinylcyclohexene system I must be mobile and must readily isomerize to II.

(c) Isomerization must not proceed exclusively to III, unless III can rearrange to II under the conditions of the second diene addition to IV.

(d) Further isomerization of III to cyclohexadiene derivatives must not be considerable for these add dieneophiles as well.

Farmer and Warren found that _trans_-1,3,5-hexatriene combined with maleic anhydride at 100° C to give in quantitative yield, 6-ethylidene-1,2,3,6-tetrahydrophthalic anhydride.

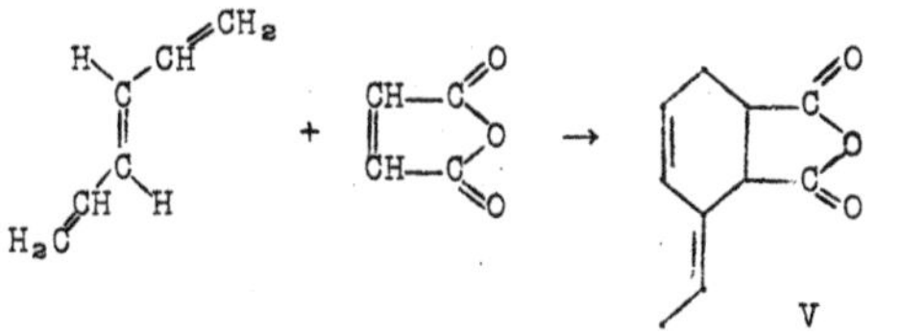
CH₂
H
CH
C
C
CH
H
H₂C
CH—C
O
CH—C
O
O
O
O
O
C
C

(exclusively)

V

Butz was able to add 4-acetoxy-2,5-toluquinone to hexatriene and more recently to cyclohexene and cyclopentane derivatives. The reactions occur at low temperatures and give good yields. Thus condition (a) seems fulfilled. Information on (b) (c) and (d) is very meager. If the reaction investigated by Farmer and Warren was general and unavoidable, the synthesis might as well be abandoned. Thus an investigation with hexatriene and 1,4-naphthoquinone was undertaken by Butz to study this tendency.

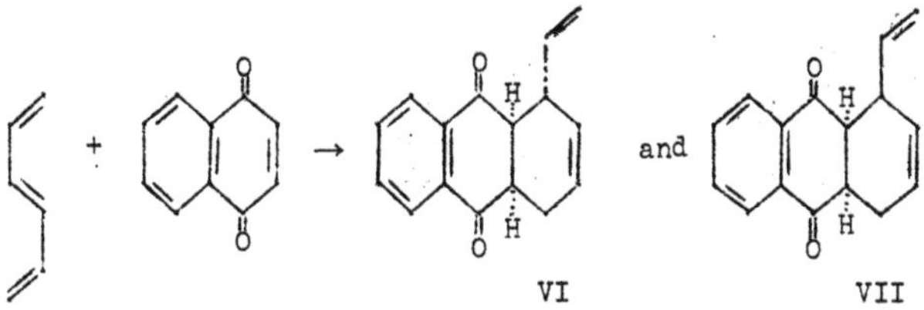
O
O
O
H
H
O
and
O
H
H
O

VI VII

<u>cis</u>-1-vinyl-1,4,4a,9a-tetrahydro-9,10-anthraquinone and <u>trans</u>-

The 1,4-naphthoquinone was chosen because it readily reacts with dienes to give tetrahydroanthraquinones which are easily oxidized to characterizable anthraquinones. It is also symmetrical about the reactive double bond and leads only to two primary adducts VI or VII.

Thus 1,3,5-hexatriene and 1,4-naphthoquinone were heated in a sealed tube for six hours at 50° C. A liquid in 70% and a solid in 25% yield (m.p. 134-6° C) were obtained, both having the empirical formula $C_{16}H_{14}O_2$. That the liquid was a mixture of the <u>cis</u> and <u>trans</u>-tetrahydroanthraquinones VI and VII was shown in the following way.

(Liquid) ($C_{16}H_{14}O_2$) (VI + VII) $\xrightarrow[\text{KOH}]{\text{EtOH}}$ red enolate $\xrightarrow{\text{air}}$ yellow solid $C_{16}H_{10}O_2$ m.p. 162-4° C.

X is the new compound 1 vinyl-9,10-anthraquinone

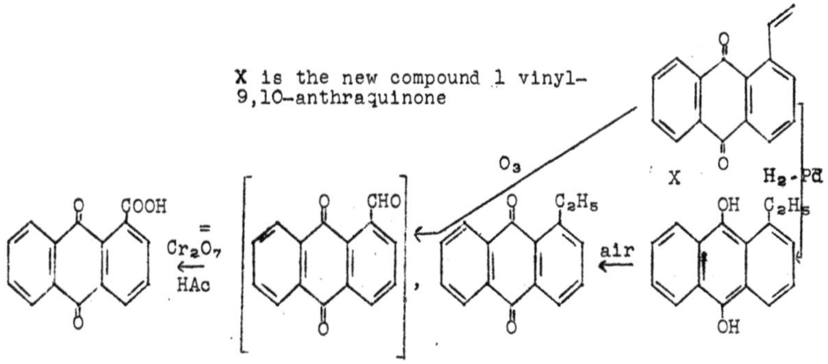
O₃
X
H₂-Pd
COOH
Cr₂O₇
HAc
CHO
C₂H₅
air
OH
C₂H₅
OH

Thus in contrast with the reaction of 1,3,5-hexatriene and maleic anhydride at 100° C, which gave the ethylidene compound V quantitatively, 1,3,5-hexatriene and 1,4-naphthoquinone gives largely the vinyl compounds (VI and VII).

The solid obtained in 25% yield in the same reaction was shown to be identical with a compound XI obtained by Diels and co-workers when 1,4-naphthoquinone and <u>cyclohexadiene</u> were heated.

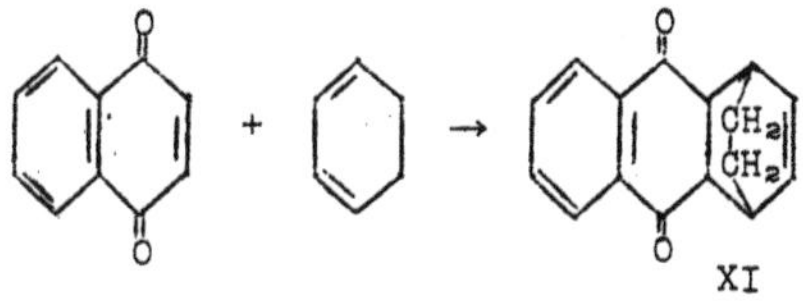
CH₂
CH₂

XI

The assumption, is that cyclohexadiene was present in the 1,3,5-hexatriene or that the conditions of the reaction were such that the 1,3,5-hexatriene isomerized to the cyclohexadiene.

Dienynes add two moles of maleic anhydride.

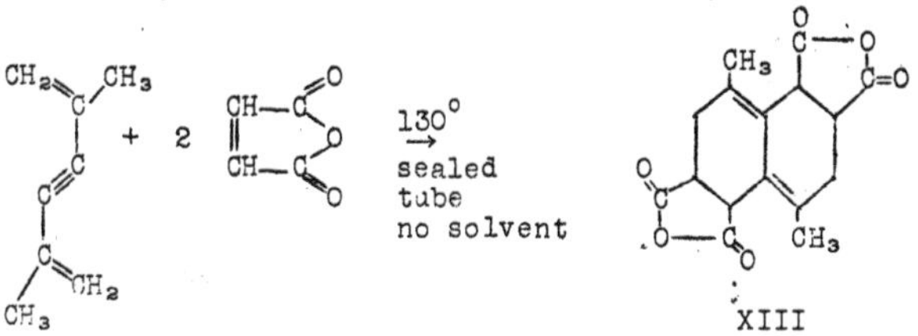
CH₂
CH₃
+ 2
CH
CH
130°
sealed
tube
no solvent
CH₃
CH₃
CH₂
CH₃

XIII

24%

Compound XIII reduces potassium permanganate, adds bromine and on treatment with hydrogen over a palladium catalyst is converted to a hydrocarbon, $C_{12}H_{12}$, XIV.

$C_{16}H_{14}O_6 \xrightarrow[Pd]{H_2} C_{12}H_{12}$

XIII → XIV

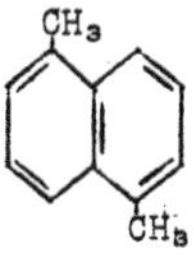

XIV

This hydrocarbon is easily sublimed, melts at 77° and forms a picrate that checks for 1,5-dimethylnaphthalene. The absorption spectra of both XIII and its tetraethyl ester show an extinction coefficient of 22,000 at a wave length of 2500 Å, characteristic of a conjugated carbon double bond system.

Butz and co-workers have reported the total synthesis of a non-benzenoid steroid. Cyclohexenylcyclopentenylacetylene is condensed with maleic anhydride at 130° C. There is no reaction at 70°.

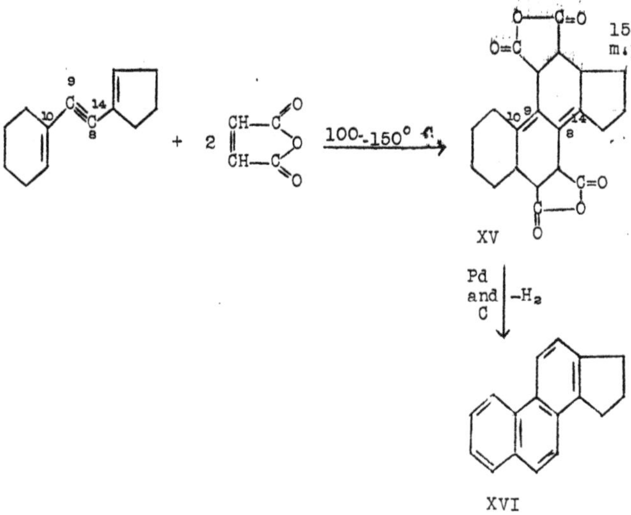

9
14
10
8
+ 2
CH—C
CH—C
O
100-150° C
10 9
14
8
C=O
O=C
Pd
and
C
-H₂

XV

15-17%
m.p. 249-51° C
with decomposition

8,9-steradiene-6,7,11,12-tetracarboxylicdianhydride.

XVI

15,16-dihydro-17-cyclopenta[a]phenanthrene. low yield

Compound XVI did not depress the melting point of an authentic specimen. The extinction coefficient of an ethanol solution of XV is 19,000, λ max = 2555Å, indicating the presence of a <u>conjugated</u> double bond system. Compound XV can be converted to a <u>sterene</u> derivative by the following series of reactions.

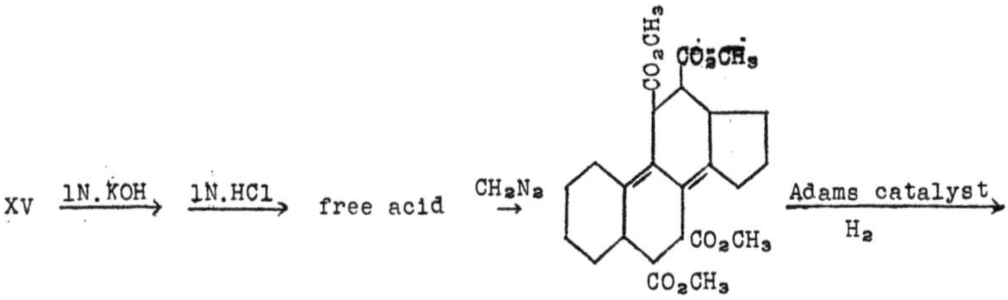
XV
1N.KOH
1N.HCl
free acid
CH_2N_2
CO_2CH_3
CO_2CH_3
CO_2CH_3
CO_2CH_3
Adams catalyst
H_2

XVa

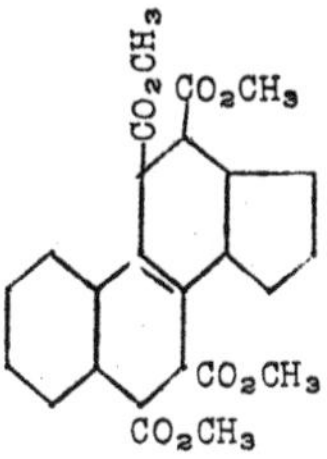
CO_2CH_3
CO_2CH_3
CO_2CH_3
CO_2CH_3

6,7,11,12-tetracarbomethoxy-
8(9)-sterene.

Compound XVII has an absorption maximum of about 2200Å with an extinction coefficient of 5000. Thus the conjugated double bond system has disappeared and XVII is suggested as the structure.

The dianhydride of a tetradecahydrochrysene-6,7,11,12-tetracarboxylic acid and a homolog with an angular methyl group has been reported by Joshel, Butz and Feldman.

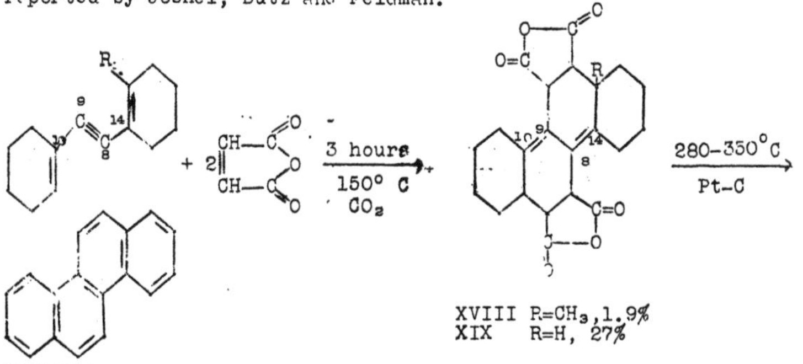
R
9
14
10
8
+ 2
3 hours
150° C
CO_2
280-350° C
Pt-C

XVIII R=CH_3, 1.9%
XIX R=H, 27%

chrysene
low yield

When compound XVIII, when R=CH_3, is pyrolyzed, chrysene is obtained, proving that the methyl group is angular, since a methylchrysene would be obtained otherwise. Absorption spectra is in accord with formulas XVIII and XIX.

The dienyne double addition reaction has been further applied to the methyl and ethyl fumarates. Dicyclohexenylacetylene and methyl or ethyl fumarate are condensed to give the <u>trans</u> chrysitadienes.

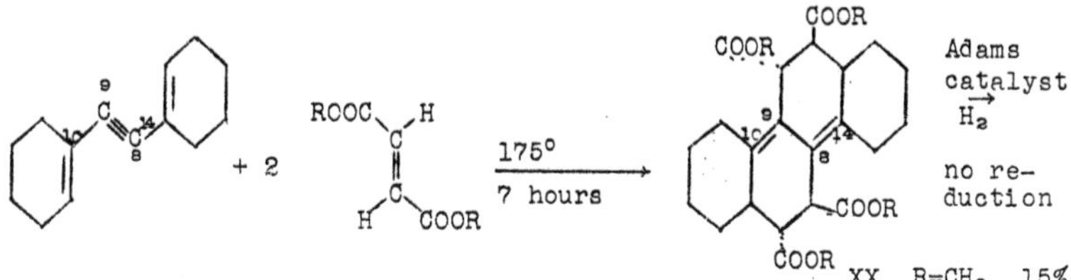

9
10
14
8
+ 2
ROOC
H
COOR
175°
7 hours
COOR
COOR
COOR
COOR
Adams
catalyst
H_2
no reduction

XX R=CH_3, 15%
XXI R=C_2H_5, 7%

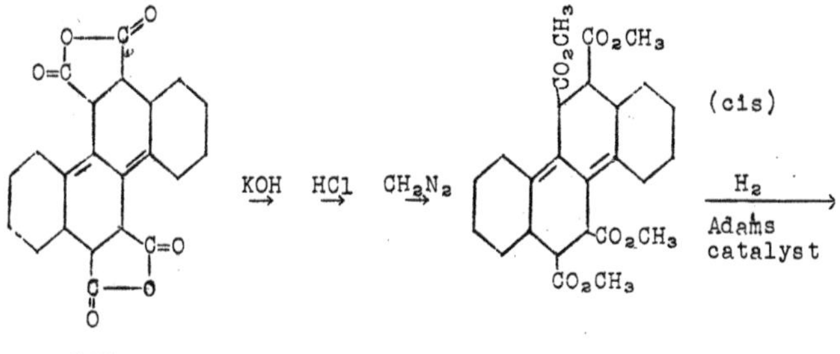

KOH
HCl
CH_2N_2
CO_2CH_3
CO_2CH_3
CO_2CH_3
CO_2CH_3
(cis)
H_2
Adams
catalyst

XIX

XXII

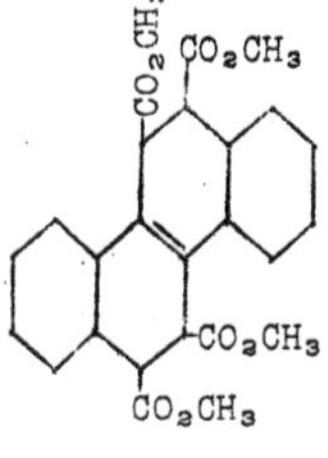

CO_2CH_3
CO_2CH_3
CO_2CH_3
CO_2CH_3

XXIII

That compound XXII is truly a stereoisomer of XX is shown, not only by a different melting point, but also by the fact that XXII readily takes up one mol. of hydrogen (just as the analogous steradiene XVa) while XX resists hydrogenation. Absorption spectra again is in accord with the formulas.

In view of the invariable presence of the 3-hydroxyl or 3-keto group in naturally occuring steroids, it is of interest that 4-methoxy-1-cyclohexenyl-1-cyclopentenylacetylene condenses with methyl fumarate to give the 3-methoxy-<u>trans</u>-6,7-<u>trans</u>-11,12-tetracarbomethoxy-8(14), 9-steradiene.

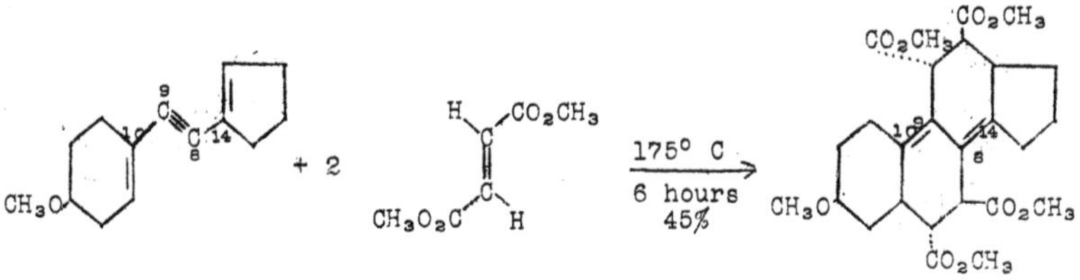
9
10
8
14
CH_3O
+ 2
H
CO_2CH_3
CH_3O_2C
H
175° C
6 hours
45%
CO_2CH_3
CO_2CH_3
10
9
14
8
CH_3O
CO_2CH_3
CO_2CH_3

XXIII

λ max. = 25[illegible] Å
ϵ = 17,000

Compound XXIII analyzed correctly but was not crystalline. Since a new asymmetric center is introduced the number of stereoisomers is doubled and this may be the reason.

Since the preceeding compounds are burdened with unwanted activating groups, the use of ethylene instead of maleic anhydride or the fumarates as the dieneophile was investigated. Under normal conditions of the Diels-Alder reaction, ethylene does not add to 1,3-dienes. However, 2,3-dimethylbutadiene, butadiene and 1,3-cyclopentadiene will condense with ethylene at the proper temperature and pressure.

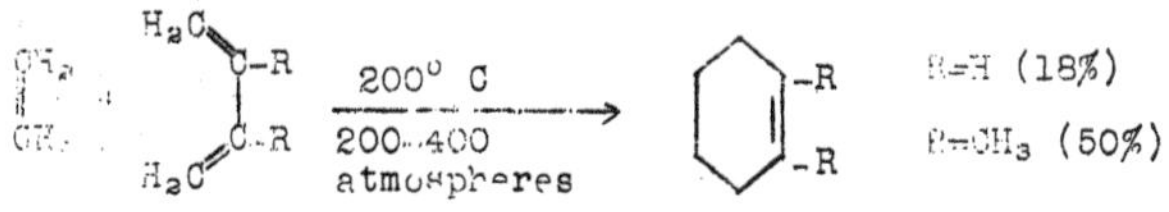
H_2C
C-R
C-R
H_2C
200° C
200-400
atmospheres
-R
-R
R=H (18%)
R=CH_3 (50%)

8130

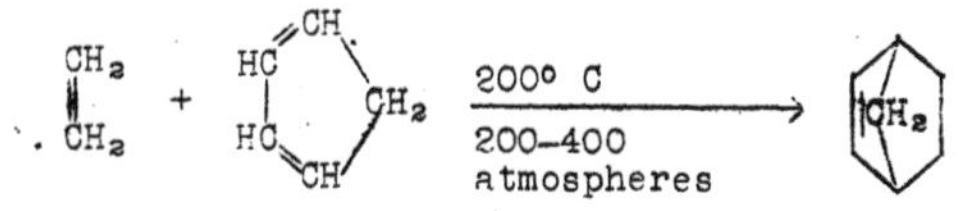
CH2
CH2
+
HC
CH
CH2
HC
CH
200° C
200-400
atmospheres
CH2

bicyclo-[2,2,1]-2-heptene 74%

Ethylene has not been added to any of the above dienyne systems as yet.

BIBLIOGRAPHY

Alder and Stein, Z. Angew. Chem., 50, 510 (1937).
Dane and Eder, Ann., 539, 207 (1939).
Cook and Lawrence, J. Chem. Soc., 58 (1938).
Meggy and Robinson, Nature, 140, 282 (1937).
Goldberg and Müller, Helv. Chim. Acta, 21, 1699 (1938).
Farmer and Warren, J. Chem. Soc., 897 (1929).
Butz, J. Am. Chem. Soc., 60, 216 (1938).
Butz, Butz, and Gaddis, J. Org. Chem., 5, 171 (1940).
Diels and Alder, Ber., 62, 2359 (1929).
Butz, Gaddis, Butz and Davis, J. Org. Chem., 5, 379 (1940).
Butz, Gaddis, Butz and Davis, J. Am. Chem. Soc., 62, 995 (1940).
Butz and Joshel, ibid., 63, 3344 (1941).
Joshel, Butz and Feldman, ibid., 63, 3348 (1941).
Butz and Joshel, ibid., 64, 1311 (1942).
Joshel and Butz, ibid., 63, 3350 (1941).

Reported by J. W. Mecorney
January 20, 1943.

THE 1,4-DIKETONE CYCLIZATION

Wilds, in a recent publication in the Journal of the American Chemical Society, has indicated the various means by which 1,4-diketones might cyclize and the experimental conditions under which these cyclizations take place. A report on this publication and the literature development of this topic is the purpose of this seminar paper.

From 1883, when Paal first synthesized and studied the reactions of 1,4-diketones to the present time, there have only been scattered reports in the literature upon the use of this condensation. Early work indicated only the formation of furan derivatives in which there were no vital interests. Thus Paal observed that the action of alcohol potash on acetonylacetophenone, I, lead to the formation of 2-methyl-5-phenylfuran, II.

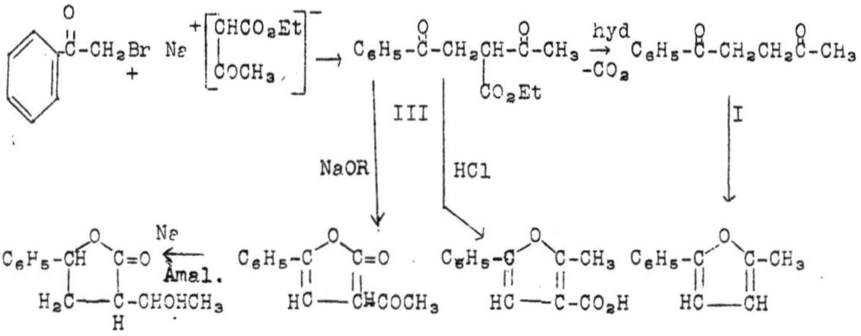

Na
CHCO₂Et
COCH₃
C₆H₅-C-CH₂CH-C-CH₃
CO₂Et
hyd
-CO₂
C₆H₅-C-CH₂CH₂C-CH₃
III
I
NaOR
HCl
Na
Amal.
C₆H₅-CH C=O
H₂C—C-CHOHCH₃
C₆H₅-C C=O
HC—CHCOCH₃
C₆H₅-C C-CH₃
HC—C-CO₂H
C₆H₅-C C-CH₃
HC—CH
VI
V
IV
II

Borsche further investigated the reaction and indicated that the products obtained were a function of the condensing agent. He treated the intermediate ester, III, with HCl and obtained the furan-1-carboxylic acid, IV; sodium alkoxide, however, led to the formation of a ketolactone, V.

Further evidence contributing to the ketolactone formation was obtained by Weltner, who, using $NaOC_2H_5$ and then Na amalgam in alcohol, obtained the corresponding reduction product of V, namely, VI as shown above.

Blaise, noting that only furan type of derivatives had been reported, proposed that another manner of condensation was possible leading to the formation of a cyclopentenone (VII); that involving the elimination of water between the carbonyl group and a methyl, or methylene, group, Thus

CH₂
C₂H₅-C
CH₂
O
H₂
CH₃-C
C=O
CH₂
C₂H₅-C
CH₂
CH₃-C
C=O

VII

This revived interest in the condensation. The synthesis of steroids and hormones require the formation of a cyclopentane ring and this reaction might be a possible approach.

With this in mind, Weidlich and his cowarkers, in their attempt to synthesize equilenin, started with 2-bromotetralone and condensed it with acetoacetic ester.

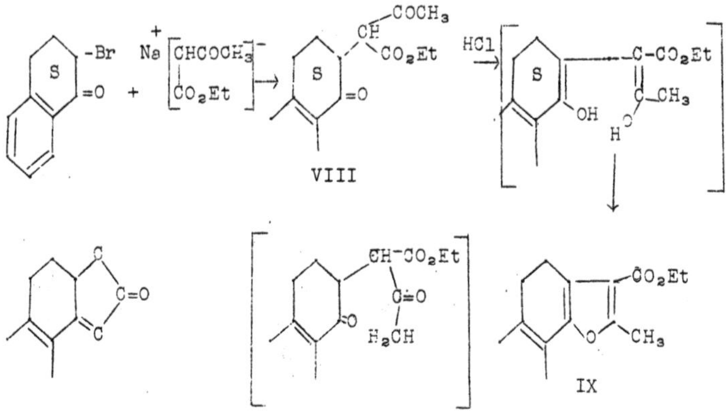
-Br
=O
Na
$CHCOCH_3$
CO_2Et
S
$COCH_3$
CH
CO_2Et
VIII
HCl
$C-CO_2Et$
$C-CH_3$
OH
H
C=O
$CH-CO_2Et$
C=O
H_2CH
CO_2Et
CH_3
IX

They then attempted to cyclize the resulting ester, VIII, to the corresponding cyclopentenone derivative, but without success. Berg, some ten years earlier in 1931, had shown that this same ester, VIII, underwent a typical furan condensation with hydrochloric acid, yielding 1-methyl-2-carbethoxy-7,8-dihydronaphthaleno-(1,2-b)-furan, IX.

The synthesis of cyclohexenone derivatives have also been accomplished. Robinson, while studying possible syntheses of sterols, containing the angular methyl group, condensed β-chloroethyl methyl ketone with 2-methylcyclohexanone, using sodium, and obtained the analogous reaction, namely,

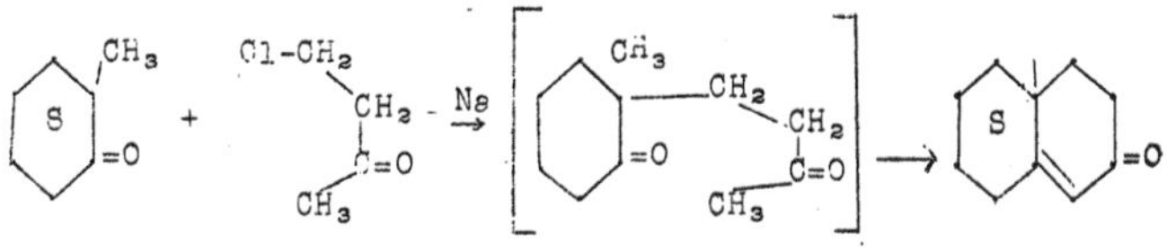
CH_3
S
=O
$Cl-CH_2$
CH_2
C=O
CH_3
Na
CH_2
=O

Another unique application, X to X[1] of this type of 1,4-diketonic cyclization to the synthesis of hormones was used by Goldberg and Muller in their preparation of 15-methyl-15-dehydro-x-norestrone, as indicated by the following series:

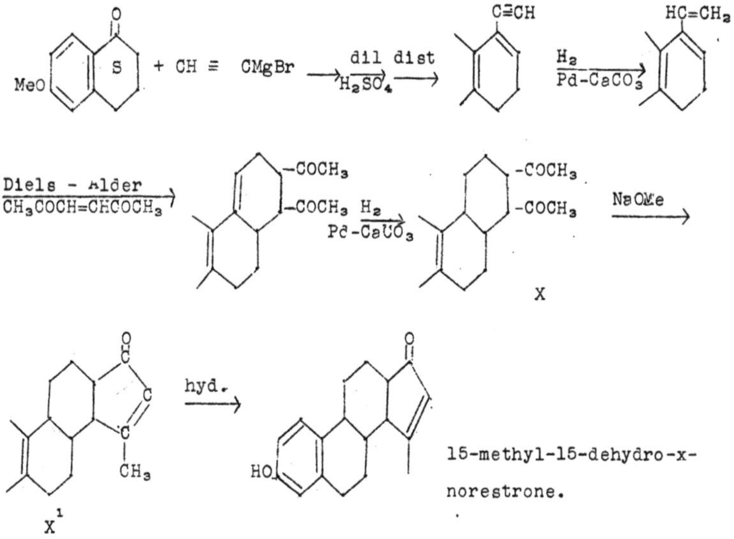
MeO
S
+ CH ≡ CMgBr
dil dist
H_2SO_4
C≡CH
H_2
Pd-$CaCO_3$
HC=CH_2
Diels - Alder
$CH_3COCH=CHCOCH_3$
-$COCH_3$
-$COCH_3$
H_2
Pd-$CaCO_3$
NaOMe
X
hyd.
CH_3
HO
X[1]
15-methyl-15-dehydro-x-
norestrone.

In considering now the work of Wilds, we find that there are esentially three products to be expected from the cyclization of an ester such as XI, produced in 85% yield from acetoacetic ester and 2-bromo-1,2,3,4-tetrahydrophenathrene-1-one:

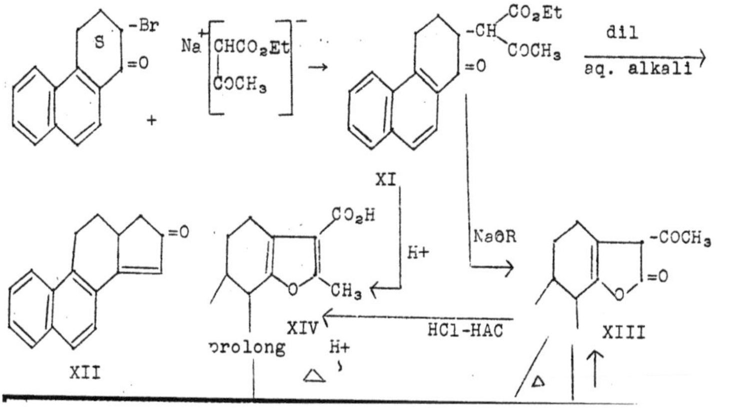
S
-Br
=O
+
Na^+ [$CHCO_2Et$ $COCH_3$]$^-$
-CH
CO_2Et
$COCH_3$
dil
aq. alkali
XI
=O
CO_2H
H+
NaOR
-$COCH_3$
CH_3
XIV
HCl-HAC
XIII
XII
prolong
H+
△

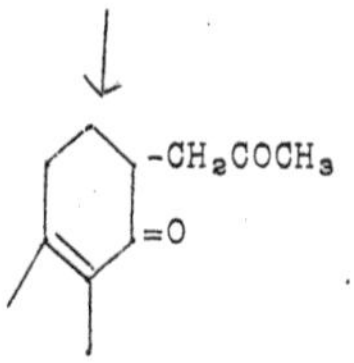
-CH2COCH3
=O

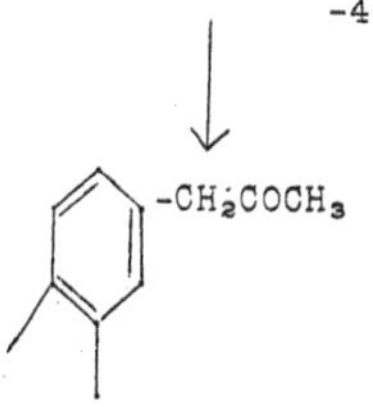
-CH2COCH3

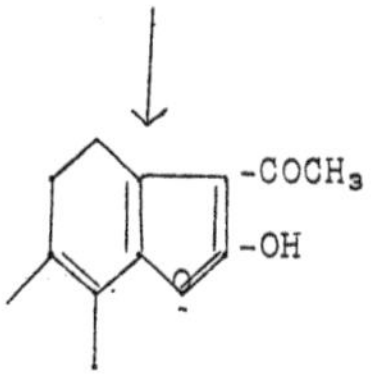
-COCH3
-OH

XVI XVII XV

These three compounds are indicated above as XII, the cyclopentenone derivative, formed by the elimination of water between the ketonic group and the methyl group with subsequent hydrolysis of the ester and decarboxylation; XIII, the ketolactone, or more correctly, XV, a hydrofuran (this was indicated by virtue of the fact that the compound in question was unreactive to alcoholic KOH, thus indicating a furan rather than a lactone), formed from the ester XI by the lactonization of the carbethoxy group with the enolic form of the keto group, followed by enolization to form XV; and thirdly, XIV, a furan-1-carboxylic acid, formed by the elimination of water between the enolic forms of the two ketone groups.

The experimental conditions leading to the formation of each of these products from the ester, XI, are indicated as follows: the cyclopentanone, XII, by the action of dilute aqueous alkali;, the hydroxyfuran, XV, by the presence of sodium alkoxide; and the furan-1-carboxylic acid, XIV, by a strong mineral acid. Prolonged heating of either XIII or XIV in the presence of an acid resulted merely in a reaction corresponding to the hydrolysis and decarboxylation of the ester, XI, and gave rise to the diketone, XVI.

The structures of all of these compounds were amply proven by esterification experiments. For example, XIV, a typical acid, would be expected to esterify readily with either diazomethane or methanol-hydrochloric acid yielding the same product. This was found to be so. On the other hand, XIII, the hydroxyfuran, gave no methyl ester with diazomethane, as would be expected. Furthermore, XIII reacted with alcohol-hydrochloric acid, but hydrolysis of the resulting ester did not regenerate the starting material, XIII. Instead, XIV was formed indicating that the hydroxyfuran, XIII, had undergone rearrangement during esterification to its isomer, XIV, the furan-1-carboxylic acid.

A different approach was used in the proof of cyclopentenone, XII. Through a Clemmensen reduction, followed by dehydrogenation, it was converted to the known 1,2-cyclopentenophenanthrene. The exact position of the double bond in this compound, XII, was not definitely proven, but rather inferred from the ease of hydrogenation in catalytic hydrogenation experiments.

One curious reaction might be pointed out in conclusion. The hydroxyfuran of the type XIII was found to lose carbon dioxide and aromatize to a phenanthrene derivative, XVI. This phenanthrene nucleus was justly proven by hydrogenation experiments. Product XVI is noted to be isomeric with the cyclopentenone, XII.

BIBLIOGRAPHY

Wilds, J. Amer. Chem. Soc., 64, 1421 (1942).
Paal, Ber. 16, 2865 (1883), 17, 2756 (1884).
Weltner, ibid., 17, 66 (1883).
Borsche and coworkers, ibid., 39, 1809, 1922; 41, 190
Weidlich, ibid., 72, 1590, (1941).
Blaise, Compt. rend., 158, 708.
Robinson, J. Chem. Soc., 1937, 53; 1938, 1994.
Goldberg and Muller, Helv. Chim. Acta, 23, 831 (1940)
Ebel, ibid., 12, 16 (1929)
Smith and Carlin, J. Amer. Chem. Soc., 64, 435, 524 (1942).

Reported by S. S. Drake
January 20, 1943

HYDROGEN FLUORIDE AS A CONDENSING AGENT

The first reported use of hydrogen fluoride as a condensing agent was in 1938 when Simons and Archer alkylated benzene with several olefins and tertiary alkyl chlorides. Since then its uses have rapidly expanded, and it has been found to promote polymerization, alkylation, acylation, certain rearrangements, and some unusual ring closures. It has been found that reactions involving oxygen-containing compounds require more hydrogen fluoride, since there is a great tendency for the formation of addition compounds.

Polymerization by hydrogen fluoride has found little application as it is almost uncontrollable. Under proper conditions almost any olefinic compound, and many carbonyl compounds are polymerized.

Alkylation of aromatic and olefinic compounds has been accomplished using alkyl halides, olefins, alcohols, esters, ethers, and strained ring compounds. Hydrogen fluoride has been used successfully for compounds which are as difficult to alkylate as benzoic acid. The reason for this success is probably that it is possible to use high temperatures for long periods without tar formation. On the other hand, even such sensitive compounds as hydroquinone can be alkylated smoothly at low temperatures.

Olefins are the best alkylating agents, but alcohols and the corresponding ethers give yields which are almost as high. Halides usually give somewhat lower yields.

One of the first alkylations accomplished through the use of hydrogen fluoride is a typical example of the use of olefins.

$$CH_3CH{=}CH_2 + C_6H_6 \xrightarrow[0^\circ]{HF} C_6H_5CH(CH_3)_2 \qquad 84\% \text{ yield}$$

In these reactions hydrogen fluoride behaves in a manner similar to sulfuric acid. If the concentrated acid is used in the alkylation of phenol by diisobutylene, a cleaved product (p-t-butylphenol) is obtained; 70% hydrogen fluoride gives fairly good yields of the uncleaved p-t-octylphenol.

The mechanism of alkylation by olefins is not known. Since olefins react more readily than the corresponding fluorides, it seems unlikely that the halogen compounds are intermediates.

Alkylation by cyclopropane gives n-propyl derivatives exclusively. Thus, when cyclopropane is bubbled through a benzene hydrogen fluoride mixture at 0°, a 40% yield of n-propyl benzene is obtained, in addition to smaller amounts of di-n-propyl-(20%) and tri-n-propyl-(3%) benzenes.

Secondary and tertiary alcohols or the corresponding ethers usually give excellent results as alkylating agents in the presence of hydrogen fluoride at room temperatures. Primary compounds require a temperature of 100° before appreciable yields are obtained. A typical example is the alkylation of benzene by <u>t</u>-butyl alcohol.

$$(CH_3)_3COH + C_6H_6 \xrightarrow[\text{room temp.}]{HF} \begin{cases} 40\% \text{ t-butyl benzene} \\ 50\% \text{ di-t-butyl benzene} \end{cases}$$

Higher temperatures favor more highly alkylated compounds.

A very unusual and interesting example is the alkylation of hydroquinone by isopropyl alcohol where the products depend upon the amount of alcohol used.

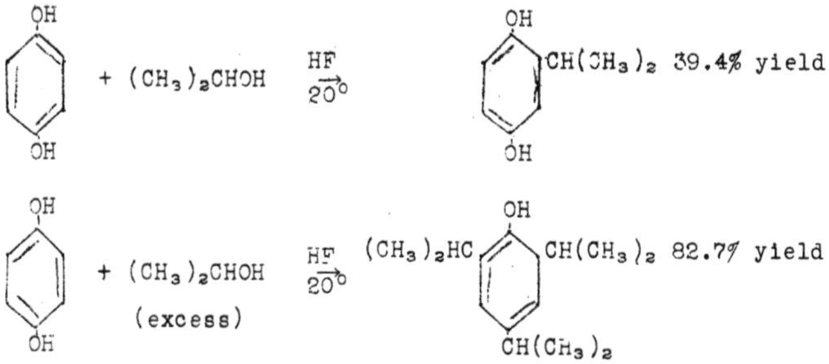
OH
OH
+ (CH3)2CHOH
HF
20°
OH
CH(CH3)2 39.4% yield
OH
OH
OH
+ (CH3)2CHOH
(excess)
HF
20°
OH
(CH3)2HC
CH(CH3)2 82.7% yield
CH(CH3)2

Hydrogen fluoride is also an effective agent for the alkylation of bemzene by esters, and the yields are usually good. Benzyl acetate gives a 75% yield of diphenylmethane and butyl acetate a 60% yield of <u>s</u>-butylbenzene.

$$C_6H_6 + CH_3COOCH_2CH_2CH_2CH_3 \xrightarrow[0°]{HF} C_6H_5CH(CH_3)CH_2CH_3$$

This rearrangement to the secondary alkyl group is typical of primary esters, alcohols, ethers and halides.

Simons has proposed an ionic mechanism for this reaction, which depends upon the fact that hydrogen fluoride is a very powerful ionizing solvent:

$$RO_2R' + 2HF \rightarrow RCO_2H_2^+ + R'^+ + 2F^-.$$

The positive alkyl ion, of course, could give a rearranged product when combining with an aromatic compound.

Although tertiary halides are very satisfactory alkylating agents at 0° in the presence of hydrogen fluoride, secondary halides require somewhat higher temperatures, and primary halides have so far failed to give appreciable yields below 100°. A new and inter-

esting condensation involving a tertiary halide and ethyl furoate has been reported to occur in the presence of hydrogen fluoride.

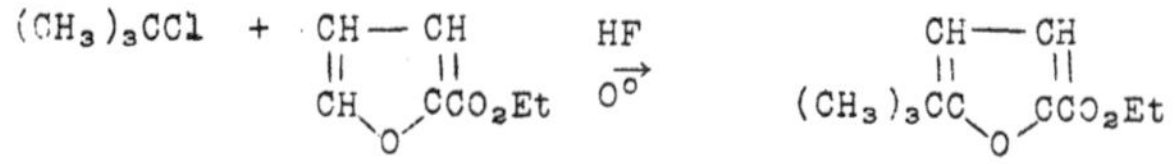

Acylation has proved to be one of the most useful of the reactions catalyzed by hydrogen fluoride. It may be accomplished with carboxylic acids, acid halides, acid anhydrides, or esters, although esters give low yields due to the fact that alkylation occurs more readily. The acids react as readily as the acid halides, which is a great contrast to the action of aluminum chloride, and, obviously, is a great advantage in many syntheses. Acylation usually goes smoothly at 80° - 100°, even when the reactants are of commercial grades. In many cases the hydrogen fluoride method is far superior to the aluminum chloride and sulfuric acid methods.

Acylation has found special use in synthesizing new ring systems especially since hydrogen fluoride, in some instances, acylates in different positions from those acylated by aluminum chloride. Fieser and Hershberg report the following unusual example, the first known instance in which acylation goes to the position indicated:

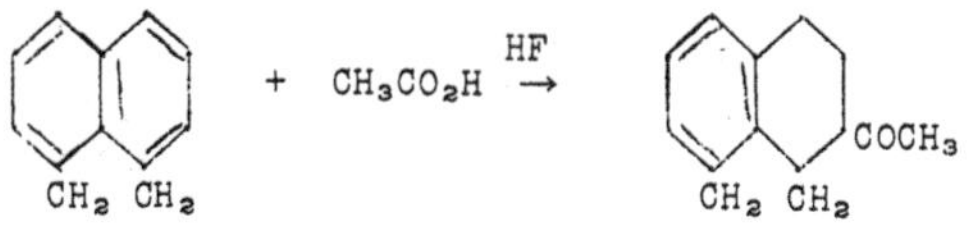

25% yield

Hydrogen fluoride has been found to be a very general agent for the cyclization of γ-arylbutyric and β-arylpropionic acids. The yields are usually very satisfactory; γ-phenylbutyric acid, for instance, gives a 92% yield of α-hydrindone in 73% yield from hydrocinnamic acid. An interesting ring closure which cannot be accomplished by other methods is used in the synthesis of perylene (II) from 1,10-trimethylene - 9 hydroxyphenanthrene (I) and acrolein:

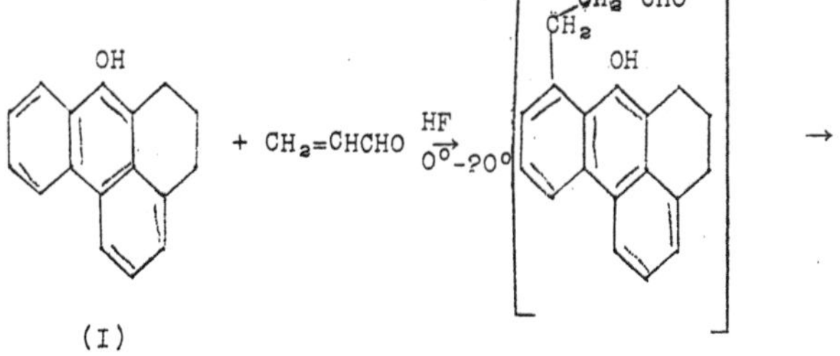

(I)

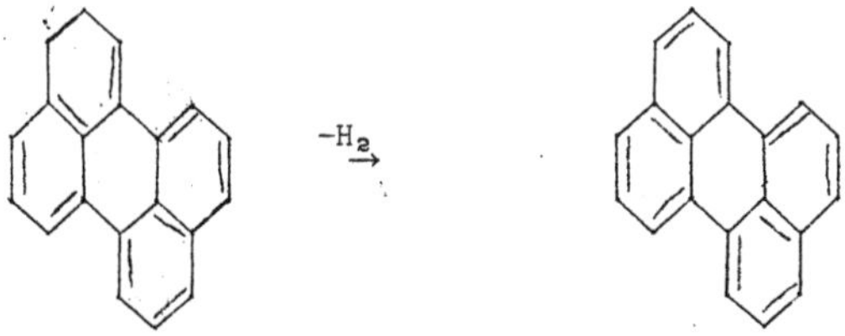

(II)

Thus far little success has been attained in bringing about rearrangements in the presence of hydrogen fluoride. Interchange of alkyl groups (Jacobsen rearrangement) does not occur. The Beckmann rearrangement goes in fairly good yields (72% in the case of the oxime of benzophenone). The Fries reaction does not occur at room temperatures. At 100° the reaction is far from smooth, but low yields of the para rearrangement product are obtained.

Another condensation which promises to be useful is the formation of acids from alcohols, secondary or tertiary halides and carbon monoxide in a copper bomb at 100° - 160°. This method was reported to give fairly good yields in some cases but in many only tars were obtained. The reaction fails to go unless a small amount of water or methanol is present, and with primary alcohols a rearranged product is obtained; 1-propanol, for instance, gives a 28% yield of isobutyric acid.

Bibliography

Simons and Archer, J. Am. Chem. Soc., 60, 986 (1938)
Simons and Archer, J. Am. Chem. Soc., 60, 2953 (1938)
Simons, Archer, and Adams, ibid., 60, 2955 (1938)
Simons, Archer, and Passino, ibid., 60, 2956 (1938)
Calcott, Tinker, and Weinmayr, ibid., 61, 949, 1010 (1939)
Fieser and Hershberg, ibid., 61, 1272 (1939)
Simons, Archer, and Randall, ibid., 61, 1795, 1821 (1939); 62, 485 (194
Simons, Ind. Eng. Chem., 32, 178 (1940)
Simons and Archer, J. Am. Chem. Soc., 62, 451, 1623 (1940)
Simons and Passino, ibid., 62, 1624 (1940)
Simons and Bassler, ibid., 63, 880 (1941)
Spraeuer and Simons, ibid., 64, 648 (1942)
Simons and Werner, ibid., 64, 1356 (1942)

Reported by John E. Wilson

January 20, 1943

SYNTHETIC RUBBER

The subject of butadiene, its mechanisms of polymerization, and synthetic rubbers in general was reviewed in three successive semesters a few years ago. Two of these reports dealt with polymerization mechanisms alone while the third and last was concerned almost exclusively with the commercial manufacture of various types of rubber.

In view of present circumstances occasioned by our war with Japan this country has found it necessary to turn to chemistry for its supply of rubber goods. And Chemistry, forced by the urgency of immediate production, has temporarily shelved development of new ideas concerning the practicality of polymers as rubber-like materials. The majority of recent advances in synthetic rubber then, have been made in the enlargement of manufacturing facilities, and developments in the actual chemistry of rubber have been left in the laboratory, to await further investigation when we have won the war.

A seminar at this time, consequently, should include a sizable quantity of production figures, economic appraisals, and manufacturing difficulties, as well as information about new methods of favoring desired mechanisms and obtaining certain physical characteristics in a product. This is such a report.

Until our entry into the war 97% of the rubber used in this country was obtained from the East Indies. Now, with this source no longer accessible, we find ourselves with an ordinary year's supply in our stock pile, 350,000 tons of reclaimed rubber possible for this year and an equal amount for next year, and 30,000 tons of imported rubber from South America and Liberia. This total quantity is not enough to meet the demands of the Army, Navy, and Air Forces, to say nothing of civilian needs.

The only hope is that the chemist and engineer can synthesize good rubber at the rate of 500,000 or even 1,000,000 tons a year within a very short time. In 1939 we produced 2,400 tons of synthetic rubber, in 1940, 5,000 tons, 12,000 in 1941, and by the end of this year we will have turned out 30,000 tons. Obviously the goal which has been set is a monumental one.

Only two types of synthetic rubber have been manufactured in any quantity previous to this emergency. Du Pont has been producing neoprene for some time now, but its high cost has limited its output to the specialty market where unusual oil resistance was desired. Expansion of neoprene plants, however, has proceeded rapidly. Thiokol, developed by Dow, several years before the war, was considered as an inferior elastic product until its use as re-tread material for tires was discovered. It now assumes great importance in the rubber emergency.

These two elastic polymers, however, are only a relatively small item in the huge program planned for the manufacture of rubber substitutes. Two others, butyl rubber and buna-S, are counted on for a substantially larger quantity of rubber-like material. Official plans for the eventual production figures for synthetic rubber are tabulated below:

Neoprene.................	40,000 tons
Thiokol..................	30,000 tons
Buna-S...................	700,000 tons
Butyl Rubber.............	130,000 tons
total......	900,000 tons

Buna-S is the type of synthetic rubber which has been selected by the government for a concentrated production drive. It is made from butadiene and styrene. The great problem is the production of butadiene, for there is a ready source of styrene made by the Dow Company.

Five methods for making butadiene have been considered, all of which have been developed at least as far as the pilot plant stage. These include -

1. Thermal cracking.
   This process, it is planned, will yield 108,000 tons of butadiene. The Baruch committee has recommended that provisions be made for the production of an additional 100,000 tons by this method. The raw material is naphtha and light oil feed stock.

2. Catalytic cracking of butane to butylene and thence to butadiene.
   340,000 tons will be synthesized by this method, most of it by Standard Oil of New Jersey.

3. Ethyl alcohol process.
   Carbide and Carbon Chemicals Corp. has fostered advancement of this method which is expected to yield 220,000 tons.

4. Butylene glycol process.
   The possibilities of this substance as a source of butadiene are as yet not completely investigated and so figures are not available. It is planned to construct a plant in the middle west to produce 27,000 tons either by this or the ethyl alcohol method. Construction date is set six months in the future to allow time for a fair comparison of the two processes.

5

5. Butyl alcohol method.
   Considerable experimental work with this process has indicated its unattractiveness as a source of butadiene. The overall yield is low and there are many other war needs served by butanol.

Butyl rubber is primarily the product of the Standard Oil Co. of New Jersey. The monomers from which it is made are isobutylene and isoprene, the latter being present to the extent of 1.5 to 4%. The presence of the diolefin unit in the polymer serves only as a site of double bonds which allow the substance to be vulcanized. After vulcanization butyl rubber is very resistant to oxidative deterioration, and so to aging.

Tires made wholly of butyl rubber have given 20,000 miles of se-

vice at speeds under 40 m.p.h. Such experimental products contained but 1.4 of diolefin and it is thought that a higher percentage of diolefin will serve even better.

Butyl rubber is an excellent barrier to passage of gas...so that it might be used in the manufacture of inner tubes.

The cost of plants for making rubber is not a matter of history and therefore cannot be set down here. Estimates vary greatly, but in our national emergency relatively little significance has been attached to any such expense. Much more attention has been devoted to the availability of strategic material needed for construction, the time required to construct such plants, and the assured availability of suitable raw materials. All of these things will determine the success of our rubber program. The last of the 24 proposed butadiene plants is scheduled for completion next October, ten months from now.

After the plants have been built there will still be new problems to iron out. The production of buna-S from a material such as ethyl alcohol involves three steps, none of which have heretofore been attempted on such a scale as is planned. Butadiene must be prepared, it must be polymerized, and this product must be processed. There are encouraging probabilities that the overall yield of buna-S from these three steps will exceed official plans by 10 or 15%, but there still remain the possibilities of unseen obstacles.

The Baruch Committee, acting at the instigation of the president, has listed several suggestions concerning a new synthetic rubber industry. Its recommendations call for an eventual annual output of over 1,100,000 tons of synthetic rubber, as listed below:

Neoprene................	69,000 tons
Thiokol.................	60,000 tons
Buna-S..................	845,000 tons
Butyl Rubber............	130,000 tons

The synthesis of large quantities of rubber like material can become economically practical only if the monomers needed for the syntheses are available in large quantities. This was one of the factors which has been considered in the program undertaken by the government.

The monomers needed for the Buna-S type are butadiene and styrene. At the present time butadiene is being produced by only three of the methods outlined above, (1) thermal cracking, (2) catalytic cracking (3) and from ethyl alcohol.

Thermal cracking consists in heating certain petroleum fractions, chiefly the naptha and light oil fractions, to a temperature of 1300-14000°F, at a low pressure. One advantage of this process is that idle petroleum apparatus can be used, also equipment which originally operated at high pressure but has become unsafe for operation at high pressure can be used at a low pressure in this process. The % yield of butadiene by this method is small but the availability of raw material and suitable apparatus justify its use.

Catalytic cracking of petroleum products also fits in with the

petroleum refining apparatus very well. The best charging stock for this process would be butylene. If butylene is not available then butane must first be dehydrogenated to butylene and this in turn to butadiene. This would constitute a "two stage process" while the first would be a "one stage" process. Actually the petroleum units which are being used are flexible enough that either of the methods may be used. This is advantageous because either can be used as needed in conjunction with the production of high octane gasoline and also isobutylene for the butyl rubber program. It is impossible to discuss in detail all the modifications which can be used to convert the butane-butylene fraction of petroleum into butadiene. In general the process may be represented by one or both of the steps represented in equations (1) and (2).

$$n\text{-}C_4H_{10} \xrightarrow[\text{Al}_2\text{O}_3\text{-Cr}_2\text{O}_3 \text{ Atmospheric pressure}]{500\text{-}750^\circ C} CH_3CH = CHCH_3 + H_2 \quad (35\text{-}90\% \text{ yield}) \quad (1)$$

$$CH_3CH = CHCH_3 \xrightarrow[\text{with or without catalyst, atmospheric pressure}]{500\text{-}600^\circ C} CH_2 = CH\text{-}CH = CH_2 \quad (30\% \text{ yield}) \quad (2)$$

The synthesis of butadiene from alcohol is presumably by some modification of the Lebedev process used in Russia.

$$2CH_3CH_2OH \rightarrow CH_2 = CHCH = CH_2 + 2H_2O + H_2$$

According to the above reaction, the only byproducts are water and hydrogen. However other byproducts are formed such as 2-butene, ethylene, acetaldehyde and butyl alcohol. By 1939 the Russians had improved the process to give 70% conversion to butadiene according to the above equation. The process works best at slightly reduced pressure (630 mm.), at 400-450°C, using a catalyst composed of a mixture of aluminum and zinc oxides, or uranium oxide. Mixtures of the hydrosilicates or oxides of aluminum with salts or oxides of manganese are also claimed to be good catalysts.

Another method which may be used for converting alcohol to butadiene is shown by the following equations.

$$CH_3CH_2OH \xrightarrow{-H_2} CH_3C(=O)\text{-}H \xrightarrow{NaOH} CH_3CH(OH)CH_2CHO \xrightarrow{+H_2} CH_3CH(OH)\text{-}CH_2CH_2OH \xrightarrow{-H_2O} CH_2 = CH\text{-}CH = CH_2$$

A very pure grade of butadiene can be made from 2,3-butylene glycol by first converting the glycol to the diacetate, which is pyrolyzed at about 500°C with the production of high yield (88%) of butadiene. The glycol is obtained by fermentation of grain. The only difficulties encountered in going through the diacetate step are those which are inherent from working with concentrated acetic acid. Pyrolysis of the glycol itself does not give satisfactory yields of the diene because of competing side reactions. By the diacetate procedure it is estimated that 7 to 7.5 lbs. of butadiene may be obtained from

a bushel of corn.

Styrene of sufficient purity for copolymerization is now being produced by the Dow Chemical Co. It is made by Friedel-Crafts condensation between benzene and ethylene to ethyl benzene, using aluminum chloride as the catalyst. The ethyl benzene is then cracked at 800-950°C in the presence of super-heated steam. Small quantities of phenylacetylene and polyethylenes are obtained and must be separated by the fractional distillation before the styrene is used for copolymerization. When distilled, the monomer tends to polymerize and must be stabilized by adding a small amount of sulfur.

Other methods which have been proposed for the production of ethylbenzene are:

$$(1)\quad C_6H_6 + C_2H_5OH \xrightarrow[600^{\circ}F,\ 250\ \text{lbs.}]{H_3PO_4\ \text{catalyst}} C_6H_5CH_2CH_3 + H_2O$$

$$(2)\quad C_6H_6 + CH_2 = CH_2 \xrightarrow[\text{same as (1)}]{} C_6H_5CH_2CH_3$$

Both of these methods require benzene of higher purity than the Dow process. They operate at higher temperatures and pressures and requires recycling of the alcohol and ethylene in order to obtain high yields.

The term "Butyl" rubber does not necessarily apply only to copolymers of isobutylene and isoprene but also to copolymers of isobutylene with other diolefins such as butadiene. Isobutylene is obtained from cracking petroleum for making high octane gasoline. Isoprene is needed in very small quantities compared to isobutylene (1-5%) and may be obtained from the thermal cracking of turpentine. Isoprene can also be synthesized from acetone and acetylene.

$$CH_3\overset{O}{\overset{\|}{C}}CH_3 + CH\equiv CH \xrightarrow{NaNH_2} CH_3-\underset{OH}{\overset{CH_3}{\overset{|}{\underset{|}{C}}}}-C\equiv CH \xrightarrow{H_2} CH_3-\underset{OH}{\overset{CH_3}{\overset{|}{\underset{|}{C}}}}-CH=CH_2 \xrightarrow{-H_2O} CH_2=\underset{CH_3}{\underset{|}{C}}-CH=CH_2$$

For the synthesis of chloroprene to make neoprene, du Pont uses the following method.

$$2CH\equiv CH \xrightarrow[NH_4Cl]{Cu_2Cl_2} CH\equiv C-CH=CH_2 \xrightarrow[CuCl_2]{HCl} CH_2=\underset{Cl}{\underset{|}{C}}-CH=CH_2$$

Although the government plan does not mention such useful synthetic materials as Perbunan, Hycar OR, Chemigum, etc., they will still be produced in increasingly larger quantities as they are needed. Some of the more common synthetic rubbers are listed below.

1. Buna or Perbunan - Butadiene + acrylonitrile (Standard Oil)
2. Chemigum - Butadiene + probably acrylonitrile (Goodyear)
3. Hycar OR - Butadiene + probably acrylonitrile (Goodrich)
4. Vistanex - polyisobutylene (Standard Oil)
5. Vinylite - vinylacetate + vinyl chloride (Carbide and Carbon)
6. Koroseal - polyvinylchloride (Goodrich)
7. Flamenol - polyvinylchloride (General Electric)
8. Saran - olyvinylidenechloride (Dow Chemical Co.)
9. Thiokol A - Ethylene chloride + NaS_4 (Dow Chemical Co.)
10. Thiokol B - Dichloroethyl ether + NaS_4 (Dow Chemical Co.)
11. Polyphenylene ethylene (A.X.F.) (U.S. Rubber Co.)

Polymerization of the various monomers is generally carried out by the emulsion method. For example, the monomers, emulsified in water with the aid of emulsifying agents such as ammonium oleate or sodium alkylnaphthylene sulfonates, are treated with a polymerizing agent such as an organic peroxide or a perborate, and the mixture heated slightly (40-50°) or allowed to polymerize at room temperature in some cases. After about twenty-four hours, the product is a synthetic rubber latex which is coagulated very much as natural rubber latex by the addition of dilute acids or salt solutions. The coagulated material is then washed and dried in the same manner as natural rubber. If there is any tendency for polymerization to continue after drying a small amount (2%) of phenyl β - naphthylamine is milled into the rubber before it is ready for use. If a platicizer is to be used, it is added to the emulsion before polymerization is started. An example of this type is Koroseal in which equal parts of polyvinyl chloride and tricresyl phosphate are emulsified.

Most of the synthetic rubbers, like natural rubber, must be vulcanized in order to convert them into products of practical use. The butadiene rubbers are vulcanized with sulfur, and other compounding ingredients are added to give the finished products good aging and other desirable properties. Neoprene, however, is vulcanized with magnesium oxide.

The other compounding ingredients include such things as, accelerators of vulcanization, antioxidants; plasticizers, reinforcing and filling agents, coloring agents, odorants, etc.

Sulfur is the most common vulcanizing agent but some organic sulfur compounds are also being used, such as tetramethyl thuiram disulfide $(CH_3)_2NCS\text{-}S\text{-}S\text{-}SCN(CH_3)_2$.

Merceptobenzotniazolc (Captax) and its oxidation product, benzothiazyl disulfide (Altex) are the most commonly used accelerators.

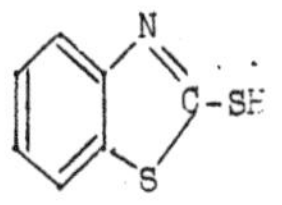

Captax

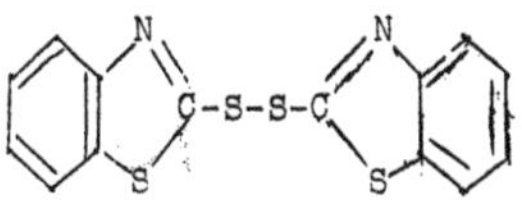

Altex

The effect of the antioxidant is to prevent further polymerization as well as to stabilize the polymer toward oxidation by air. Phenyl-β-naphtyhlamine is nearly always used for this purpose. Polymerized trimethyldihydroquinoline (Agerite resin D) is sometimes added for resistance to heat aging.

Synthetic rubbers generally require larger proportions of plasticizers and softeners than natural rubber. Some of the substances being used are the dialkyl phthalates, triaryl phosphates, soft coal tar, alkylnaphthalenes, aldol-α-naphthylamine resins, and cumarone-indene resins.

Filling agents can be carbon black, zinc oxide and other inert inorganic substances. The use of coloring agents is limited because most synthetic rubber is compounded with carbon black. Coloring agents are usually inorganic pigments.

The following ratios give an idea of the amount of various ingredients which are added to the coagulated rubber latex while it is being processed; based on 100 parts of synthetic rubber.

Sulfur 1-3 parts
Antioxidants 1-2 parts
Plasticizers and softeners 3-5 parts(for the treads) and up to 50 parts for mechanical goods.
Coloring agents - depend on the intensity of color desired. Their use is limited since most of the synthetic rubbers require carbon blacks to bring out their optimum properties.
Fillers - such as carbon black may be added in varying proportions. Rubber for tires contains as much 40% carbon black.

REFERENCES

Brooks, Ind. Eng. Chem., 34, 798 (1942)
Grosse and Ipatieff, ibid., 32, 268 (1940)
Cramer, ibid., 34, 243 (1942)
Fisher, ibid., 34, 1382 (1942)
Dow, ibid., 34, 1267 (1942)
Elder, ibid., 34, 1260 (1942)
A. Talalay and L. Talalay, Rubber Chem. Tech., 15, 403 (1942)
Howorth and Baldwin, Ind. Eng. Chem., 34, 1301 (1942)

Reported by Joseph F. Shekleton and F. W. Wyman
January 20, 1943

SEMINAR REPORTS

II Semester 1942-43

SYNTHETIC LACTONES RELATED TO THE CARDIAC AGLYCONES

The chemistry of the cardiac drugs has been very adequately reviewed in the literature (1), and Phillips, in an organic seminar report a year ago (2), discussed recent advances in the field up to that time. However, current success in the synthesis of compounds closely related to the cardiac aglycones makes it desirable to review this latest work.

It is usual to subdivide the aglycones, on the basis of the structure of the side chain, into the "digitalis - strophanthus group", characterized by the αβ-unsaturated -γ- lactone carrying the cyclopentanophenanthrene ring system as a substituent on the β-carbon atom (type formula I), and the "squill-toad venom group", in which the side chain is an α-pyrone substituted in the 5-position (type formula II)

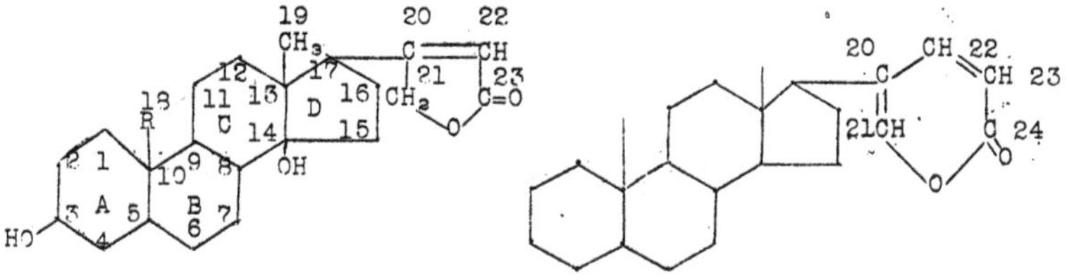

19
20
22
CH_3
C
CH
21
23
18
R
12
11
13
17
16
D
C
14
15
CH_2
C=O
O
2
1
10
9
8
OH
3
A
5
B
7
6
4
HO
20
CH
22
CH 23
C
21CH
C 24
O
O

I II

I. Aglycone	OH groups	C_3-OH/R	RingsA/B	R
(a) Digitoxigenin	3,14	trans	cis	CH_3
(b) Thevetigenin	3,14	cis	cis	CH_3
(c) Uzarigenin	3,14	cis	trans	CH_3
(d) Strophanthidin	3,5,14	trans	cis	CHO

The desire to confirm the proposed formulas by synthetic means and the search for synthetic products to supplement or supplant the natural cardiac drugs have spurred research in several laboratories; Ranganathan (3) in India, Ruzicka and Reichstein in Switzerland, and Elderfield in this country have undertaken the preparation of lactones related to the "digitalis - strophanthus group". The following methods have been employed in the synthesis of β-substituted-αβ-unsaturated-γ-lactones, or $\Delta^{\alpha\beta}$-butenolides.

The first method discovered by Elderfield and his coworkers was used for the preparation of the β-phenyl-, β-cyclohexyl-, β-cyclopentyl, and β-n-butyl-$\Delta^{\alpha\beta}$-butenolides (4,5) and employed halides as starting materials. The appropriate Grignard reagent reacts with methoxyacetonitrile to yield the ω-methoxymethyl ketone. This undergoes a Reformatsky reaction with zinc and ethyl bromoacetate to give a glycol-ether-ester, which in turn yields the unsaturated lactone (IV) when heated with hydrogen bromide in acetic acid, as outlined below.

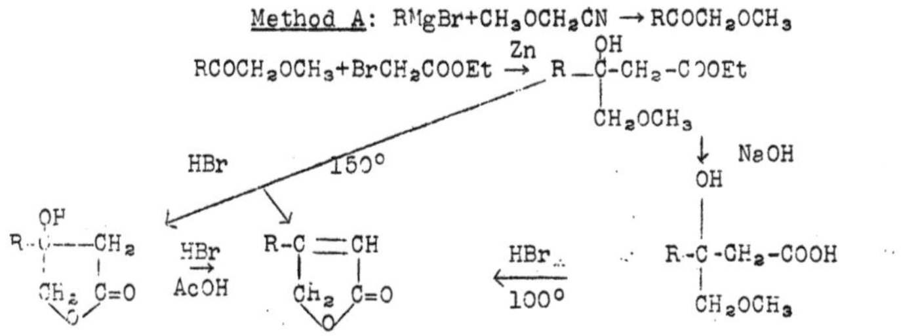

III IV

Proof that the dehydration of the β-hydroxy-lactone (III) produces an <u>αβ</u>-unsaturated lactone is to be found in the ultraviolet absorption spectrum, the exaltation of the molecular refraction, and the synthesis of the identical compound by an unambiguous method (C) of synthesis developed later.

The second method was developed approximately simultaneously in the laboratories of Elderfield (6,7) and Ruzicka (8), both proceeding from methods of preparation of diazomethyl ketones and acetoxymethyl ketones developed earlier by Reichstein and coworkers (9,10, 11). The reaction, with modifications developed later, proceeds as follows, using carboxylic acids as starting materials.

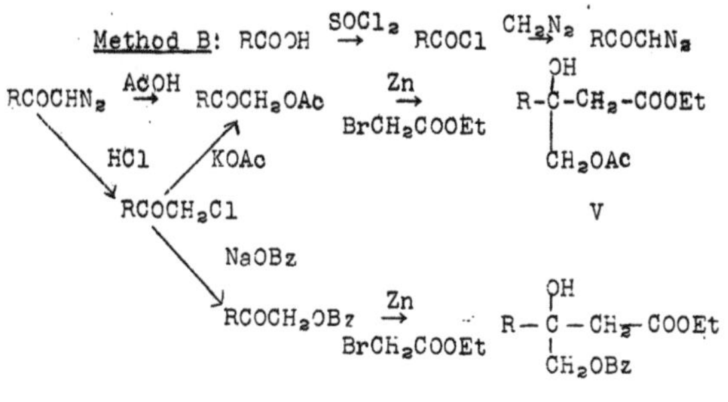

VI

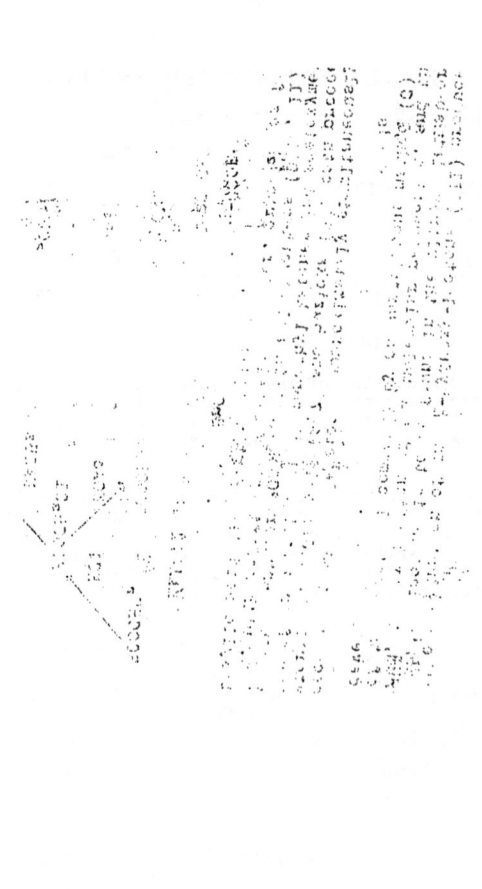

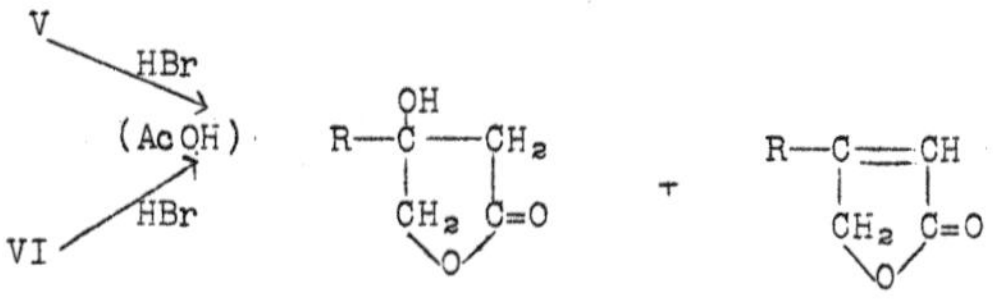

$$
\begin{array}{c}
V \\
\;\;\;\;\searrow \text{HBr} \\
\text{(AcOH)} \\
\;\;\;\;\nearrow \text{HBr} \\
VI
\end{array}
\qquad
R\!-\!\underset{\underset{O}{\overset{|}{CH_2}}\!\!\diagdown\!\!\underset{}{C=O}}{\overset{\overset{OH}{|}}{C}}\!-\!CH_2
\qquad + \qquad
R\!-\!\underset{\underset{O}{\overset{}{CH_2}}\!\!\diagdown\!\!\underset{}{C=O}}{\overset{}{C}}\!=\!CH
$$

III   IV

Compounds III and IV can be separated by chromatographic adsorption
on activated alumina, a technique used extensively throughout this
research. Compound III dehydrates to IV on further refluxing with
HBr in glacial AcOH, or on refluxing with acetic anhydride, with or
without a trace of pyridine (12,13).

The desirability of an unequivocal method of synthesizing model
αβ-unsaturated lactones was fully realized; Elderfield's workers provided just such a synthesis in the phenyl series (14), which was extended by Ruzicka (15) to △αβ-butenolides substituted in the β-position by the steroid nucleus. The reaction depends upon the selenium
dioxide oxidation of a methyl group, activated by an αβ-unsaturated
ester grouping, to a primary alcohol, followed by ring closure; the
method can be outlined as applied to ethyl-β-methyl cinnamate as
starting material.

Method C:

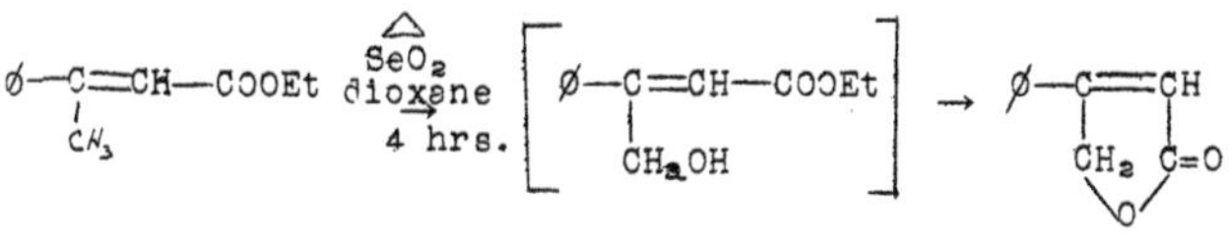

$$
\phi\!-\!\underset{\overset{|}{CH_3}}{C}\!=\!CH\!-\!COOEt
\;\;\overset{\overset{\triangle}{SeO_2}}{\underset{\underset{4\ hrs.}{dioxane}}{}}\;\;
\left[\; \phi\!-\!\underset{\overset{|}{CH_2OH}}{C}\!=\!CH\!-\!COOEt \;\right]
\;\rightarrow\;
\phi\!-\!\underset{\underset{O}{\overset{}{CH_2}}\!\!\diagdown\!\!\underset{}{C=O}}{\overset{}{C}}\!=\!CH
$$

IVa.

This unambiguous method produced compounds identical with those obtained using methods A and B.

The most recent development in synthetic methods of preparation
of △αβ-butenolides is one which utilizes the readily available methyl
ketones as starting materials (26). Using cyclohexyl-methyl ketone
and ethyl dichloroacetate, a Darzens' condensation (27) with magnesium amalgam yields the expected product, VII, which on heating with
HBr in glacial AcOH is acetolyzed, dehydrated and ring-closed to
α-chloro-β-cyclohexyl butyrolactone (VIII). This, on refluxing with
potassium acetate in glacial AcOH, yields β-cyclohexyl-△αβ-butenol-
ide (IVb).

Method D:

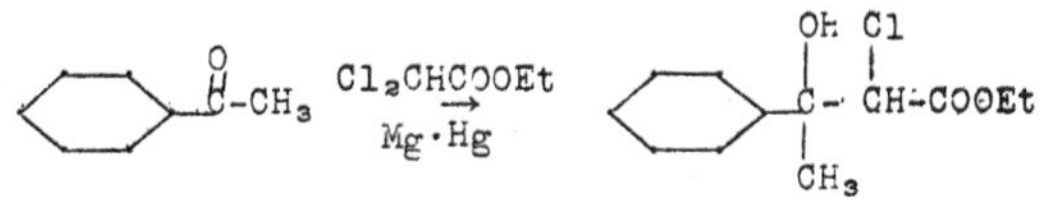

VII

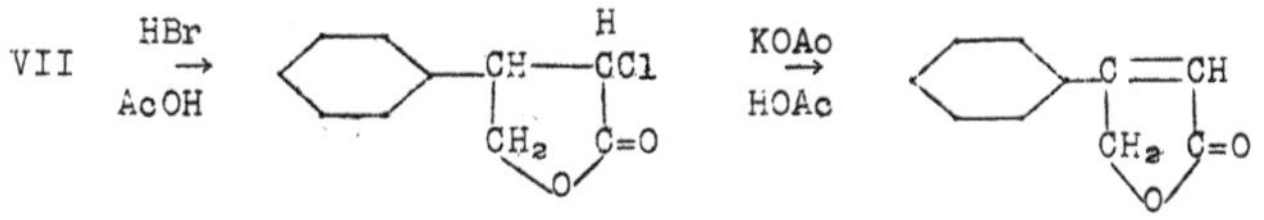

VIII IVb

The properties of the final product correspond exactly to those of the cyclohexyl compound synthesized by methods A and B, and the mixed melting points of derivatives obtained from hydrolysis products furnished direct proof that identical compounds result from all these methods.

In the last two years, a large number of β-substituted-$\Delta^{\alpha\beta}$-butenolides have been synthesized by the four methods outlined, as may be seen from the table at the end of this review. Those αβ-unsaturated lactones substituted with steroid nuclei have ultraviolet absorption spectra with maxima at approximately 2200A° and log =4.3 just as do the naturally occurring aglycones, strophanthidin and periplogenin. The first indirect comparison of synthetic and natural steroid lactones was described by Ruzicka (18), who compared synthetic saturated isomers of 3(β)-acetoxy-21-hydroxy-norallocholenic acid lactone (IX)

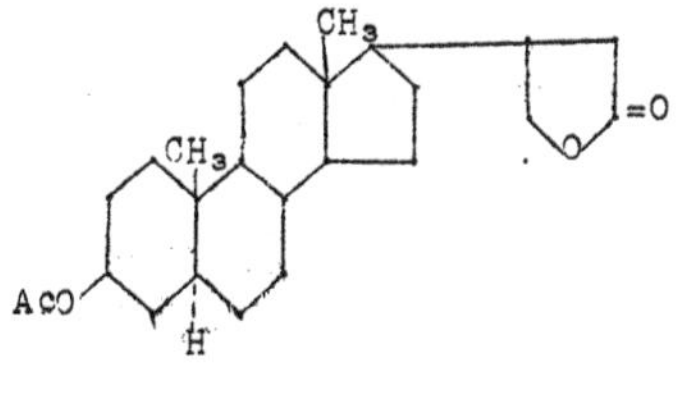

IX

with the isomers, α_1- and α_2-tetrahydroanhydrouzarigenin, obtained by Tschesche (25) on reduction of anhydrouzarigenin:

Compound	Ruzicka MP	Ruzicka $[\alpha]_D$	Tschesche MP	Tschesche $[\alpha]_D$
Isomer α_2	203-204°	+19°	205°	+20.2°
Isomer α_1	243.°	+5.9°	248°	+3.9°

While the agreement in physical properties is certainly suggestive, no mixed melting points were obtained, and it remained for Elderfield (20) and coworkers to establish a direct correlation between lactones prepared from bile acids and the cardiac aglycones themselves, with the carbon skeleton of the side chain of the latter intact. This was accomplished by proof of the identity of catalytically reduced synthetic 3,14-bisdesoxythevetigenin (XI) with hexahydrodianhydrothevetigenin (from naturally occurring digitoxigenin).

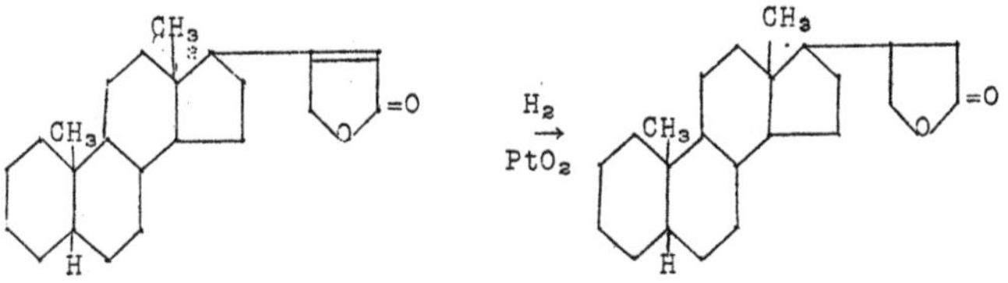

CH3
CH3
=O
O
H
H2
PtO2
CH3
CH3
=O
O
H

X XI

The melting points and specific rotations of compounds from the two sources were found to be identical, and the melting point was not depressed when the compounds were mixed.

The preparation of 5-substituted-α-pyrones as analogs of the "squill-toad venom group" of cardiac aglycones has proceeded more slowly than the synthesis of the $\Delta^{\alpha\beta}$-butenolides, in fact, the 5-methyl- and 5-ethyl-α -pyrones are the only analogs reported in the literature to date (4,16,17). The method of synthesis of 5-ethyl-α-pyrone proceeds as follows: condensation of ethyl-$\Delta^{\alpha\beta}$-hexenoate (XII) with ethyl oxalate in the presence of potassium ethylate gives diethyl 1-ethyl-2-hexene-5-one-1,6-dioate (XIII). Elderfield's workers have found recently that pyridine greatly increases the yield in this reaction. Compound XIII, after hydrolysis of the ester groups, is heated with HBr in glacial AcOH to yield 5-ethyl-6-carboxy-α-pyrone (XIV), which on distillation with copper powder yields the desired 5-ethyl-α-pyrone (XV).

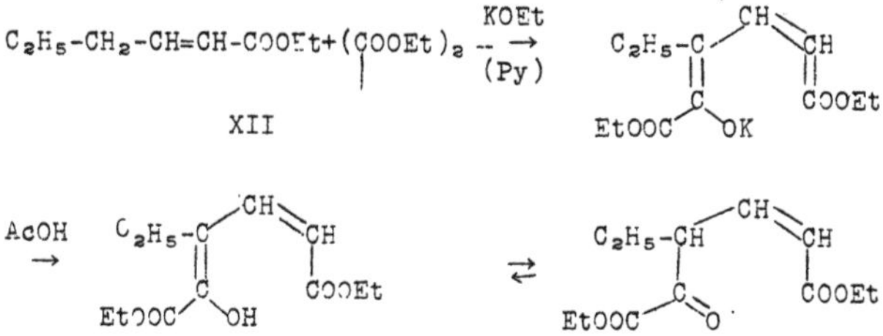

C2H5-CH2-CH=CH-COOEt+(COOEt)2
KOEt
(Py)
C2H5-C
CH
CH
COOEt
C
EtOOC
OK
AcOH
C2H5-C
CH
CH
COOEt
C
EtOOC
OH
C2H5-CH
CH
CH
COOEt
C
EtOOC
O

XII

XIII

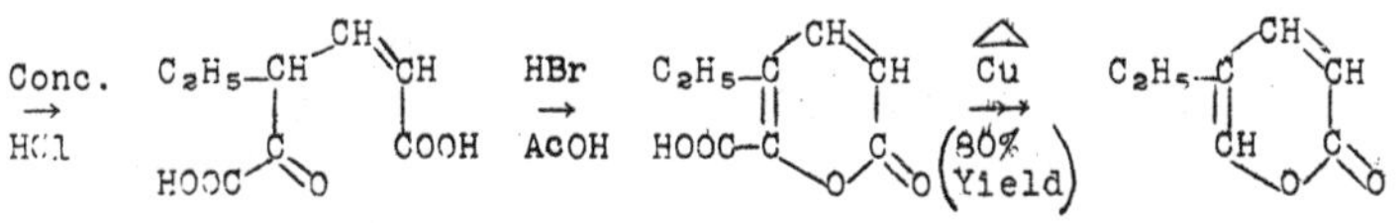

XIV XV

The ultraviolet spectrum of XV corresponds exactly with that of bufotalidin, the naturally occurring cardiac aglycone from toad venom. In contrast to the behavior of γ-alkyl-substituted crotonic esters, γ-cyclohexyl-crotonic ester and ethyl oxalate in the presence of potassium ethylate and pyridine react to form the α-condensation product XVI (28).

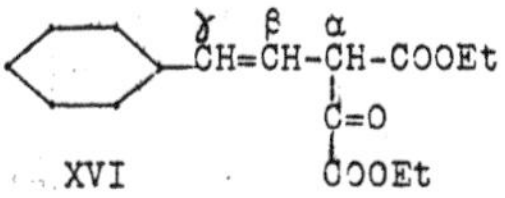

XVI

This was proved by hydrogenating the condensation product and subsequently hydrolyzing it to the malic acid, which was identical with an authentic sample of α-hydroxy-α'-(2-cyclohexyl-ethyl)-succinic acid. The occurrence of this α-condensation seems to rule out synthesis of α-pyrones with an alicyclic substituent in the 5-position according to the method projected above.

So much for the synthesis of substances related to the cardiac aglycones. The pharmacological activity of cardiac drugs is determined by frog and cat tests, of which the latter is the more indicative and important. In the cardiac aglycones, it has been determined that the unsaturated lactone ring is important for activity; rupture or reduction of the ring results in nearly complete loss of cardiac action (29); stereochemical configuration of the steroid nucleus also seems important, since digitoxigenin and thevetigenin (rings A/B-trans) cis) are active and uzarigenin (rings A/B-trans) is not. The aglycones occur in nature as glycosides, and in every case the parent glycoside is more powerful on the heart than the aglycone alone (30,31), probably because of the influence of the sugar molecule on the water-solubility and diffusibility of the material.

Recently, synthetic glycosides of strophanthidin (32) and glucosides of digoxigenin and digitoxigenin (33) have been tested pharmacologically with the following results: assayed in cats and frogs, strophanthidin acetate, strophanthidin β-d-glucoside, β-d-xyloside, and β-l-arabinoside all prove more potent than strophanthidin; they have digitalis - like action in the same range as cymarin, naturally-occurring cymaroside of strophanthidin. The synthetic "digitalis glucosides" have more than twice the activity in cats of the corresponding natural glycosides. The syntheses of the above glycosides of strophanthidin, digoxigenin, and digitoxigenin are based on the original Königs-Knorr condensation method with slight modifications (34,35); final deacetylation was accomplished by hydrolysis with barium methylate, and overall yields were between 15 and 30% of theory.

Finally, it might be worthwhile mentioning some of the compounds which have shown suggestive cardiac activity, that is, have caused systolic contraction in the frog heart upon injection of adequate doses. They are: methyl and ethyl coumalinate (36), βγ-angelica lactone, crotonolactone γ-acetic acid and its methyl ester (37), and l-ascorbic acid (an αβ-unsaturated γ-lactone in its enol form) (38).

Synthetic lactones and pyrones arranged in chronological order of appearance in the literature.

α-Pyrones

Compound	Method of Preparation	Ref.	Cardiac Activity
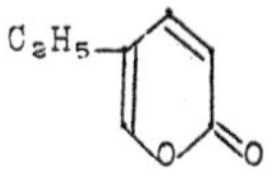			
5-ethyl-α-pyrone		4,16,17	
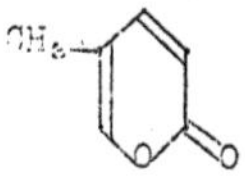			
5-methyl-α-pyrone		16,17	----

$\Delta^{\alpha,\beta}$-Butenolides

Compound	Method of Preparation	Ref.	Cardiac Activity
β-phenyl-$\Delta^{\alpha\beta}$-butenolide	A	4,5	----
	B	6	
	C	14	
β-cyclohexyl-$\Delta^{\alpha\beta}$-butenolide	A	4,5	----
	B	6,13	
	D	26	
β-cyclopentyl-$\Delta^{\alpha\beta}$-butenolide	A	3,5	----
β-n-butyl-$\Delta^{\alpha\beta}$-butenolide	A	4,5	----

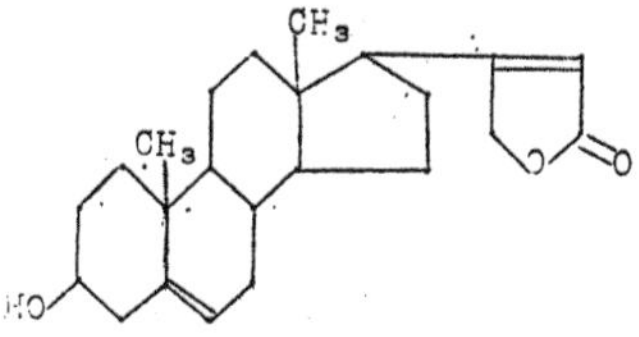

CH3
CH3
HO
O
O

$\Delta^{5,6,20,22}$-3(β),21-dihydroxy-norcholadienic acid lactone	B	8,18
$\Delta^{5,6,20,22}$-3(β)-acetoxy-21-hydroxy norcholadienic acid lactone	B	8,18

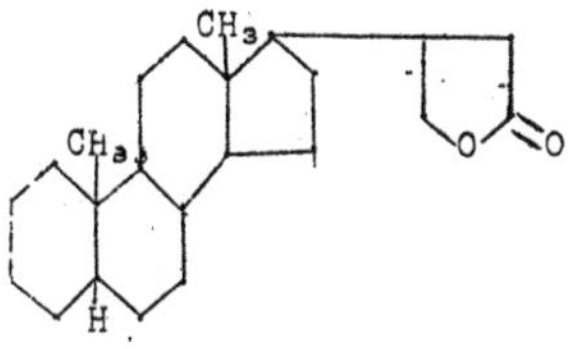
CH3
CH3
H
O
O

$\Delta^{20,22}$-21-hydroxy-norcholenic acid lactone(3,14-Bisdesoxy-thevetigenin or-digitoxigenin)	B	19,20

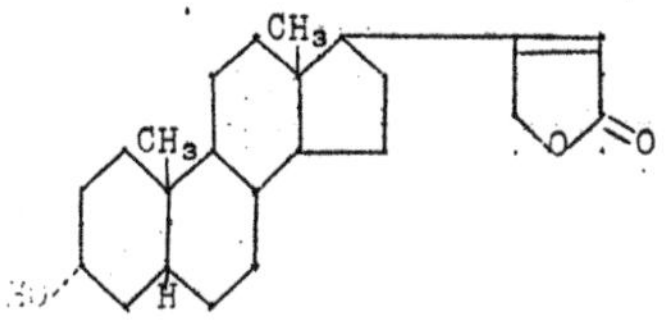
CH3
CH3
H
O
O

$\Delta^{20,22}$-3(α),21-dihydroxy-norcholenic acid lactone(14-desoxydigitoxigenin)	B	21
$\Delta^{20,22}$-3(α)-acetoxy-21-hydroxy-norcholenic acid lactone	B	21

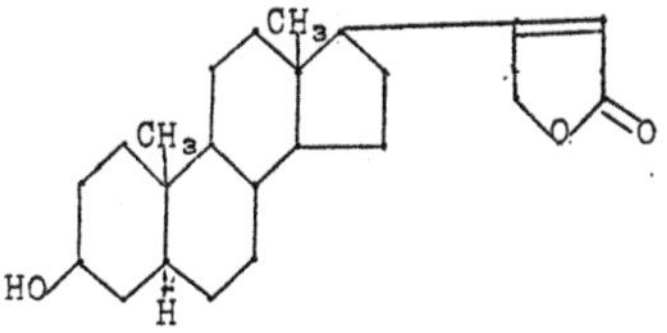
CH3
CH3
HO
H
O
O

$\Delta^{20,22}$-3(β),21-dihydroxy-norallo-cholenic acid lactone (14-desoxyuzarigenin)	B	12
$\Delta^{20,22}$-3(β)-acetoxy-21-hydroxy-norallocholenic acid lactone	B	12
	C	15

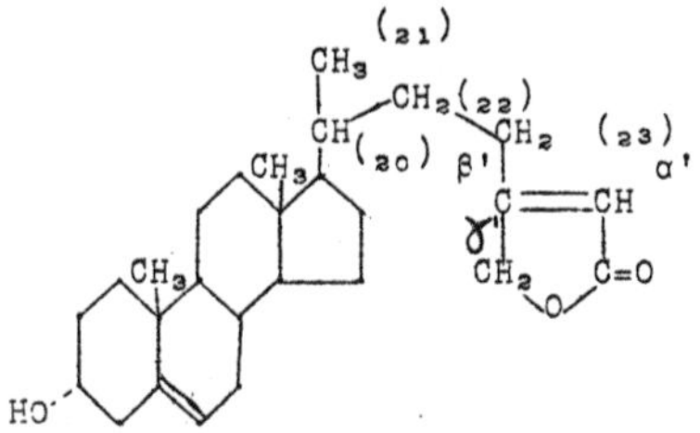

(21)
CH3
CH2 (22)
CH
CH2
(23)
(20)
β'
α'
CH3
C
CH
γ'
CH3
CH2
C=O
O
HO

β'-[3(β)-hydroxy-Δ^5-norcholenyl-(23)]-$\Delta^{\alpha'\beta'}$-butenolide	B	22
β'-[3(β)-acetoxy-Δ^5-norcholenyl-(23)]-$\Delta^{\alpha'\beta'}$-butenolide	B	22

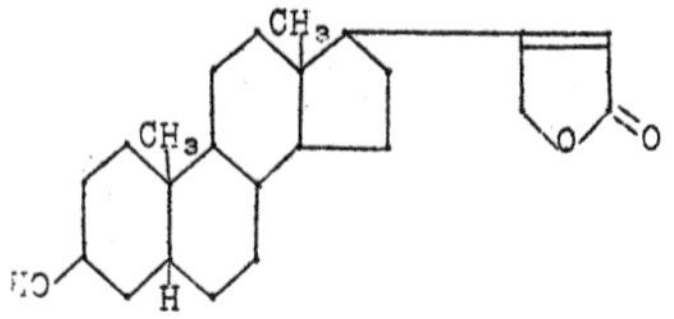

CH3
CH3
O
O
HO
H

$\Delta^{20,22}$-3(β),21-dihydroxy-norcholenic acid lactone (14-desoxythevetigenin)	B	20
$\Delta^{20,22}$-3(β)-acetoxy-21-hydroxy-norcholenic acid lactone	B	20
β-(α-naphthyl)-$\Delta^{\alpha\beta}$-butenolide	B	23
β-(β-naphthyl)-$\Delta^{\alpha\beta}$-butenolide	B	23
β-(6-hydroxy-2-naphthyl)-$\Delta^{\alpha\beta}$-butenolide	B	23
β-(decahydro-β-naphthyl)-$\Delta^{\alpha\beta}$-butenolide	B	23
β-(1-hydrindenyl)-$\Delta^{\alpha\beta}$-butenolide	B	23

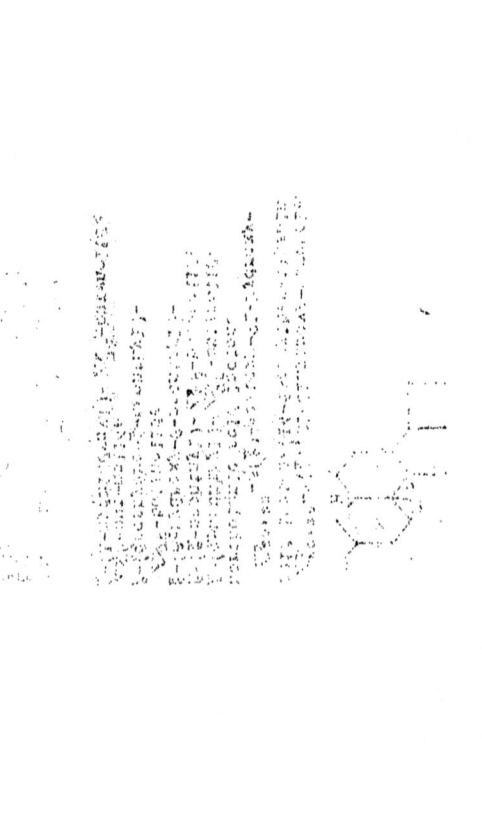

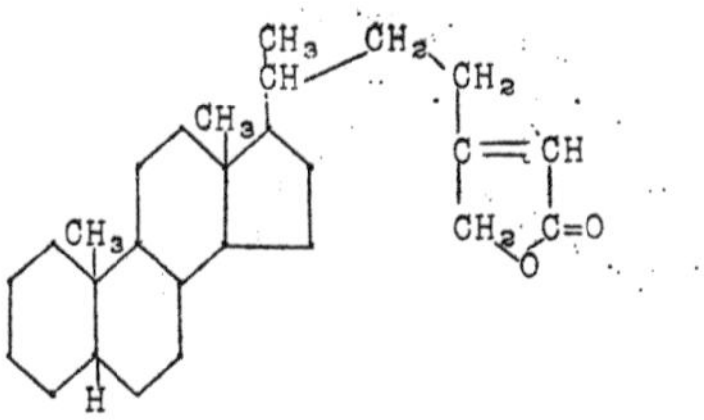

β-norcholanyl- $\Delta^{\alpha\beta}$-butenolide	B	13	---
β-(4-hydroxyphenyl)- $\Delta^{\alpha\beta}$-butenolide	A,B	24	---
β-(3-hydroxyphenyl)- $\Delta^{\alpha\beta}$-butenolide	B	24	

Bibliography

1. Fieser, Chemistry of Natural Products Related to Phenanthrene, Reinhold, N. Y. (1937), p. 256 ff.
Gilman, Organic Chemistry, Wiley, N. Y. (1942), p. 1341 ff.
Tschesche, Ergeb. Physiol. Biol. Chem. exptl. Pharmakol. 38, 31 (1936).
Elderfield, Chem. Rev. 17, 208 (1935)

2. Phillips, Organic Seminar Reports, Univ. of Ill., Jan. 7, 1942,
3. Ranganathan, Current Science 9, 458 (1940)
4. Fried, Rubin, Paist and Elderfield, Science 91, 435 (1940)
5. Rubin, Paist and Elderfield, J. Org. Chem. 6, 260 (1941)
6. Linville and Elderfield, J. Org. Chem. 6, 270 (1941)
7. Paist, Blout, Uhle and Elderfield, J. Org. Chem. 6, 273 (1941)
8. Ruzicka, Reichstein and Fürst, Helv. Chim. Acta 24, 76 (1941)
9. Steiger and Reichstein, Helv. Chim. Acta 20, 1164 (1937)
10. Reichstein and Montigel, Helv. Chim. Acta 22, 1212 (1939)
11. Reichstein and Fuchs, Helv. Chim. Acta 23, 658 (1940)
12. Ruzicka, Plattner and Fürst, Helv. Chim. Acta 25, 79 (1942)
13. Knowles, Fried and Elderfield, J. Org. Chem. 7, 383 (1942).
14. Torrey, Kuck and Elderfield, J. Org. Chem. 6, 289 (1941)
15. Ruzicka, Plattner and Potaki, Helv. Chim Acta 25, 425 (1942)
16. Fried and Elderfield, J. Org. Chem. 6, 566 (1941)
17. Fried and Elderfield, J. Org. Chem. 6, 577 (1941)
18. Ruzicka, Plattner and Fürst, Helv. Chim. Acta 24, 716 (1941)
19. Linville, Fried and Elderfield, Science 94, 284 (1941)
20. Fried, Linville and Elderfield, J. Org. Chem. 7, 362 (1942)
21. Ruzicka, Plattner and Balla, Helv. Chim. Acta 25, 65 (1942)
22. Ruzicka, Plattner and Heusser, Helv. Chim. Acta 25, 435 (1942)
23. Knowles, Kuck and Elderfield, J. Org. Chem. 7, 374 (1942)
24. Marshall, Kuck and Elderfield, J. Org. Chem. 7, 444 (1942)
25. Tschesche and Bohle, Ber. 68, 2252 (1935).

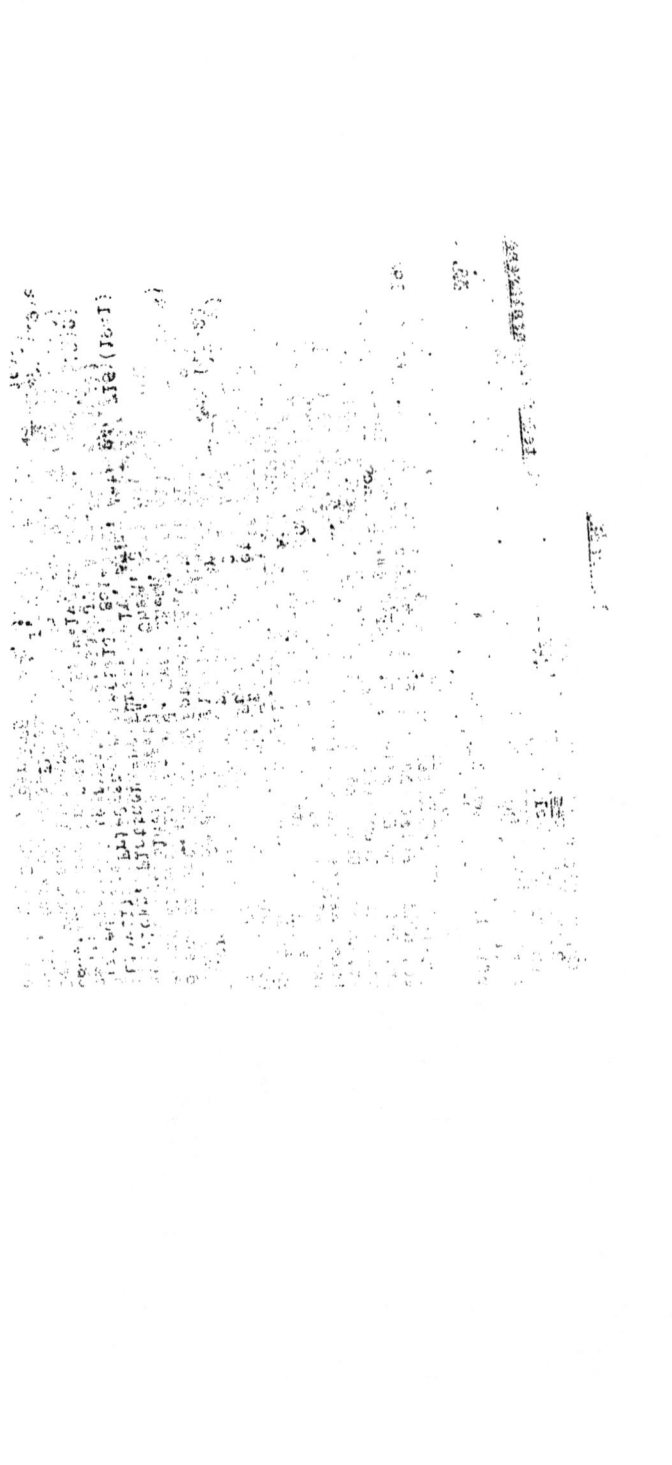

26. Blout and Elderfield (in press)
27. Darzens, Compt. rend., 203, 1374 (1936)
28. Blout, Fried and Elderfield (in press)
29. Chen and Elderfield, J. Pharmacol. 70, 338 (1940)
30. Chen, Chen and Anderson, J. Am. Pharm. Assoc. 25, 579 (1936)
31. Chen, Robbins and Worth, J. Am. Pharm. Assoc. 27, 189 (1938)
32. Chen and Elderfield, J. Pharmacol. 76, 81 (1942)
33. Chen, Fried and Elderfield (in press)
34. Uhle and Elderfield (in press)
35. Fried and Elderfield (in press)
36. Chen, Steldt, Fried and Elderfield, J. Pharmacol. 74, 381 (1942)
37. Krayer, Mendez, deEspanés and Linstead, J. Pharmacol. 74, 372 (1942)
38. Linstead and Krayer, Science 95, 332 (1942)

Reported by Nelson J. Leonard
February 10, 1943

THE CHEMISTRY OF ACRYLONITRILE

Although acrylonitrile has been known for a good many years, it has only recently assumed importance. Since it is now used in the synthetic rubber industry, its commercial production is currently being rapidly expanded. The rubber substitutes Chemigum, Perbunan (Buna N) and Perbunan Extra are copolymers of acrylonitrile with butadiene. Acrylonitrile, a colorless liquid which boils at 77°, possesses, as would be expected, an extremely reactive double bond due to its conjugation with the nitrile group. In some respects its reactions parallel those of acrylic ester; however, its versatility greatly exceeds that of the latter. Our knowledge of the chemistry of acrylonitrile has been greatly extended by the work of Bruson, carried on in the Laboratories of Röhm and Haas. It is the reactions, exclusive of polymerization, of acrylonitrile that are to be discussed in this seminar report.

Reactions with Amines

According to the patent literature (1) many primary and secondary amines, both aliphatic and aromatic, react with acrylonitrile to give substituted β-cyanoethylamines. For example:

$$CH_3CH_2CH_2CH_2NH_2 + CH_2{=}CHCN \rightarrow CH_3CH_2CH_2CH_2NHCH_2CH_2CN$$

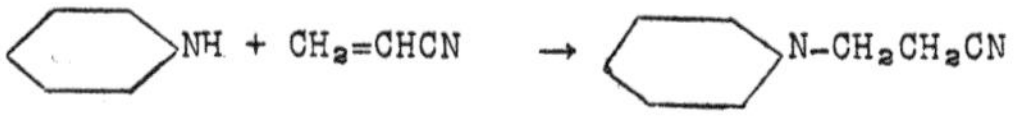

NH + CH₂=CHCN →
N-CH₂CH₂CN

$$NH_3 + CH_2{=}CHCN \rightarrow NH_2CH_2CH_2CN \rightarrow NH(CH_2CH_2CN)_2$$

Since the resulting cyanides can be hydrolyzed to acids, reduced to amines, or converted to esters a wide variety of products may be obtained. That the reaction may be more than a paper patent is indicated by the known fact that ethyl acrylate will react with amines in an exactly analogous manner (2a). For instance, it has been shown in this laboratory that merely heating a mixture of the ethyl ester of l-proline and ethyl acrylate on the steam bath for 12 hours results in the formation of ethyl-β-N-(2-carbethoxypyrralidyl)-propionate in 93% yield(2b).

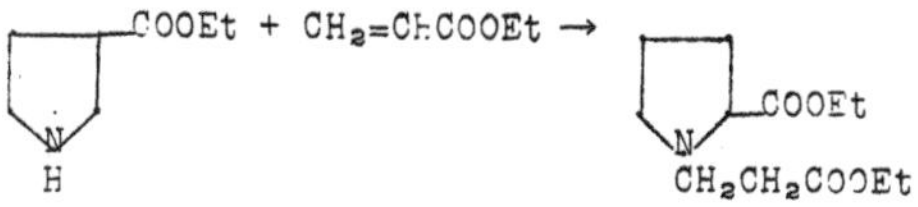

COOEt + CH₂=CHCOOEt →
N
H
COOEt
N
CH₂CH₂COOEt

With Water, Alcohols, Phenols and Oximes

Bruson(3) has shown that two molecules of acrylonitrile condense with one of water to form bis-(β-cyanoethyl) ether, although

$$2CH_2=CHCN + H_2O \xrightarrow{NaOH} O(CH_2CH_2CN)_2$$

the same product can be obtained in a much better yield (91%) by the action of acrylonitrile on ethylene cyanohydrin.

Many polyhydric alcohols such as propylene glycol, trimethylene glycol, glycerol, pentamethylene glycol, etc., undergo complete cyanoethylation in 80-90% yields (3). As an example, the reaction with ethylene glycol is given below.

$$\begin{matrix} CH_2OH \\ | \\ CH_2OH \end{matrix} + 2\ CH_2=CHCN \longrightarrow \begin{matrix} CH_2OCH_2CH_2CN \\ | \\ CH_2OCH_2CH_2CN \end{matrix}$$

Many monohydric unsaturated alcohols, especially those with β,γ-unsaturation react practically quantitatively with acrylonitrile(4).

Numerous phenols are also cyanoethylated, according to patent claims(5) to form β-cyanoethyl ethers. Langley and Adams (6) found that acrylonitrile reacts with resorcinol in the presence of zinc chloride and an ethereal hydrogen chloride solution, not to give a ketone (Hoesch reaction) but instead to give I.

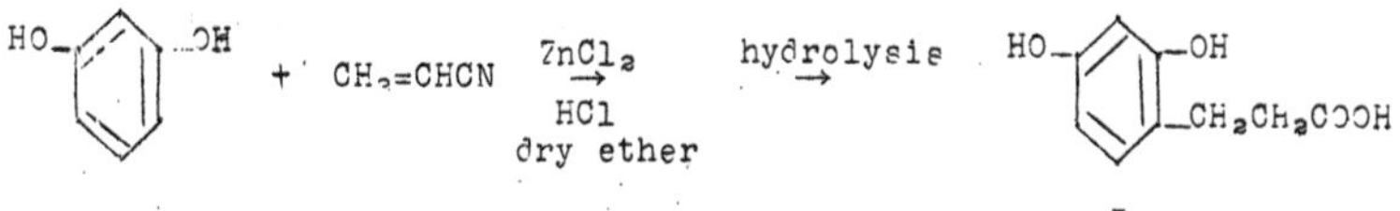
HO
OH
+ CH₂=CHCN
ZnCl₂
HCl
dry ether
hydrolysis
HO
OH
CH₂CH₂COOH

I

Both ketoximes and aldoximes readily react with acrylonitrile to form oximino-O-cyanoethyl ethers (3),

$$C_6H_5-\underset{\displaystyle CH_3}{C}=NOH + CH_2=CH-CN \longrightarrow C_6H_5-\underset{\displaystyle CH_3}{C}=N-O-CH_2CH_2CN$$

```
HC—CH                                       HC—CH
||  ||                                      ||  ||
HC  C-CH=NOH + CH2=CHCN        ——>          HC  C-CH=N-O-CH2CH2CN
  \O/                                         \O/
```

The condensations of acrylonitrile with water, alcohols and oximes are effected by the use of small amounts of alkaline reagents such as sodium or potassium hydroxide, sodium methylate, and trimethylbenzylammonium hydroxide ("Triton B"). The reaction with phenols is accomplished at elevated temperatures and pressures with alkali.

With Halogens

Everything on halogenation of acrylonitrile appears to be in the patent literature (7) and consequently merits only brief mention here. Chlorination in the vapor phase over activated charcoal allegedly produces α-chloroacrylonitrile. If the chlorination occurs in an aqueous medium at 20-30°, α,β-dichloropropionitrile and α-chloro-β-hydroxypropionitrile result.

With Hydrogen Sulfide

A patent has been issued on the addition of hydrogen sulfide to acrylonitrile to give bis-(β-cyanoethyl)-thio ether (8).

With Diazonium Salts

Koelsch (9) has recently shown that several diazonium salts react with acrylonitrile to give derivatives of α-chlorohydrocinnamonitrile in yields from 34-48%. Methyl acrylate couples in

$$ArN_2Cl + CH_2{=}CHCN \rightarrow ArCH_2CHClCN{+}N_2$$

a similar manner to give the corresponding ester. Since the resulting α-chloro derivatives can be dehydrohalogenated readily by boiling for five minutes with diethylaniline, the method may be of some importance in the preparation of various substituted cinnamic acids.

Michael Type Condensations with Acrylonitrile

Bruson (3.10,11,12) has demonstrated that acrylonitrile undergoes a direct Michael reaction with a large number of compounds which possess active methylene or methenyl groups. An effective catalyst for these reactions is trimethylbenzylammonium hydroxide which is employed in the form of an aqueous 40% solution (Triton B). In some cases sodium or potassium methylate, 30% methanolic potassium hydroxide, or even aqueous 40% sodium hydroxide are effective, but the solubility of "Triton B" and its high degree of alkalinity renders it particularly effective where the other alkalies either fail to initiate the reaction at all or to give good yields. A large number of the Michael condensations to be discussed take place at room temperature with the evolution of heat. In order to prevent excessive polymerization of the acrylonitrile and to allow the reaction to proceed smoothly an inert diluent such as dioxane or tert.butanol is recommended. This is especially necessary when the reactive methylene compound is a solid.

Cyanoethylation of Methylene Groups Activated by Carbon-Carbon Unsaturation

In the presence of a catalytic amount of "Triton B" acrylonitrile readily condenses with fluorene, II, to yield bis-9,9-(β-cyanoethyl) fluorene, III. Anthrone gives the dicyanoethylated product IV (10).

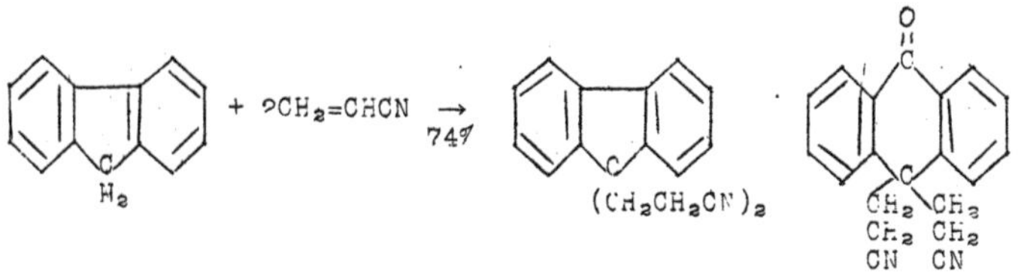

II III IV

The reactions of acrylonitrile with cyclopentadiene are of interest. In the absence of a catalyst these two substances undergo a Diels-Alder condensation as illustrated in the following equation.

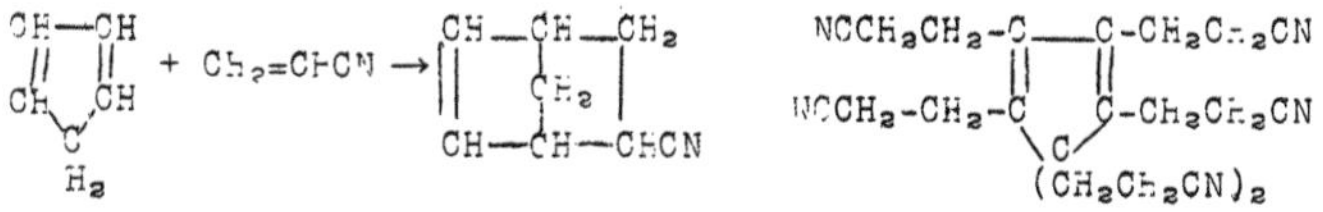

V VI

However, in the presence of "Triton B" the formation of the Diels-Alder type of adduct is so far repressed that each of the six hydrogen atoms in the cyclopentadiene molecule adds to acrylonitrile to form the crystalline hexacyanoethylation product VI, accompanied by a mixture of lower poly-cyanoethylated products. This unexpected reaction of cyclopentadiene may be explained in one of two possible ways: by assuming that a resonating system of double bonds make all of the methylene and methenyl hydrogen atoms equally active, or by a mechanism involving a shift of the residual methylene hydrogen atom to a contiguous carbon atom as soon as the first cyanoethyl group is introduced thus forming new reactive methylene groups successively around the cycle as each cyanolthyl group enters. Interruption of the resonating system, as for example by dimerization of the cyclopentadiene or by adduct formation with the loss of a double bond, result in products that will no longer react with acrylonitrile. Thus, neither dicyclopentadiene nor the adduct V will react.

Attempts to replace acrylonitrile with methyl or ethyl acrylate in the above condensations fail, although these substances form a Diels-Alder adduct with cyclopentadiene regardless of whether "Triton B" is used as a catalyst or not.

With Ketones

Aromatic methyl ketones of the type $ArCOCH_3$, as exemplified by acetophenone, and its nuclear substituted derivatives and 2-acetylnaphthalene, readily undergo the Michael reaction to give tricyanoethation products (11).

$$Ar\underset{O}{C}CH_3 + 3CH_2{=}CHCN \longrightarrow Ar{-}\underset{O}{C}C(CH_2CH_2CN)_3$$

Acetomesitylene reacts more sluggishly than the other ketones but gives a 30% yield of the tricyanide. Propiophenone, $C_6H_5COCH_2CH_3$, and desoxybenzoin dicyanoethylate on the methylene carbon adjacent to the carbonyl group. Cyclic ketones such as α-tetralone, o-methylcyclohexanone and cyclopentanone can be cyanoethylated to form compounds VII, VIII, and IX, respectively.

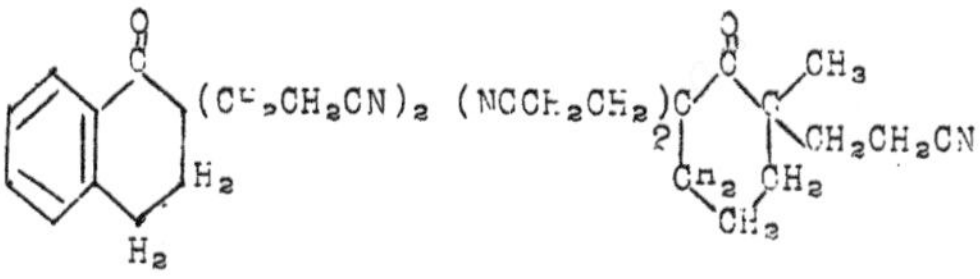

VII VIII

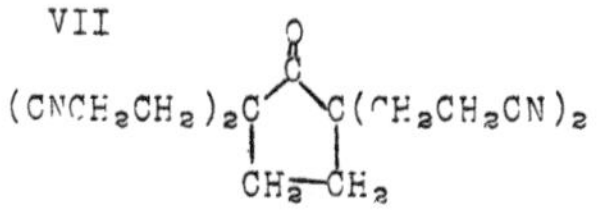

IX

Although cyclic ketones possessing the $-CH_2-\overset{O}{C}-CH_2-$ group are tetracyanoethylated, straight chain ketones such as dibenzyl ketone and diethyl ketone undergo condensation with only three molecules of acrylonitrile, X.

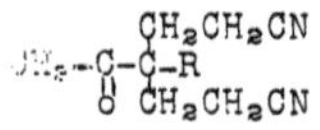

XI ($R{=}CH_3, C_2H_5, n{-}C_5H_{11}$)

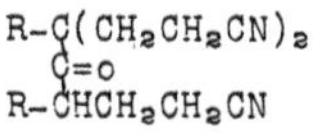

X ($R{=}CH_3$ or C_6H_5)

In general, saturated ketones of the type $CH_3\overset{O}{C}-CH_2-$ react with two molecules of acrylonitrile both of which enter on the methylene group rather than on the methyl group. Thus methyl ethyl, methyl propyl and methyl n-hexyl ketones give γ-acetyl- γ-alkyl pimelonitriles, XI. Acetoacetic ester gives a crystalline product in which two β-cyanoethyl groups are on the methylene carbon.

It should be noted that attempts to utilize α-methyl acrylonitrile in the above reactions were unsuccessful. Furthermore, it was found that certain ketones such as camphor, isophorone and diisobutyl ketone could not be cyanoethylated.

With Alpha, Beta-Unsaturated Ketones and Nitriles

In the presence of trimethylbenzylammonium hydroxide mesityl oxide reacts with acrylonitrile to produce two products: the crystalline γ-acetyl-γ-isopropenylpimelonitrile, XII (75% yield), and the liquid monocyanoethylation product, 1-(β-cyanoethyl)-1-isopropylidene acetone, XIII, (10-15% yield). Pure XIII could be converted to XII by further cyanoethylation (12).

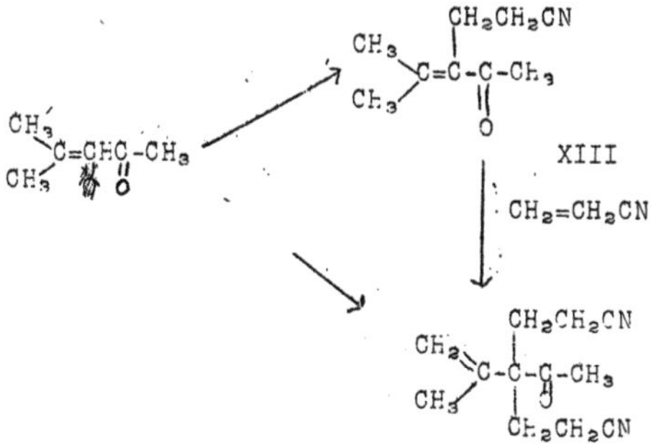
CH₂CH₂CN
CH₃
C=C-C-CH₃
CH₃
O
CH₃
C=CHC-CH₃
CH₃
O
CH₂=CH₂CN
CH₂CH₂CN
CH₂
C-C-C-CH₃
CH₃
O
CH₂CH₂CN

XIII

XII

To explain the formation of XII, Bruson assumed that the unsaturated ketone existed in a three-carbon desmotropic equilibrium and that the powerful acceptor characteristics of acrylonitrile

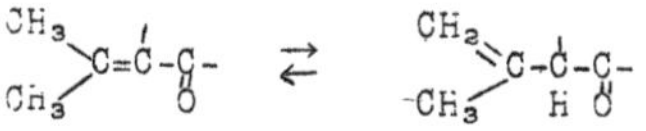
CH₃
C=C-C-
CH₃
O
⇄
CH₂
C-C-C-
CH₃
H O

continually shifted this equilibrium to the right.

α,β-Unsaturated nitriles, which are known to form three-carbon desmotropic systems, react with acrylonitrile in an analagous manner. Thus, either crotononitrile or allyl cyanide react to give the two products XIV and XV.

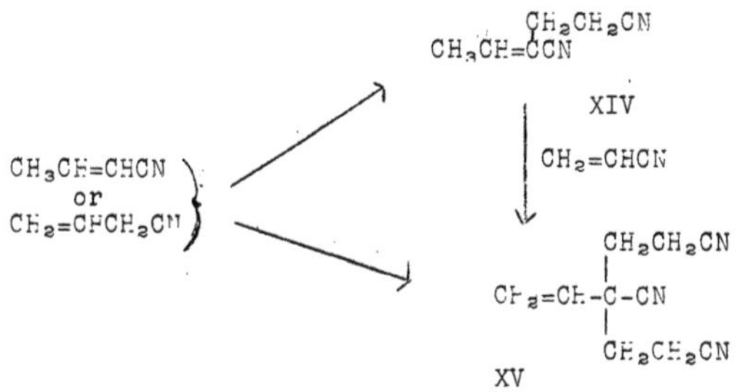
CH3CH=CHCN
or
CH2=CHCH2CN
CH2CH2CN
CH3CH=CCN
XIV
CH2=CHCN
CH2CH2CN
CH2=CH-C-CN
CH2CH2CN
XV

Neither cinnamonitrile nor α-methylacrylonitrile will react with acrylonitrile in the presence of "Triton B", possibly because these compounds do not possess a mobile hydrogen atom capable of a 1,3-shift.

<u>With Other Compounds Which Possess Active Hydrogen</u>

The only nitro compound which Bruson reported studying was nitromethane. This reacts vigorously with three equivalents of acrylonitrile to give XVI (3). Examples of each of the following

$$(NCCH_2CH_2)_3C{-}NO_2$$

XVI

types of compounds undergo a Michael condensation with acrylonitrile: aryl-CH_2CN, $ROOCCH_2CN$, $H_2NCOCH_2CONH_2$, $CH_2(COOR)_2$, $R^1CH(COOR)_2$, $NCCH_2CONH_2$ and aryl-$CH_2SO_2NH_2$. In each case dicyanoethylation occurs, the yields vary from 40-90%.

<u>Bibliography</u>

(1) I. G. Farbenindustrie A.G., British Patents 404,744 (1934) 457, [illegible] (1936); Hoffmann and Jacobi, U.S. Patents 1,992,615 (1935), 2,017,537(1935), German Patent 598,185(1934).
(2) [illegible] Filippi and Galter, Monatsh; <u>51</u>, 253-266(1929);
[illegible] Adams and Carmack, Unpublished Work at the Univ. of Illinois.
(3) Bruson and Riener, J. Am. Chem. Soc., <u>65</u>, 23(1943)
(4) Bruson, U.S. Patent 2,280,790(1942); 2,280,791(1942); 2,280,[illegible](1942)
(5) [illegible] Ger. Patent [illegible] (1939)
(6) Langley and Adams, J. Am. Chem. Soc., <u>44</u>, 2320 (1922)
(7) Long, U.S. Patent 2,2[illegible],363(1941); D'Ianni, U.S. Patent 2,[illegible],360 (1941); Lichty, U.S. Patent 2,231,838 (1941)
(8) Keyssner, U. S. Patent 2,163,176 (1939)
(9) Koelsch, J. Am. Chem. Soc., <u>65</u>, 57 (1943)
(10) Bruson, J. Am. Chem. Soc., <u>64</u>, 2457 (1942)
(11) Bruson and Riener, ibid., <u>64</u>, 2850 (1942)
(12) Bruson and Riener, ibid., <u>65</u>, 18 (1943)

Reported by John E. Mahan
February 10,1943

THE ALDEHYDE AMIDE CONDENSATION

The best known aldehyde-amide condensations are those that take place between urea and formaldehyde to give polymeric compounds. However, aldehydes can be condensed with amides to give bimolecular compounds. It is with the latter type of condensation that this report is concerned.

Although the aldehyde-amide condensation has been known for over 70 years, a great deal of investigational work still remains to be completed. In general, the reaction of aldehydes with amides to form simple products may be classified into three groups.

$$(1)\quad RCHO + 2\ HNHCOR' \rightarrow RCH(NHCOR')_2 + H_2O$$

$$(2)\quad RCHO + HNHCOR' \rightarrow RCH(OH)NHCOR'$$

$$(3)\quad RCHO + H_2NCOR' \rightarrow RCH{=}NCOR' + H_2O$$

Reactions of type (1) are by far the most numerous and may be classified as a general reaction with types (2) and (3) as the exceptions. Usually when an aldehyde and an amide are heated together at 130-140°C. for several hours without a solvent and with or without a condensing agent, alkylidene- or arylidene-diamides are produced.

$$HCHO + 2\ HNHCOCH_3 \rightarrow HC\begin{cases} NHCOCH_3 \\ NHCOCH_3 \end{cases}$$

Either aliphatic or aromatic amides or aldehydes can be used and the yields of product obtained vary widely. Although the experimental data available is not sufficient to lead to accurate prediction of the course of type (1) reactions, the following generalizations may be put forth tentatively.

(1) Long chain aliphatic amides give the best yields.

(2) Aromatic aldehydes give higher yields than aliphatic aldehydes.

(3) m- and p-substituted aromatic aldehydes give higher yields than o-substituted aldehydes.

Condensations of type (2) are likely to occur when the aldehyde is substituted with negative groups. For example, butyrchloral and acetamide condense to form butyrchloralacetamide.

$$CH_3CHClCCl_2CHO + HNHCOCH_3 \rightarrow CH_3CHClCCl_2C\begin{cases} OH \\ NHCOCH_3 \end{cases}$$

Chloral and bromal behave similarly. More stringent conditions and an excess of the amide bring about condensations of type (1) even with these aldehydes. If the type (2) products are treated with mild dehydrating agents such as acetic anhydride or phthalic anhydride, then water is eliminated and type (3) products are obtained.

$$CCl_3CH(OH)NHCOH \rightarrow CCl_3CH{=}NCOH$$

Reactions of type (3) occur directly when an aromatic aldehyde containing a phenolic group is condensed with an amide. Salicylaldehyde, for example, condenses with acetamide to produce nearly quantitative yields of o-hydroxybenzylidenaecetamide and an excess of the amide does not affect the nature of the final product.

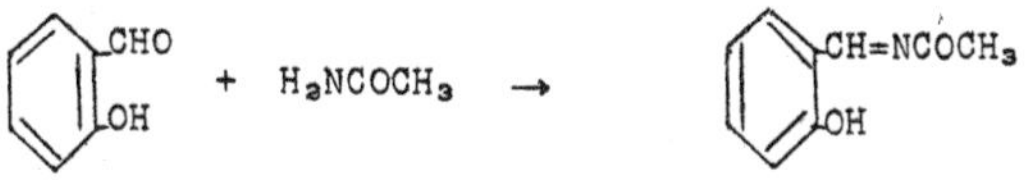
CHO
OH
+ H2NCOCH3
CH=NCOCH3
OH

m- and p-Hydroxybenzaldehydes behave in an analogous manner. Condensing agents such as pyridine usually increase the yield of type (3) reactions. The products obtained, in contrast to those obtained by type (1) reactions, are more difficult to purify, tend to be glassy instead of crystalline, do not give sharp melting points but generally decompose before reaching it, and are obtained in greatest yield from o- and p-substituted aldehydes and least from m-substituted aldehydes. It is interesting to note that if the phenolic group is methylated then the methoxybenzaldehydes produced undergo the normal type (1) condensation with two molecules of the amide.

An interesting reaction occurs when o-hydroxyarylamides are allowed to react with aldehydes. Salicylamide and m-nitrobenzaldehyde, for example, react and presumably an intermediate is formed which undergoes ring closure to form 2-(m-nitrophenyl-)-dihydro-1,3-benzoxazine-4-one.

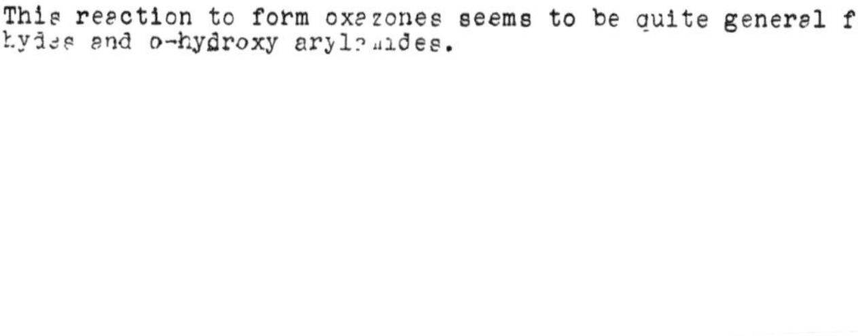
CHO
NO2
+
H2NCO
HO
CH=NCO
HO
NO2
H
N
CH
C=O
O
C
C
NO2

This reaction to form oxazones seems to be quite general for aldehydes and o-hydroxy arylamides.

TABLE I - Yield of Nitrobenzylidenediamide

Amide	Nitrobenzaldehydes % o-	% m-	% p-
Form-	40	68	68
Acet-	48	60	55
Propion-	49	67	54
n-Butyr-	65	85	72
n-Hept-	97	97	97
Benz-	54	80	80
Phenylacet-	47	97	97

TABLE II - Yield of Methoxybenzylidenediamide

Amide	Methoxybenzaldehydes % o-	% m-	% p-
Acet-	49	54	54
Benz-	58	52	47
Phenylacet-	56	51.5	56
Propion-	38	51.5	40

TABLE III - Yield of Hydroxybenzylideneamide

Amide	Hydroxybenzaldehydes % m-	% p-
Form-	0.0	63
Acet-	0.0	92
Propion-	27.5	84
Benz-	58.0	83
Phenylacet-	40.0	81

TABLE IV - Typical Alkyl- and Aryldiacetamides

Aldehyde	Product -diacetamide	Melting Point C.	Yield %
Form-	methylene-	197.5-198	54
Acet-	ethylidene-	180	44
Propion-	propylidene-	190-190.5	7.5
Butyr-	butylidene-	189	11.5
Isovaler-	2-Me-4-butylidene	184	26
Hept-	heptylidene-	171-2	6.5
Benz-	benzylidene-	238	48
Cinnam-	cinnamylidene-	234	52
p-Tolyl-	p-tolylidene-	274	27
Isobutyr-	2-Me-propylidene-	216	25
Capron-	hexylidene-	145	8
2,4-di-Nitrobenz-	none	-	0.0

Bibliography

th, Ann. 154, 72, (1870).
huster, ibid., p.80.
dicus, Ann. Ch. Pharm., clvii, 44.
ffmann and V. Meyer, Ber. 25, 209-13, (1892).
low, ibid., 26, 1972-4, (1893).
ich, Monatsh. 25, 933-42, (1904).
ich, ibid., p.966.
Titherley, J.C.S., 91, 1419, (1907).
Hicks, ibid., 97, 1032, (1910).
ucka and C.Rogl. Ber. 59B, 756-62, (1926).
aser and S. Frisch, Arch. Pharm. 266, 103-116, (1928).
Noyes and D. B. Forman, J.A.C.S., 55, 3493, (1933).
Hann, J. Wash. Acad. Sci., 24, 124-6, (1934).
ntsev and Benevolenskaya, J. Gen. Chem. (USSR), 7, 2361-4, (193[illegible]).
andya and T.S.Sodhi, I, Proc.Ind.Acad.Sci., 7A, 361-8, (1938).
Mehra and Pandya, II, ibid., p.376-80.
and Pandya, III, ibid., 9A, 508, (1939).
and Pandya, IV, ibid., 10A, 279, (1939).
nzur and Pandya, V, ibid., p.282.
and Pandya, VI, ibid., p.285.
a and G. Varghese, VII, ibid., 14A, 18, (1941).
a and Varghese, VIII, ibid., p.25.
Ittyerah and Pandya, IX, ibid., 15A, 6, (1942).
rah and Pandya, X, ibid., p.258.
duMont and G. Ratzel, Ber. 72B, 1500-5, (1939).

ted by R. D. Lipscomb
ary 17, 1943

RELATION OF SPATIAL CONFIGURATION TO PHARMACOLOGICAL ACTIVITY

Last February Dr. Hamlin presented a seminar report which dealt with the correlation between structure and physiological activity of various position isomers and homologues of compounds belonging to the ephedrine-epinephrine series. It is the purpose of this seminar to point out a few of the quantitative and qualitative physiological variations that have been reported for optically active molecules and to present some of the explanations proferred which attempt to account for these differences.

Diastereoisomeric compounds are encountered everywhere, the best known and most intensively studied illustration being, perhaps, the simple sugars. The fact that differences in physiological response are produced upon administration of diastereoisomers is not surprising in view of the chemical and physical differences existant between them. No proof is available yet to indicate how these properties may affect physiological activity and, consequently, the following information may at first glance appear to be heterogeneous and unrelated; it is given primarily to emphasize the fact that diastereoisomers are not identical compounds.

1. Light Absorption: Meso and the active tartaric acids show different absorption bands.[1]

2. Adsorption: From solutions of the same concentration $Al(OH)_3$ will absorb approximately 50% more of the dl-tartaric acid than of the meso isomer.[2]

3. Crystalline Structure of Solids: The first example to come to kind is the mirror image relationship of the $Na-NH_4$-tartrates which were first studied by Pasteur.[3]

4. Vapor Pressure of Liquids: The (-)menthyl esters of dl-2-methyl butanoic acid and of dl-2-methoxy propionic acid were vaporized through a 60 plate column. In each case the esters of the dextrorotatory acids came over first.[4]

5. Solubilities in Optically Active Solvents: On shaking together (-)carvone, water and dl-mandelic acid, the aqueous layer became dextrorotatory because of the unequal distribution of the (+) and (-)acids between the water and the optically active organic solvent.[5]

6. Rates of Reaction: A mixture of (+)neoisomenthol and (+)neomenthol could be separated into its component fractions by means of partial esterification with 3,5,-dinitrobenzoyl chloride. (+)Neoisomenthol reacts more rapidly and could thus be separated as its esters

7. Reactions with Enzymes: Since enzymes are optically active, organic, biocatalysts, it is to be expected that the rate and extent of reaction with steoisomers should differ. Thus α-methyl-d-glucoside is hydrolyzed by yeast and β-methyl-d-glucoside is hydrolyzed by emulsin.[7] Enzymes are active not only in degradation but also in synthetic reactions. Thus it has been found that o-methylcyclohexanone is reduced by hops to a dextrorotatory alcohol.[8] Benzaldehyde and hydrogen cyanide in the presence of emulsin form optically active mandelonitrile.

From the above observations it appears that neither the enzyme nor the substrate alone determines the resultant of their reactions; an enzyme may select one isomer of one pair, the other of another, while on the other hand one enzyme may destroy the dextrorotatory component, another the levorotatory antipode of a racemic mixture. In some cases each isomer is destroyed at the same rate: for example, racemic lactic acid subjected to certain ferments is slowly destroyed without developing any optical activity.[9]

It is interesting to note that the same variation occurs in the case of the combination of substances of known constitution. Thus a levorotatory alkaloid may be separated from its mirror image by the insolubility of its salt with an optically active acid, while with another acid the solubility may be reversed and with a third the difference may be too small to afford any separation.

In view of the differences in physical and chemical properties of diasterioisomers a difference in pharmacological response is not unexpected. Quantitative physiological differences between enantiomorphs may probably be attributed to the diastereoisomeric complexes formed by "protoplasmic" reaction with the abundant optically active components in the tissues.

Thus, in view of the chemical and physical differences between diastereoisomers and the quantitative, physiological variations demonstrable upon the administration of antimers, a correlation of physicochemical properties of spatial isomers with pharmacological activity is to be expected. However, before such correlation may be established it will be necessary to procure more quantitative data than are now available, and so far as conditions permit the studies should be made with compounds having small variations in chemical structure and producing comparable physiological effects. There is, perhaps, no group of compounds which lends itself better for such examination than the epinephrine-ephedrine series.

In this seminar report the following compounds will be considered:

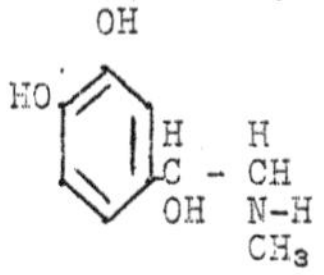

Epinephrine
I

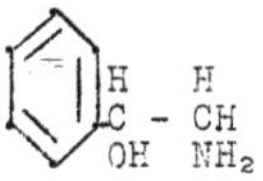

Phenylethanolamine
II

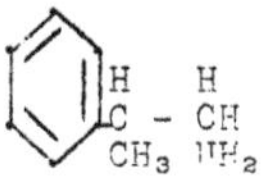

β-Phenylpropylamine
III

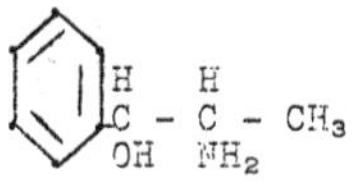

Propadrine
IV

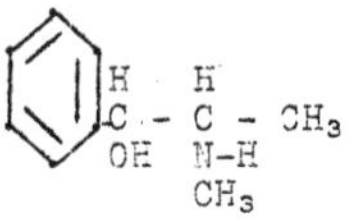

Ephedrine
V

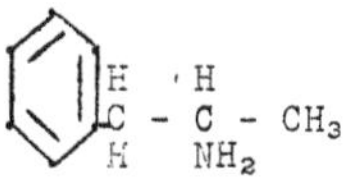

Benzedrine
VI

The above series of compounds are commonly referred to as "sympathomimetic" amines because they have the property of causing physiological effects very similar to those caused by stimulation of the sympathetic nervous system. The characteristic rise in blood pressure produced upon the administration of these compounds is used as the basis for the quantitative comparison of the physiological activity of these "pressor" drugs.

An interesting correlation exists between the solubility of the mandelates of the isomeric ephedrines and their effect on blood pressure. The salts increase in solubility in the following sequence:

(-)Ephedrine(-)mandelate
(+) " (-) "
(+)ψ " (-) "
(-)ψ " (-) "

The pressor action of the ephedrine bases decreases in the same order.[10] Thus (-)ephedrine forms the least soluble (-)mandelate and is the most effective pressor of the four bases.

The resolution of compounds II, III, IV, and VI by means of (-)mandelic acid and a subsequent study of the solubility of the amine salts formed was carried out to discover if there was any consistency in this relationship.[18] It was shown that the less soluble salts of (-)mandelic acid from each pair of enantiomorphous bases were: (-)propadrine, (+)ψ-propadrine, (-)phenylethanolamine, (-)β-phenylpropylamine and (+)benzedrine. The spatial relationships of the asymmetric carbons in the compounds studied show a definite correlation to the solubility of their salts. Thus in those pressor amines possessing only one asymmetric center the less soluble (-)mandelate is formed by that base whose phenyl-bearing carbon atom is levorotatory (Ex.II, III) or whose carbinamine carbon is dextrorotatory (Ex.V). Propadrine and ephedrine each possess two centers of asymmetry. It has been established by the work of Nagai[11], Leithe,[12] Freudenberg[13] and others that in (-)ephedrine and (-)propadrine the hydroxyl-bearing carbon is levorotatory while the carbinamine carbon is dextrorotatory. In (+)ψ-propadrine and (+)ψ-ephedrine both asymmetric centers are exerting their effect in the same direction. Thus of the four possible isomers of these two bases the least soluble (-)mandelate is found to be that one answering the specifications mentioned before; namely that the carbinamine carbon be dextrorotatory while the hydroxyl-bearing carbon has a levorotation. In the event both centers of asymmetry are of similar sign it is the dextrorotatory isomer which forms the less soluble (-)mandelate.

Pharmacodynamic information about the sympathomimetic amines studied shows that of the enantiomorphous pairs (-)propadrine, (-)ephedrine, (+)benzedrine, (+)ψ-propadrine, and (+)ψ-ephedrine are the more active. Neither quantitative nor qualitative comparisons in the physiological behavior of the optical isomers of phenylethanolamine or β-phenylpropylamine is available. However from the analogies with the compounds described above the prediction is ventured that in each case the levorotatory isomer will prove to be the more active. In epinephrine, where the hydroxyl-bearing carbon is asymmetric, the levorotating isomer is 15 times more effective than the dextrorotatory in causing an elevation in the blood pressure. In both phenylethanolamine and β-phenylpropylamine the corresponding

carbon atoms are asymmetric.

The relative configuration about the asymmetric centers in the amines of the ephedrine series has not been established conclusively. According to the work of Freudenberg[13] the hydroxyl-bearing carbon atom of ephedrine belongs to the d-series when it is levorotatory. He based his conclusions on the synthesis of (-)ephedrine from d(-)mandelic acid:

$$C_6H_5\text{-CHOH-COOH} \xrightarrow{} C_6H_5\text{CHOH-CONH}_2 \xrightarrow{CH_3MgI} C_6H_5\text{CHOH-COCH}_3$$

d(-)mandelic acid

$$C_6H_5\text{CHOH-COCH}_3 \xrightarrow{(H)\ CH_3NH_2} (-)\text{Ephedrine}$$

Leithe assigned the relative configuration of the carbinamine carbon atom to the d-series. His conclusions are based upon a steric comparison of benzedrine (of which the (+)isomer has the same relative configuration as (+)desoxyephedrine) with the optically active forms of α-phenethylamine. Freudenberg and Nikolai[14] synthesized (+)desoxyephedrine from l(+)alanine and thus placed the methylamino-bearing carbon in the l-series:

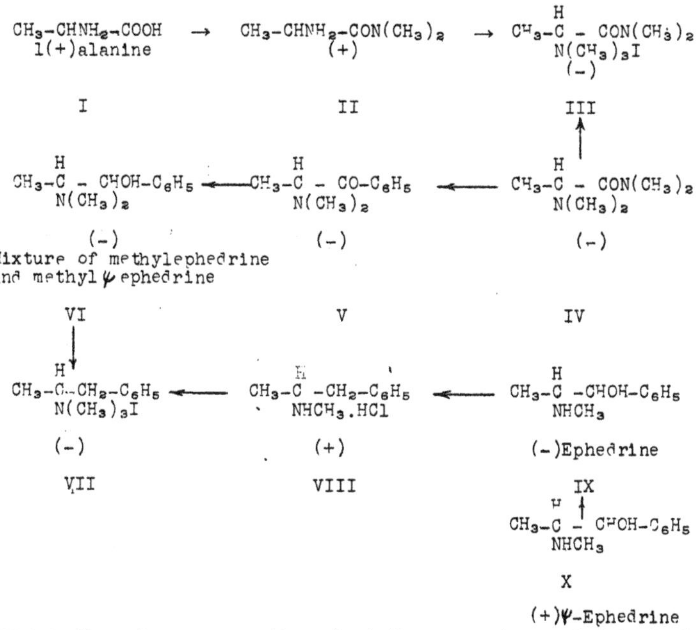
CH3-CHNH2-COOH
l(+)alanine
I
CH3-CHNH2-CON(CH3)2
(+)
II
H
CH3-C - CON(CH3)2
N(CH3)3I
(-)
III
H
CH3-C - CON(CH3)2
N(CH3)2
(-)
IV
H
CH3-C - CO-C6H5
N(CH3)2
(-)
V
H
CH3-C - CHOH-C6H5
N(CH3)2
(-)
Mixture of methylephedrine
and methyl ψ ephedrine
VI
H
CH3-C-CH2-C6H5
N(CH3)3I
(-)
VII
H
CH3-C -CH2-C6H5
NHCH3.HCl
(+)
VIII
H
CH3-C -CHOH-C6H5
NHCH3
(-)Ephedrine
IX
H
CH3-C - CHOH-C6H5
NHCH3
X
(+)ψ-Ephedrine

However there is some question about the correctness of this designation as may be seen from the following:

COOH
H_2N
H
CH_3
$CH_2C_6H_5$
H_2N
H
CH_3

1(+)alanine → A

COOH
H_2N
H
$CH_2C_6H_5$
CH_3
H_2N
H
$CH_2C_6H_5$

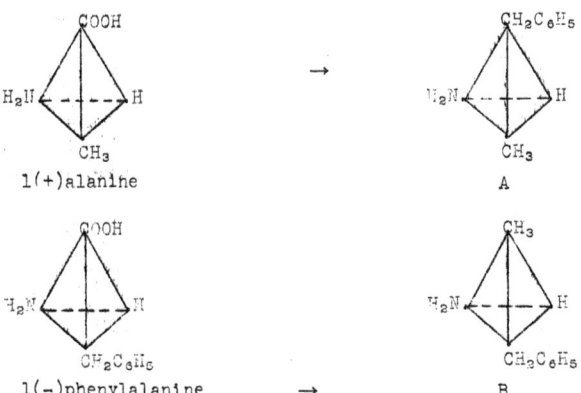

1(-)phenylalanine → B

By starting with l(+)alanine the authors obviously obtained the ephedrine whose relative configuration is represented by space formula A. Had they started with l(-)phenylalanine and reduced the carboxyl to a methyl they would have obtained the isomer represented by space formula B. A and B are mirror images, for an observer standing at position H, figure A, would find the order CH_3-NH_2-$CH_2C_6H_5$ to be counter-clockwise, whereas in figure B the same order is clockwise. In other words A has the relative configuration of l-alanine or d-phenylalanine, depending on which amino acid is taken as the reference standard.

From these results one is led to believe that the optimum configuration for activity is found in those isomers in which the phenyl-bearing carbon atom, if asymmetric, is levorotatory (d-series if referred to phenylalanine, l-series if referred to alanine).

Many investigators have postulated theories for the difference in physiological activity between enantiomorphs.[15-17] Cushny postulated that the optical isomers combine with another optically active component in the body and thus form two diastereoisomers. Since the diastereoisomers would be entirely different compounds they could have different biological properties. As opposed to this, followers of the Meyer-Overton theory supposed that the differences in pharmacological action were due to physical changes brought about in the cells, such as solubility and diffusibility.

Easson and Stedman have elaborated an alternative view according to which there appears to be no reason for differentiating between molecular dissymmetry and structure in regard to the manner in which they influence physiological activity. They consider that the difference in physiological activity of optical isomers may frequently be ascribed to circumstances which are identical with those which cause different symmetrical molecules to exhibit different physiological activities. For example, consider the case of the

following symmetrical, structural isomers:

$C_6H_5-CH_2-NHCH_3$ (I) $C_6H_5-CH_2-CH_2-NH_2$ (II)

The pressor action of II is very much greater than I. According to the proponents of this theory the molecular arrangement of the compound is the factor responsible for difference in physiological response of structural isomers. Their contention is that molecular dissymmetry and the optical activity which accompanies it are merely accidental adjuncts of different molecular arrangements.

The theoretical basis for their suggestion starts from the postulate that a drug is attached to its specific receptor in the tissues in such a manner that a considerable proportion of the drug molecule is involved. If an asymmetric carbon atom is involved three of the groups present may be concerned in the process.

As an example, the epinephrines are represented by models III and IV:

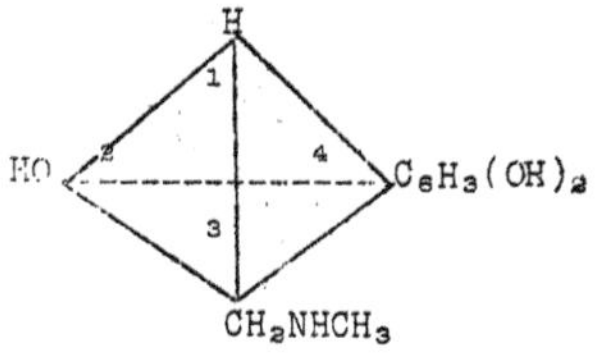
H
1
HO
2
4
$C_6H_3(OH)_2$
3
CH_2NHCH_3

III

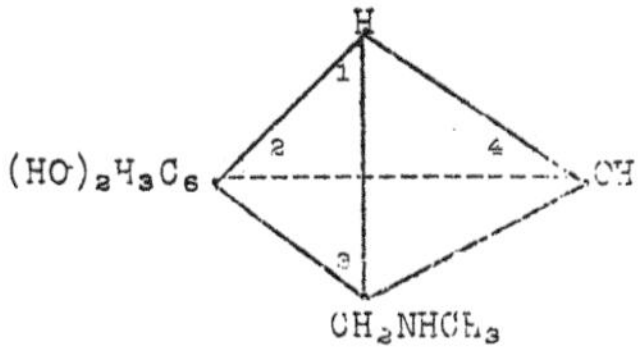
H
1
$(HO)_2H_3C_6$
2
4
OH
3
CH_2NHCH_3

IV

2
4
3

V

Figure V represents diagrammatically the surface of the specific receptor in the tissues. For the drug molecule to produce a maximum physiological effect it must be attached to the receptor in such a manner that the groups 2,3,4 in the drug coincide with 2,3,4 in the receptor. Such coincidence can only occur with the more active levo isomer. If the hydrogen at position 1 were replaced by hydroxyl the resulting symmetrical compound should still have the same activity as (-)epinephrine since the base of the tetrahedron remains unchanged. The acquisition of such a compound was not feasible. Hence another test was applied to prove the validity of their theory using (+)epinephrine. The hydroxyl at position 4 in (+)epinephrine cannot be concerned with the attachment of the drug to its receptor. This attachment is brought about by the face, 1,2,3, and the smaller physiological activity of the isomeride is attributed to the less perfect combination which results. By substituting

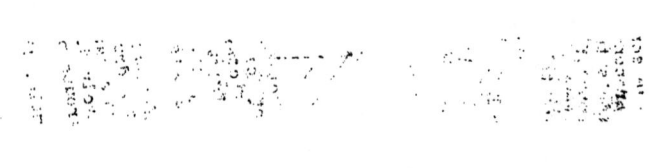

2!

a hydrogen atom for the hydroxyl in (+)epinephrine a symmetrical compound is obtained which, they theorized, should possess the same affinity for the receptor. The fact that this symmetrical compound, 3,4-dihydroxy-β-phenylethylmethylamine (epinine), does possess the same physiological activity bears out their hypothesis very well.

BIBLIOGRAPHY

(1) Kuhn, Freudenberg and Wolf, Ber., 63, 2379, (1930).
(2) Dumanski and Jakowlew. Kolloid Z., 48, 155, (1929).
(3) Pasteur, L. , "Recherches sur la dissymmetrie moleculaire des produits organiques naturels", Paris (1861).
(4) Bailey and Hass, J.A.C.S., 63, 1969, (1941)
Hass, Personal Communication to W. H. Hartung, Professor of Pharmaceutical Chemistry, University of Maryland.
(5) Schroer, Ber., 65, 966, (1932)
(6) Huckel and Niggemeyer, Ber., 72, 1354, (1939)
(7) Fischer, E., Z.physiol. Chem., 26, 60, (1898).
(8) Akamatsu, Biochem.Z., 142, 188, (1923).
(9) Ehrlich, F., Biochem.Z., 63, 379, (1914).
(10) Chen, K.K., Wu, Chang-Kong, and Henriksen, E., Journ. Pharm and Exp.Ther., 36, 363-400, (1929).
(11) Nagai, N., J. Pharm. Soc.Japan 121, 200, (1932)
(12) Leithe, W., Ber., 64B, 2827-32, (1931).
(13) Freudenberg,K., Schoffel,E., and Braun,E., J.A.C.S., 54, 234-236, (1932).
(14) Freudenberg,K., and Nikolai,F., Ann., 510, 223, (1934)
(15) Cushny, A.R., "Biological Relations of Optically Isomeric Substances", Williams and Wilkins, Baltimore, (1926).
(16) For a general review of the theories of the difference in physiological activity of antipodes see Cushny. (See 15).
(17) Easson, L.H., and Stedman, E., Biochem.J., 27, 1257-66, (1933).
(18) Hartung,W.H., and coworkers, University of Maryland, School of Pharmacy (1942).

Reported by:
C. I. Jarowski
February 17, 1943

THE GRAEBE-ULLMANN SYNTHESIS OF CARBAZOLES

By analogy with the thermal decomposition of o-aminobenzophenone to give fluorenone, Graebe and Ullman reasoned that o-aminodiphenylamine should give carbazole. However, it was found that the compound was stable toward heat. The conversion was found to be possible through the intermediate of 1-phenyl-1,2,3-benzotriazole which gives carbazole on distillation at 360° in almost quantitative yield.

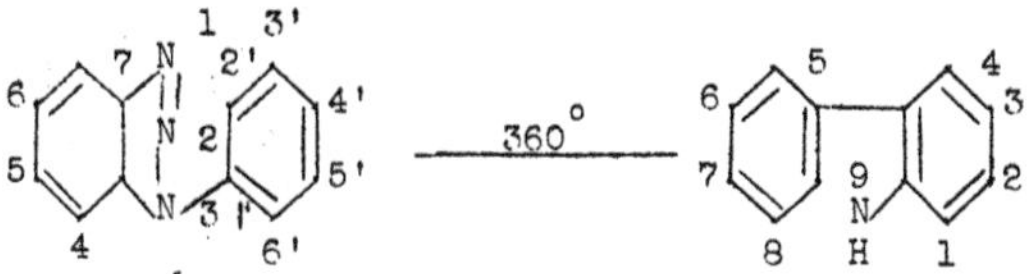
1
3'
7
N
2'
6
4'
N
2
5
5'
N
3
1'
4
6'
360°
5
4
6
3
7
9
2
N
8
H
1

1-phenyl-1,2,3-benzotriazole is readily prepared starting with o-chloronitrobenzene, since this compound reacts with aniline by refluxing in alcohol with sodium acetate or potassium carbonate to give excellent yields of 2-nitrodiphenylamine. The nitro compound is readily reduced to the corresponding amino compound. 2-Aminodiphenylamine sulfate on treatment with nitrous acid gives 1-phenyl-1,2,3-benzotriazole which precipitates out. It and its substitution products are recrystallizable from the usual organic solvents and are stable above their melting points. Many of them may be distilled at reduced pressure, but in general lose nitrogen more or less completely at their boiling points, sometimes with explosive violence.

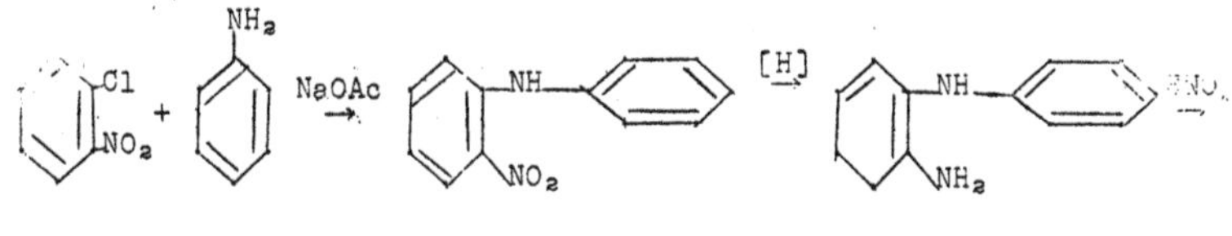
NH2
Cl
NO2
NaOAc
NH
NO2
[H]
NH
NH2

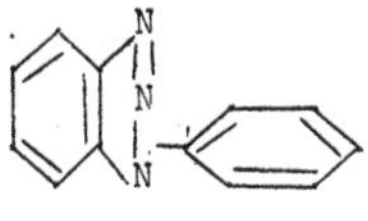
N
N
N

The value of the synthesis however lies in the preparation of substituted carbazoles, which with certain specific exceptions, are not satisfactorily obtainable in any other way. The preparation of carbazole derivatives by heating a substituted 1-phenyl-1,2,3-benzotriazole proceeds smoothly when the substituent is a saturated group such as alkyl, amino, halogen, etc. However when an unsaturated group such as nitro acetyl, or cyano is present in either benzene ring the reaction is difficult and sometimes impossible. Frequently it is advantageous to employ a 3-nitro-4-chlorobenzoic acid as the starting material. In the decomposition of the benzotriazole quick lime is used, leading to decarboxylation as well as loss of nitrogen.

Although no mechanism has been proposed in the literature it seems logical to assume that the reaction is quite analogous to the formation of biphenyls by the action of diazotates on aryl hydrocarbons.

A good example of the use of the reaction is the synthesis of 3-methylcarbazole.

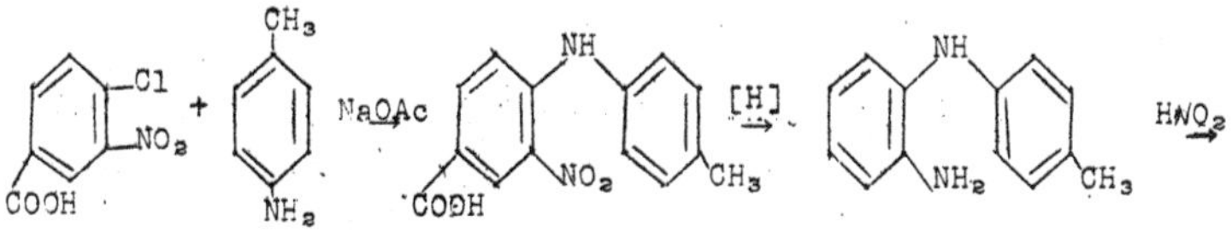
CH3
Cl
NO2
COOH
NH2
NaOAc
NH
NO2
COOH
CH3
[H]
NH
NH2
CH3
HNO2

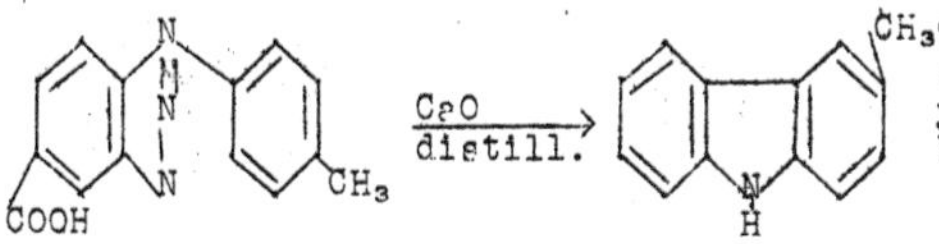
N
N
N
COOH
CH3
CaO
distill.
N
H
CH3

The over-all yield on the synthesis is about 50%, the yield on the last step being 84%.

If the substituent is in the 2' position of the benzotriazole the reaction is often somewhat more difficult. If this substituent is a methyl group acridine derivatives are formed as by-products. Thus in the preparation of 1-methylcarbazole, acridine is formed as a by-product.

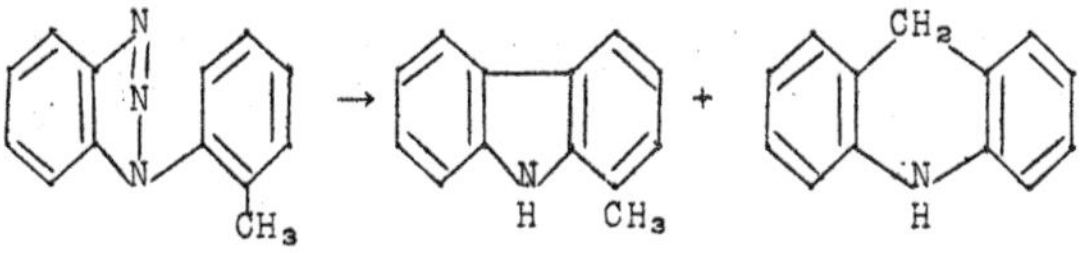
N
N
N
CH3
N
H
CH3
CH2
N
H

The reaction is still a useful preparative method however since the carbazole predominates and since the two substances are easily separated by extracting the acridine with acid.

Halogen derivatives have been prepared in the same manner. In preparing the starting material for the 3-halo compound advantage is taken of the difference of reactivity of halogens <u>ortho</u> and <u>meta</u> to a nitro group.

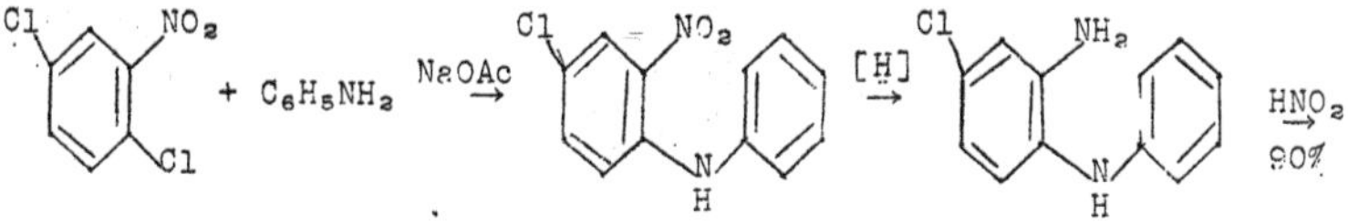
Cl
NO2
Cl
+ C6H5NH2
NaOAc
Cl
NO2
N
H
[H]
Cl
NH2
N
H
HNO2
90%

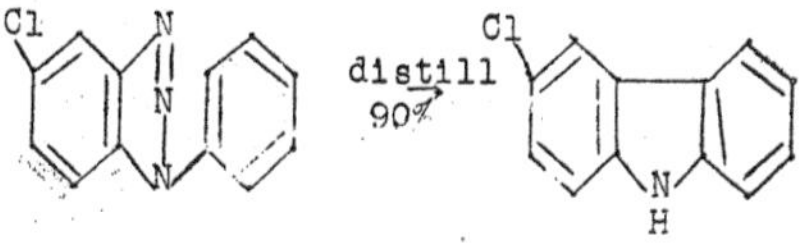
Cl
N
N
N
distill
90%
Cl
N
H

Attempts to prepare 1-bromocarbazole have led to the elimination of bromine and the formation of carbazole itself.

When unsaturated substituents are present the yields, as would be expected, are much lower. For example, the 3-acetyl compound is obtained in only 22% yield.

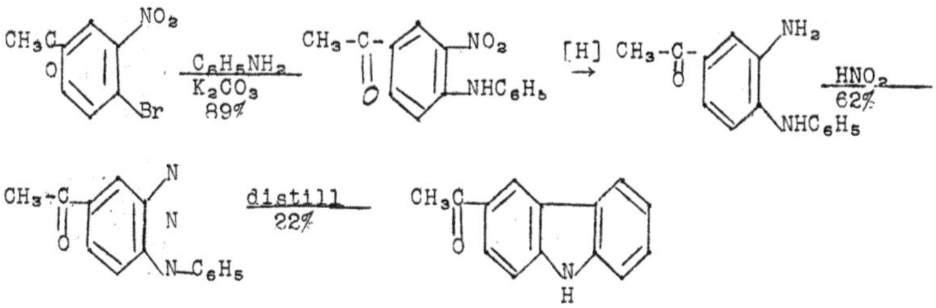
CH3C
O
NO2
Br
C6H5NH2
K2CO3
89%
CH3-C-
O
NO2
NHC6H5
[H]
CH3-C-
O
NH2
NHC6H5
HNO2
62%
CH3-C
O
N
N
N-C6H5
distill
22%
CH3C
O
N
H

However this reaction is still useful for preparing the 3-acyl derivatives since a Friedel-Crafts reaction on carbazole with acetyl chloride or benzoyl chloride gives the 3,6 disubstituted product.

The 3-nitro compound can be made in low yields but is much more readily obtained by the nitration of carbazole, which gives almost exclusively the 3-nitro compound with a trace of the 1-nitro compound. The 1-nitro compound can be made in 18% yield by the Graebe-Ullmann method, but it is more readily prepared by nitrating 3,6-dibromocarbazole and then removing the bromine atoms by reduction.

Attempts to make carbazoles with substituents on each of the benzene rings have met with varying success, For example 3-amino-6-methyl carbazole is obtained in 87.5% yield from the corresponding phenylbenzotriazole. If an amine group is substituted for the methyl only a very low yield of the diamino compound is obtained. If a bromine or a methoxyl is substituted for the methyl none of the disubstituted carbazole is obtained.

An especially interesting synthesis is that of 5-carboline from the pyridine analog of phenylbenzotriazole in almost quantitative yield.

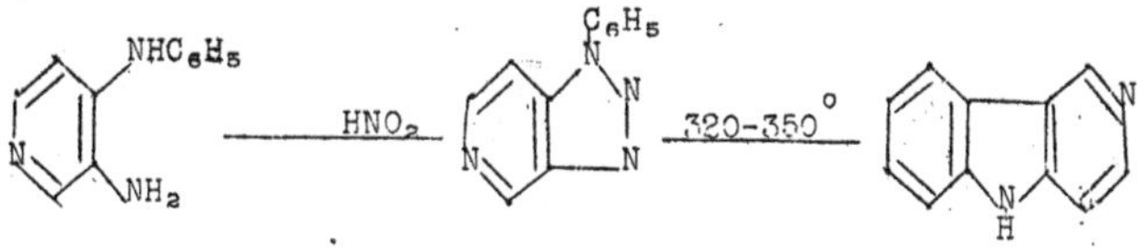
NHC6H5
N
NH2
HNO2
C6H5
N
N
N
N
320-350°
N
N
H

It found however that substituents on either ring seemed to stop the reaction completely and only unchanged starting material and resins were obtained. Similarly the quinaldinyl derivative of benzotriazole gives 5-methyl-2,3benzo- -carboline.

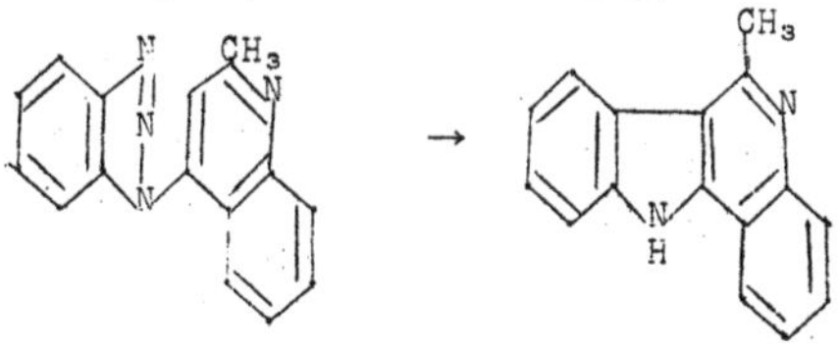
N
N
N
CH3
N
CH3
N
N
H

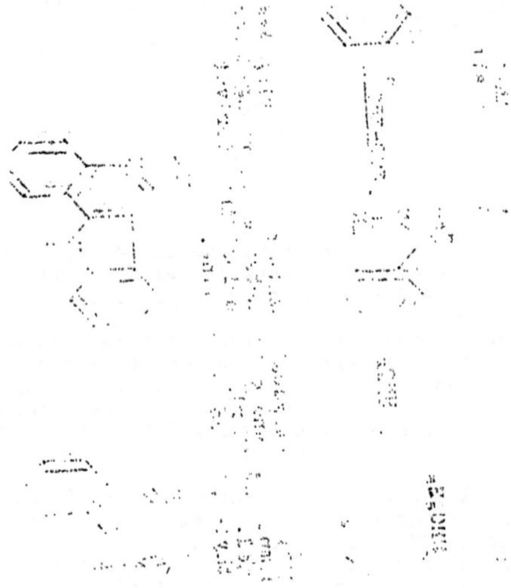

Recently improvements of yields have been made by heating the phenylbenzothiazole in solvents such as syrupy phosphoric acid, naphthalene, biphenyl, and most successfully nitrobenzene. Thus, Waldmann and Back have made 1,3-dinitrocarbazole in excellent yields from the corresponding phenylbenzotriazole by heating for thirty minutes in nitrobenzene at 300 . While Borsche was unable to close the ring without a solvent.

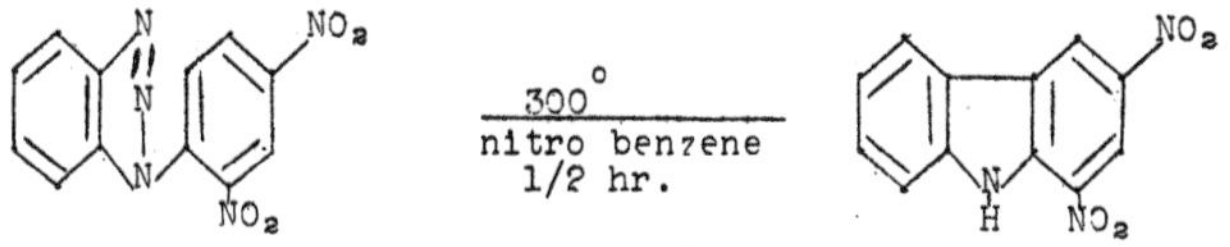
N
N
N
NO_2
NO_2
300°
nitro benzene
1/2 hr.
NO_2
N
H
NO_2

By an anlogous reaction Waldmann and Back were able to synthesize 1,9-benzacridine and some of its derivatives.

N
N
N
heat
naphthalene
NO_2
N
H

In the following table an attempt has been made to collect references to the majority of the syntheses that have been attempted by the Graebe-Ullmann method.

TABLE

Compounds Whose Preparation has been Attempted by the Graebe-Ullmann Synthesis.

Carbazoles

Substituents	Yield	Reference
Unsubstituted	almost quantitative	1,2,3
1-Methyl-	fair + acridine	2
3-Methyl-	84%	2
1,3-Dimethyl-	fair + 2-Methylacridine	2
3,7-Dimethyl-	good	3
1,2-Benzo-	good	2,3,16
1,2-Benzo-7-methyl-	70%	2,3
1-Bromo-	none only carbazole	4
2-Chloro-	84%	2
3-Chloro-	95%	2
1-Amino-	good	5
3-Amino-	fair	2,3
3-Amino-6-methyl	87.5%	6
3,6-Diamino-	very poor	6

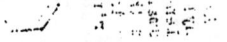

TABLE Contd.

Substituents	Yield	Reference
3-Amino-6-bromo-	none	6
3-Amino-6-methoxy-	none	6
2-Amino-7-methoxy-	15%	22
3-Cyano-	none	7
	34%	8
1-Nitro-	18%	8,9
3-Nitro-	poor	2,8
1,3-Dinitro-	excellent	16
	none	9
3-Acetyl-	22%	8
	none	7
3-Acetyl-6-ethyl-	none	12
1-Benzoyl-	none	11
3-Benzoyl-	fair	10
3,6-Dibenzoyl-	----	11
3,6-Diacetyl-	none	12
3-Chloro-6-acetamido-	none	13
3-Nitro-9-Ethyl-	----	14
	none	15
2,3-Dimethoxy-	good	21

Analogous Compounds

3-Carboline	fair	17,18
5-Carboline	almost quant.	6,19
4-Chloro-5-carboline	none	6
9-Amino-5-carboline	none	6
5-Methyl-2,3-benzo- -carboline	good	20
15-Methoxy-5-methyl-2,3-benzo- -carboline	fair	20
1,9-Benzacridine	good	16
5,7dinitro-1,9-benzacridine	excellent	16

BIBLIOGRAPHY

1) Graebe and Ullmann, Ann., 291, 16, (1896)
2) Ullmann, Ann., 332, 82, (1904)
3) Ullmann, Ber., 31, 1697, (1898)
4) Campbell and MacLean, J. Chem. Soc., 504 (1942)
5) Lindemann and Wessel, Ber., 58, 1221, (1925)
6) Bremer, Ann., 514, 2[illegible], (1934)
7) Borsche, Stackmann, and Makaroff-Semljanski, Ber., 49, 2222, (1916)
8) Preston, Tucker, and Cameron, J. Chem. Soc., 500, (1942)
9) Borsche and Rantscheff, Ann., 379, 1[illegible], (1911)
10) Hunter and Darling, J. Am. Chem. Soc., 53, 4183, (19[illegible])
11) Plant and Tomlinson, J. Chem. Soc., [illegible] (1932)
12) Plant, Rogers, and Williams, ibid, [illegible], (1935)
13) Blom, Helv. Chim. Acta, 4, 1036, (19[illegible])
14) Deletra and Ullmann, Arch. Sci. phys. nat. Geneve, 1[illegible] 78, (1904)
15) Storrie and Tucker, J. Chem. Soc., [illegible], (1931)
16) Waldmann and Back, Ann., 545, 52, (19[illegible])
17) Spath and Eiter, Ber., 73B, 719, (1940)
18) Lawson, Perkin, and Robinson, J. Chem. Soc., 125, 626, (1924)
19) Robinson and Thornley, ibid, 2169, (1924)
20) Kernach and Smith, J. Chem. Soc., 1999, (1930)
21) Hughes, Lions, Maunsell, and Wright, J. Proc. Roy. Soc. N.S. Wales, 71, 428, (1938). (C.A., 33, 613, (1939))
22) Berkenheim and Lur'e, J. Gen. Chem., (U.S.S.R.), 6, 1043, (1936) (C.A. 31, 1780, (1937)).

Reported by Zeno W. Wicks Jr. - February 24, 1943.

REACTIONS OF DIAZONIUM SALTS WITH α,β-UNSATURATED NITRILES AND CARBONYL COMPOUNDS

Meerwein and Co-workers; Marburg University, Germany
Bergmann and Weinberg; Rehovoth, Palestine
Koelsch; University of Minnesota

In 1938 it was reported that α,β-unsaturated carbonyl compounds such as aldehydes, ketones, or carboxylic acids react with diazonium salts in dilute inorganic acids, or in organic acids at elevated temperatures, in such a way that nitrogen is split off and the α-hydrogen atom of the carbonyl compound is replaced by an aryl group.

$$>C{=}CH\overset{O}{\overset{\|}{C}}R + ArN_2^{+} \rightarrow \; >C = \underset{Ar}{C}\overset{O}{\overset{\|}{C}}R + N_2 + H^{+}$$

In certain cases the arylated α,β-unsaturated compounds are obtained as their hydrogen halide addition products, and may be treated with agents that split off hydrogen halide. With certain α,β-unsaturated carboxylic acids the carbonyl group is lost as carbon dioxide during the reaction. The addition of copper or of copper salts was reported to favor the reaction, which may be carried out in strong acetic acid, aqueous alcohol, or aqueous acetone solutions, to which alkali acetates or the alkali salts of other weak acids have been added. Other solvents may also be used, although the results are not as good. The presence of negative substituents in the diazonium compound is favorable.

Meerwein and co-workers have further investigated this reaction, and have found that acetylenes behave in a manner similar to compounds containing the double bond. Their investigations led to the conclusion that acetone not only facilitates solution, but also has some special effect in the reaction, and that cupric, but not cuprous, salts act as catalysts.

Obviously these reactions have a wide variety of possible synthetic applications. There has been particular interest in using this method for the preparation of substituted stilbenes, in the search for oestrogenic and trypanocidal compounds. Bergmann and Weinberg were just beginning to investigate the uses and mechanism of these syntheses when the war interrupted their work.

Meerwein prepared several 3-arylcoumarins; the results are summarized in the following tables. In each case, except the second, the diazonium salt was added to an aqueous acetone solution of coumarin to which cupric chloride and sodium acetate had been added. In the second example, methyl cyanide was substituted for acetone.

TABLE I

	Diazonium Salt	Coumarin Derivative	Yield (%)
1.	p-$ClC_6H_4N_2Cl$	3(p-chlorophenyl)-	45.9
2.	p-$ClC_6H_4N_2Cl$	3(p-chlorophenyl)-	57.
3.	$C_6H_5N_2Cl$	3(phenyl)-	25.9
4.	β-$C_{10}H_6N_2Cl$	3(β-naphthyl)-	17.6
5.	p-$CH_3OC_6H_4N_2Cl$	3(p-methoxyphenyl)-	35.3
6.	p-$CH_3CONHC_6H_4N_2Cl$	3(p-acetaminophenyl)-	27.8
7.	p-$HO_3SC_6H_4N_2Cl$	3(p-phenyl sodium sulfonate)-	43

Stilbenes with the trans configuration are prepared by the reaction of a diazonium salt with cinnamic acid or a cinnamic acid derivative. Carbon dioxide is lost in the course of the reaction. The yields would probably be much higher if the stilbenes could be isolated readily. TABLE II shows some typical stilbene syntheses.

TABLE II

Diazonium Compound	Acid	Product	Yield (%)
$C_6H_5N_2Cl$	cinnamic	stilbene	36.5
p-$CH_3C_6H_4N_2Cl$	cinnamic	p-methylstilbene	48.8
p-$ClC_6H_4N_2Cl$	cinnamic	p-chlorostilbene	68.8
o,p-Cl_2?$C_6H_3N_2Cl$	cinnamic	o,p-dichlorostilbene	34. 4
p-$O_2NC_6H_4N_2Cl$	cinnamic	p-nitrostilbene	58.4
o-$O_2NC_6H_4N_2Cl$	cinnamic	o-nitrostilbene	43.9
p-$CH_3OC_6H_4N_2Cl$	cinnamic	p-methoxystilbene	49.2
p-$HO_3SC_6H_4N_2Cl$	cinnamic	stilbene-p-sulfonic acid	77.5
p-$ClC_6H_4N_2Cl$	p-hydroxycinnamic	p-chloro-p'-hydroxy-stilbene	56.5
p-$ClC_6H_4N_2Cl$	p-methoxycinnamic	p-chloro-p'-methoxy-stilbene	61.2

It is possible to substitute the nitrile for the ester in which case the yield of the arylated product may be increased. If the ester of the acid is used in a pyridine-acetic acid solution, the hydrogen halide addition product is obtained.

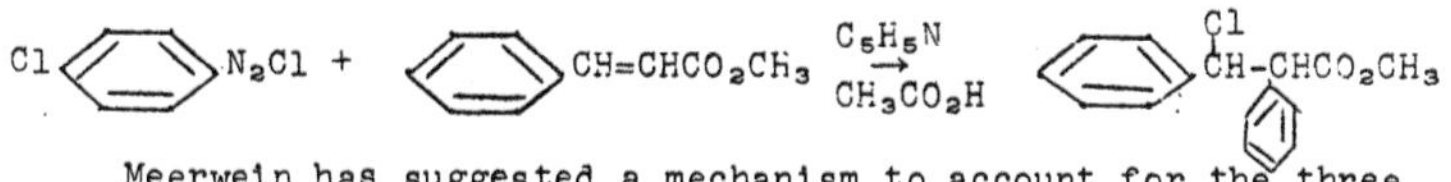

Meerwein has suggested a mechanism to account for the three types of products obtained in these reactions. He postulates the decomposition of the diazonium salt to give a positive aryl ion and free nitrogen. The positive aryl ion then unites with a tautomeric form of the carbonyl compound.

$$ArN_2^+ \rightarrow Ar^+ + N_2$$

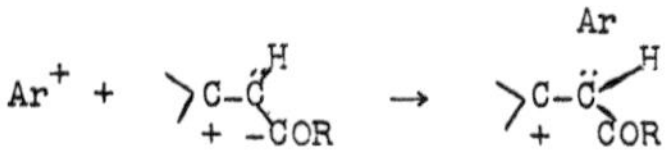

The positive ion thus formed may react in three different ways.

(a) It may lose a proton, thus reforming the double bond.

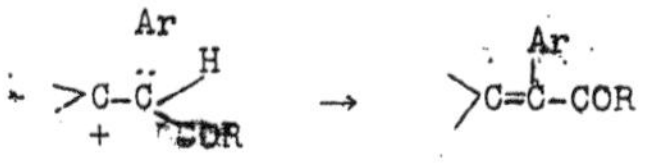

(b) It may unite with a halogen ion to give the hydrohalide.

(c) It may form an unstable lactone which loses carbon dioxide to form an unsaturated hydrocarbon.

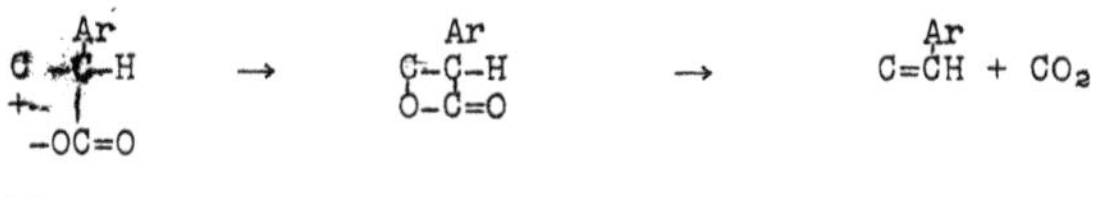

He was unable to explain what causes the reaction to go in one of these manners in preference to another.

Recently Koelsch has further studied these reactions. Meerwein and co-workers concluded that the aryl group was invariably attached to the carbon α to the carbonyl group, but the only compounds whose structures were rigorously proved were derived from α,β-unsaturated compounds already substituted in the β position. It has been shown that several reactions of diazonium salts probably involve free radicals, which led Koelsch to believe that steric factors might be more important than the direction of polarization of the α,β-unsaturation in determining the carbon atom to which the aryl group becomes attached. For this reason he investigated the reactions of diazonium salts with acrylonitrile and methyl

acrylate, and he found that in every case the product is formed through the union of an aryl group with the β carbon atom. The yields of the hydrocinnamic acids are low because of the ease with which they lose hydrogen chloride during the purification process. This appears, however, to be a promising synthesis for cinnamic acids since elimination of hydrogen chloride and hydrolysis of the nitrile group can be accomplished in good yield.

TABLE III

	Coupling Products	Yield % Pure	Yield % Crude
I	α-chlorohydrocinnamylnitrile	34	--
II	α-chloro-m-nitrohydrocinnamylnitrile	38	53
III	α-chloro-p-nitrohydrocinnamylnitrile	48	61
IV	α-chloro-p-methylhydrocinnamylnitrile	40	45
V	methyl-α-chloro-p-methylhydrocinnamate	--	23

Koelsch is still studying the effect of various substituents in the unsaturated portion of the molecule on the proportion of α- to β-arylation which it undergoes.

Bibliography

Brit. Pat. 480,617; C.A., 32, 6262 (1938)
Meerwein, Buchner, and van Emster, J. Prakt. Chem., 152, 237 (1939)
Bergmann and Weinberg, J. Org. Chem., 6, 134 (1941)
Koelsch, J. Am. Chem. Soc., 65, 57 (1943)
Haworth, Heilbron, and Hey, J. Chem. Soc., 372 (1940)
Sundholm, Seminar Abstracts 1/13/43 (3 references given)

Reported by John E. Wilson
February 24, 1943

AROMATIC IODINATION

Iodination with iodine usually involves the following equilibrium:

$$I_2 + ArH \rightarrow ArI + HI$$

There are three ways that this equilibrium can be favorably shifted. The first is by oxidizing the hydriodic acid to iodine, the regenerated iodine then reacting; the second consists in neutralizing the hydriodic acid with a suitable base; and the third, which is of little value, consists of removing the hydriodic acid by the precipitation of the iodide ion with the silver ion.

The first method is the more drastic one and is used with compounds that are difficult to iodinate. A great variety of oxidizing agents have been used. They include iodic acid, nitric acid, sulfur trioxide, potassium persulfate, nitrosulfonic acid, sodium nitrite and sulfuric acid, fuming sulfuric acid, a mixture of nitric and sulfuric acids, and chromium oxide.

Klages and Liecke succeeded in iodinating mesitylene easily in the presence of iodic acid. An aqueous solution of iodic acid was added to a mixture of iodine and mesitylene in acetic acid and this mixture was heated for fifteen minutes. A good yield of iodomesitylene was obtained.

Biphenyl was iodinated in the presence of nitric acid by Guglialmelli and Franco. Concentrated nitric acid was added dropwise to a mixture of biphenyl and iodine at 100°C. The main product was 4-iodo-4'-nitrobiphenyl.

Juvalta dissolved phthalic anhydride in fuming sulfuric acid containing 50-60% of the anhydride, heated this to 90-100°C., and added iodine cautiously. This produced the tetraiodophthalic anhydride.

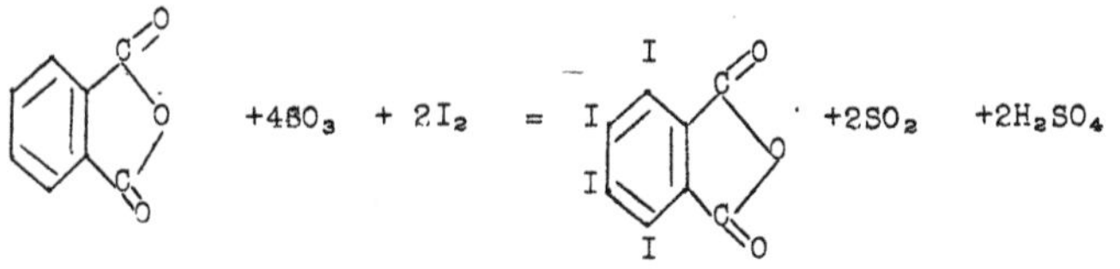

+4SO₃
+ 2I₂
=
+2SO₂
+2H₂SO₄

The most extensive investigation of this type of reaction is that of Varma in India. He has studied the bromination and iodination of a large variety of aromatic compounds using nitrosulfonic acid, sodium nitrite and sulfuric acid, and fuming nitric acid as promoters. The mixture of fuming nitric acid and nitrosulfonic acid he used was obtained by passing sulfur dioxide through the fuming nitric acid until a mixture containing about 50% of nitrosulfonic acid was obtained.

Aromatic acids are not so easily iodinated as aromatic hydrocarbons, phenols, or amines. Most of the iodo-derivatives of the aromatic acids have been obtained by the oxidation of the side-chain of the corresponding homologues of benzene or by the replacement of the amino group in aromatic acid derivatives by iodine through diazotisation. Varma found that he could iodinate benzoic acid in good yield in the presence of these promoters. Benzoic acid and iodine were dissolved in carbon tetrachloride, and the mixture heated to refluxing. Sodium nitrite and fuming sulfuric acid were added over a half an hour. This produced a 50% yield of m-iodobenzoic acid. The nitrosulfonic acid used as the promoter gave a slightly higher yield. This method fails with the three phthalic acids and with p-nitro-, p-chloro-, and o-bromobenzoic acids.

Varma iodinated benzonitrile by the direct action of potassium iodide with concentrated sulfuric acid on the nitrile.

He obtained good results in the iodination of o- and m-chlorotoluenes, but was unable to iodinate p-chlorotoluene. The chlorotoluene and iodine were dissolved in acetic acid and fuming nitric acid or nitrosulfonic acid added; then the mixture was heated for four hours. The iodo- group in both cases was oriented para to the chloro group. This may be why the p-chlorotoluene could not be iodinated.

He reported good results also on compounds like bromobenzene, p-dichlorobenzene, p-dibromobenzene, iodobenzene, benzene, toluene, the xylenes, ethylbenzene, and pseudocumene. By repeated iodination of 2,4-diiodomesitylene he received 2,4,6-triiodomesitylene using nitrosulfonic acid as the promoter. He iodinated naphthalene to α-iodonaphthalene in 40% yield using a mixture of fuming nitric and nitrosulfonic acids as the promoter.

A variety of basic reagents have been used to shift the equilibrium to the right by neutralizing the hydriodic acid formed. One of the most useful is mercuric oxide.

Recently, Hoch and Culbertson iodinated anisole using mercuric oxide as the promoter. Anisole was dissolved in absolute alcohol, mixed with mercuric oxide, and iodine added in portions with shaking. An 85% yield of the para product was obtained. Phenol, durene, carbazole, and thiophene have also been iodinated in this manner. Hodgson diiodinated o-nitrophenol in the 4,6 positions using mercuric oxide in acetic acid.

The general method for iodinating phenols and cresols is by adding iodine dissolved in potassium iodide solution to the phenol in aqueous alkali. It is quite difficult to control the iodination of phenol. Attempts to prepare 2,4-diiodophenol also gave some of the 2,4,6-triiodophenol. Körner got a preponderance of 2-iodophenol by the action of a calculated amount of iodine and iodic acid on phenol in dilute alkali.

Bordeinau reported using ammonia as the base. Varma used ammonium hydroxide as the solvent in iodinating β-naphthol with iodine in potassium iodide solution. He believes that the nascent ammonium triiodide formed by the action of iodine on ammonium hydroxide is the active iodinating agent.

[illegible]

[illegible] were adjusted [illegible]

[illegible] results [illegible] This may be [illegible] the explanation [illegible]

[illegible]

A variety of [illegible] have been used [illegible]

[illegible]

Wheeler and Liddle used calcium carbonate as the base in preparing 3-iodo-p-toluidine from p-toluidine. They reported getting an 84% yield. This is a very good method with amines. It is not necessary to use basic promoters with amines as they themselves can act as such. Recently Militzer, Smith and Evans reported iodinating some aromatic amines in dilute acetic acid directly with iodine. However, their yields were not too good. The presence of another base might have increased the yields. They could not iodinate p-aminoacetophenone, p-nitraniline, or acetanilide under these conditions.

An example of the third type of shifting the equilibrium is that using silver nitrate with the precipitation of silver iodide. Some Russians used this method for the quantitative determination of some phenolic compounds.

Iodine monochloride is an excellent iodinating agent. Its action on aromatic compounds was first investigated by Schuetzenberger in 1861. He produced mono- and diiodobenzene by the action of this reagent on sodium benzoate. In 1864, Stenhouse obtained triiodoorcin by its action on orcin. Michael and Norton in 1880 made an important study of its action on aromatic amines and amino-acids. They used the vapors of the iodine monochloride as their reagent. These vapors were passed into a solution of acetanilide in glacial acetic acid and gave an 85% of p-iodoacetanilide. They reported excellent results in iodinating m-nitraniline, p-nitraniline, and p-toluidine. It did not iodinate aceto-p-toluide, but diiodinated the free amine. Two moles diiodinated p-aminobenzoic acid.

Kerschbaum reported iodinating 2,4-dimethylacetanilide with iodine monochloride in which concentrated hydrochloric acid was used as the solvent. This compound could not be iodinated with iodine and mercuric oxide.

Bradfield, Orton, and Roberts prepared an excellent iodinating reagent by adding sodium iodide to dichloramine-T in glacial acetic acid. Aniline with one equivalent gives p-iodoaniline; with three squivalents, triiodoaniline.Acetanilide, heated with two equivalents for fifteen minutes gives an 80% yield of p-iodoacetanilide.

Le Fevre and Ganguly added iodine dissolved in ether to an ethereal solution obtained by mixing 3-bromocymene, ether, and magnesium until the color of the iodine persisted. This produced a 15% yield of 3-iodocymene.

Bibliography

Klages and Liecke, J. Prakt. Chem. 2 61, 311
Guglialmelli and Franco, Anales asoc. quin Argentina, 19, 5-33, (1931)
Juvalta, D.R.P. 50177
Varma and co-workers, J. Ind.Chem. Soc., 7, 503-4, (1930), 10, 593-8, (1933) 11,203-4, (1934), 13,31-33, 187-8, (1936), 13. 192-3, (1936)

Koch and Culbertson, Proc. Iowa Acad. Sci., 47, 265-6, (1940)
Korner, A., 137, 213
Bordeinau, Ann. sci. univ. Jassy, 20, 131-8 (1935)
Wheeler and Liddle, Amer. Ch. Jour., 42, 441, (1909)
Militzer, Smith, and Evans, J. A. C. S. 63, 436, (1941)
Stenhouse, J. Chem. Soc. 2, 2, 327
Michael and Norton, Amer. Ch. Jour., 1, 255, (1879-80)
Kerschbaum, Ber., 11, 107; 28, 2799
Bradfield, Orton, and Roberts, J. Chem. Soc., 1928, 782-5
Le Fevre and Ganguly, ibid., 1934 1697-9

Reported by N. K. Sundholm
March 3, 1943

THE SCHMIDT REACTION

Prior to 1923 the only products which had been isolated from the decomposition of hydrazoic acid were nitrogen and ammonia. About this time K. F. Schmidt discovered that an aqueous solution of hydrazoic acid in contact with concentrated sulfuric acid decomposed upon warming yielding nitrogen and hydroxylamine. If the hydrazoic acid were first dissolved in benzene or chloroform and then stirred and warmed with concentrated sulfuric acid, hydrazine sulfate was the main product. When dissolved in benzene, aniline sulfate was also a product. Schmidt explained these transformations on the basis of the decomposition of hydrazoic acid to nitrogen and an "imine residue". The "imine residue" would then react as follows:

$$HN_3 \rightarrow N_2 + NH \qquad NH + NH \rightarrow HNNH$$

$$H_2O + NH \rightarrow NH_2OH \qquad HNNH + HNNH \rightarrow {}_2HNNH_2 + N_2$$

$$C_6H_6 + NH \rightarrow C_6H_5NH_2$$

Since these initial discoveries, the method of carrying out the Schmidt reaction has been more or less standardized. The hydrazoic acid is usually dissolved in chloroform and added to a concentrated sulfuric acid solution of the reacting compound.

If the reaction is carried out in the presence of a carbonyl compound the decomposition occurs more rapidly and at a lower temperature. Use of approximately equimolecular quantities of carbonyl compound and acid led to the following results:

$$RCOR \xrightarrow[H_2SO_4]{HN_3} RCONHR + N_2 \qquad RCHO \xrightarrow[H_2SO_4]{HN_3} RCN + N_2 + H_2O$$

$$RCHO \xrightarrow[H_2SO_4]{HN_3} RNHCHO + N_2$$

In order to account for the observed products Schmidt postulated the following mechanism which involves a univalent nitrogen atom. This type of substance is no longer in favor but was at one time frequently employed in several molecular rearrangements.

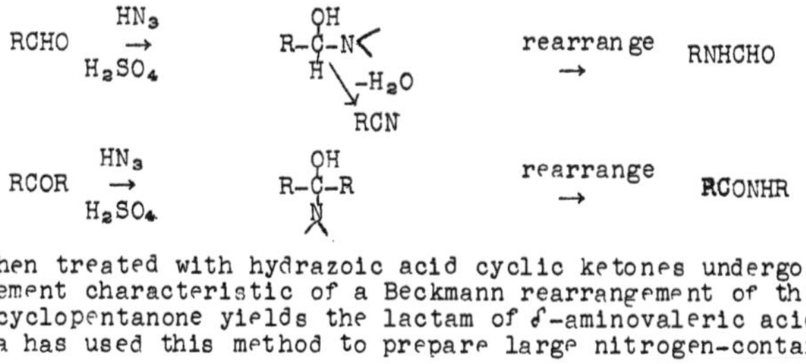

RCHO
HN3
H2SO4
OH
R-C-N
H
-H2O
RCN
rearrange
RNHCHO
RCOR
HN3
H2SO4
OH
R-C-R
N
rearrange
RCONHR

When treated with hydrazoic acid cyclic ketones undergo ring enlargement characteristic of a Beckmann rearrangement of the oxime. Thus, cyclopentanone yields the lactam of δ-aminovaleric acid. Ruzicka has used this method to prepare large nitrogen-containing rings.

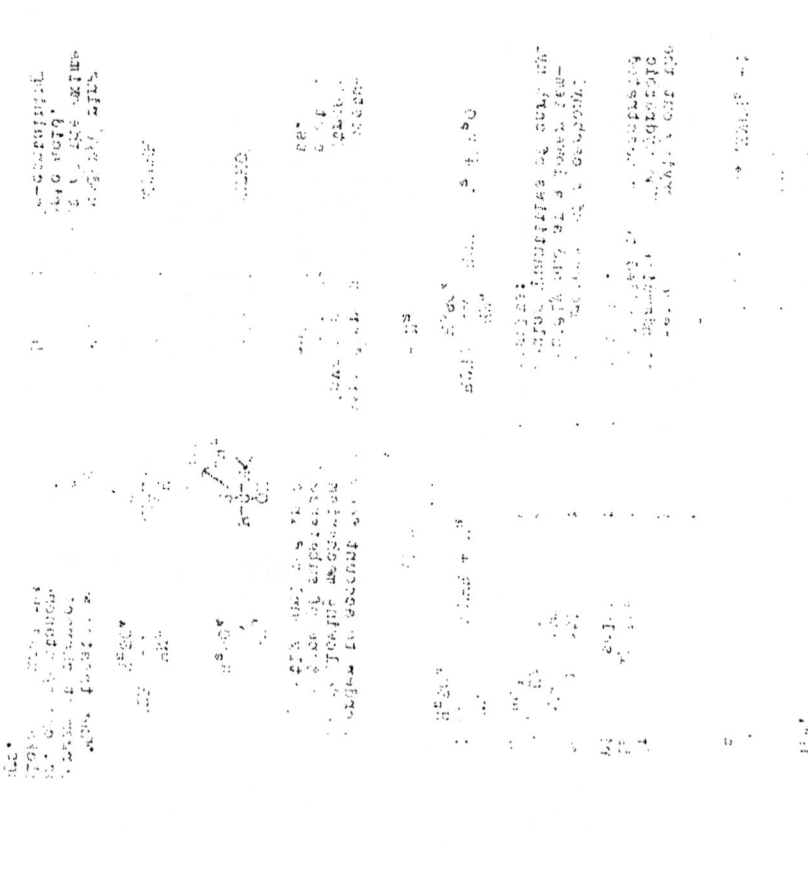

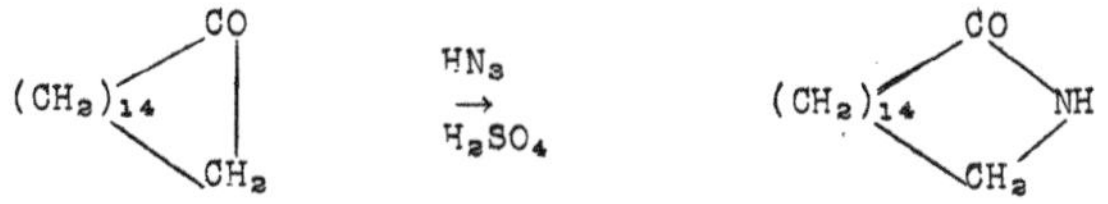

Treatment of ketones with excess acid leads to the formation of tetrazoles. An example follows.

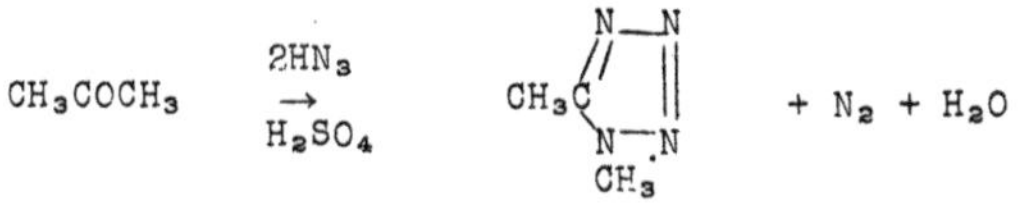

A limited application of the Schmidt reaction is the preparation of β-arylethylamides which are useful in the Bischler-Napieralski synthesis of isoquinoline derivatives. Ring closure is usually effected by phosphoric anhydride in boiling xylene or tetralin.

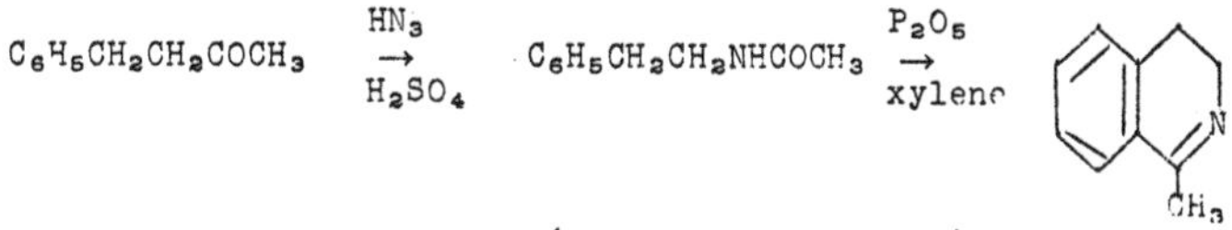

The presence of a double bond (as in benzalacetone) leads to a derivative of methyl amine rather than a derivative of acetamide.

$$C_6H_5CH{=}CHCOCH_3 \xrightarrow[H_2SO_4]{HN_3} C_6H_5CH{=}CHCONHCH_3$$

β-Ketoesters, when treated with hydrazoic acid followed by hydrolysis, produce α-amino acids in good yield. α-Aminobutyric acid, α-aminoisoamylacetic acid, dibenzylaminoacetic acid, glycine, leucine, and phenylalanine have been prepared in yields of 80-98%.

$$CH_3CO\overset{R}{C}HCO_2Et \xrightarrow[H_2SO_4]{HN_3} CH_3CONH\overset{R}{C}HCO_2Et \xrightarrow{H_2O} R\overset{NH_2}{C}HCOOH$$

The products obtained from ketones suggests an oxime formation followed by a Beckmann rearrangement. Spielman and Austin have cited evidence in opposition to such formation at least in the case of benzil. Treatment of benzil in the Schmidt manner yields benzoylphenylurea together with traces of oxanilide.

$$C_6H_5COCOC_6H_5 \xrightarrow[H_2SO_4]{HN_3} C_6H_5NHCOCOC_6H_5 \xrightarrow{HN_3} (CONHC_6H_5)_2 \text{ (traces)}$$

$$C_6H_5NHCOCOC_6H_5 \xrightarrow{HN_3} C_6H_5NHCONHCOC_6H_5$$

(main product)

In order to account for the product, it seems likely that the initial rupture must be between a phenyl group and a carbonyl group since this is the only ketonic intermediate. The formation of some oxanilide is interpreted as evidence of this rupture. The possibility of the formation of a dioxime is ruled out because the dioxime which would yield the product is γ-benzildioxime. This dioxime is stable toward sulfuric acid and mixtures of sulfuric and hydrazoic acids.

The most useful aspect of the Schmidt reaction involves the replacement of a carboxyl group by an amino group. The reaction is thought to be a modified Curtius rearrangement.

$$RCH_2CO_2H \xrightarrow[H_2SO_4]{HN_3} RCH_2CON_3 \rightarrow RCH_2NCO \rightarrow RCH_2NH_2$$

The yield of amine from the acid is usually good. Some yields of over 90% have been reported. von Braun has applied this method in elucidating some structures in the camphor series and in the determination of the structures of some constituents of crude petroleum. von Braun has also shown that optically active substances undergo the Schmidt reaction with retention of configuration.

$$\underset{\underline{d}}{\overset{C_6H_5CH_2}{\underset{CH_3}{>}}CHCO_2H} \xrightarrow[H_2SO_4]{HN_3} \underset{\underline{d}}{\overset{C_6H_5CH_2}{\underset{CH_3}{>}}CHNH_2}$$

Oesterlin has carried out many Schmidt reactions upon carboxylic acids. He found that α-amino acids, o-aminocarboxylic acids and malonic acid were unattacked under normal conditions. The effect of the double bond was observed in that cinnamic acid yielded phenylacetaldehyde probably through the intermediate formation of β-aminostyrene.

$$C_6H_5CH{=}CHCO_2H \xrightarrow[H_2SO_4]{HN_3} C_6H_5CH{=}CHNH_2 \rightarrow C_6H_5CH_2CH{=}NH \rightarrow C_6H_5CH_2CHO$$

Adamson has taken advantage of the stability of a carboxyl group situated alpha to an amino group to prepare some of the basic amino acids. α-Aminopimelic acid yields dl lysine and α-aminoadipic acid yields dl-ornithine. The yield in each case is 75%. If 2-carbethoxycyclohexanone is used as a starting material, dl-lysine is obtained in 60% yield.

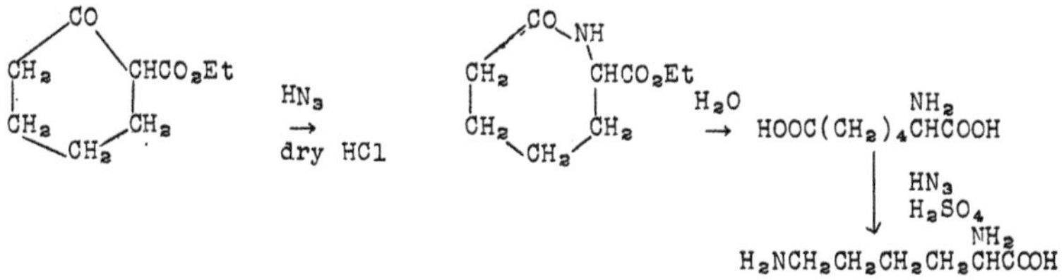

Dibasic acids containing more than four carbon atoms give good yields of diamine. Adipic acid gives putrescine in yields of 80%. Cyclobutane-1,2-dicarboxylic acid leads to 1,2-diaminocyclobutane in 35% yield.

The long chain fatty acids give better yields of amine than do the shorter ones. For example n-undecanoic acid gives rise to n-decyl amine in 90% yield and stearic acid gives heptadecyl amine in 96% yield.

The solvent for the hydrazoic acid has considerable influence upon the yields. Benzoic acid is converted to aniline in yields of 85% when chloroform or benzene is used as a solvent. If anhydrous ether is used the yield falls to 24%.

Some of the compounds which have been subjected to the Schmidt reaction are listed in Table I.

BIBLIOGRAPHY

K. F. Schmidt, Ber., 57, 704 (1924)
K. F. Schmidt, ibid., 58B, 2413 (1925)
Oesterlin, Angew. Chem., 45, 536 (1932)
Adamson and Kenner, J. Chem. Soc., 842 (1934)
Adamson, ibid., 1564 (1939)
von Braun, Ann., 490, 100 (1931)
von Braun, Ber., 66, 684 (1933)
von Braun and Kurtz, ibid., 67, 225 (1934)
Spielman and Austin, J. Am. Chem. Soc., 59, 2658 (1937)
Buchman, Reims, Skei and Schlatter, ibid., 64, 2696 (1942)
Ruzicka, Goldberg, Hurbin and Boekenoogen, Helv. Chim. Acta., 16, 1323 (1933)
Gordon, De Ath and Ellis, J. Chem. Soc., 61 (1942)

TABLE I

COMPOUND	PRODUCT	%YIELD
Acetylalanine	None	--
Adipic Acid	Putrescine	80
Alanine	None	--
α-Aminoadipic Acid	dl-Ornithine	74
α-Aminopimelic Acid	dl-Lysine	75
Benzoic Acid	Aniline	85
Benzylmalonic Acid	None	--
n-Caproic Acid	n-Amylamine	75
n-Caprylic Acid	n-Heptylamine	75
2-Carbethoxycyclohexanone	dl-Lysine	60
2-Carbethoxycyclopentanone	dl-Ornithine	42
Cinnamic Acid	Phenylacetaldehyde	43
Decanoic Acid	n-Nonylamine	75
Dibenzylacetic Acid	Dibenzylmethylamine	70
Ethyl Acetoacetate	Glycine	over 80
Ethyl Benzylacetoacetate	Phenylalanine	over 80
Ethyl Dibenzylacetoacetate	Dibenzylaminoacetic Acid	over 80
Ethyl Ethylacetoacetate	α-Aminobutyric Acid	over 80
Ethyl Isoamylacetoacetate	Isoamylaminoacetic Acid	over 80
d-Glutamic Acid	d-α,γ-Diaminobutyric Acid	42
Glycine	None	--
n-Heptanoic Acid	n-Hexylamine	75
Hexahydrobenzoic Acid	Cyclohexylamine	80
Hippuric Acid	None	--
Hydrocinnamic Acid	β-Phenylethylamine	70
Levulinic Acid	Methylamine	low
Malonic Acid	None	--
p-Methoxyhydrocinnamic Acid	p-Methoxy-β-phenylethylamine	55
o-Nitrobenzoic Acid	o-Nitroaniline	65
m-Nitrobenzoic Acid	m-Nitroaniline	65
p-Nitrobenzoic Acid	p-Nitroaniline	65
Pelargonic Acid	n-Octylamine	75
Phenylacetic Acid	Benzylamine	75
Phenylaminoacetic Acid	None	--
α-Phenyl-β-[illegible] Acid	None	--
o-Phthalic Acid	Anthranilic Acid	80
Salicylic Acid	o-Aminophenol	10
Stearic Acid	n-Heptadecylamine	96
Succinic Acid	Ethylenediamine	6
o-Toluic Acid	o-Toluidine	46
m-Toluic Acid	m-Toluidine	24
p-Toluic Acid	p-Toluidine	70
Trimethoxygallic Acid	3,4,5-Trimethoxyaniline	55
Undecanoic Acid	n-Decylamine	90

Reported by W. E. Wallace
March 3, 194[illegible]

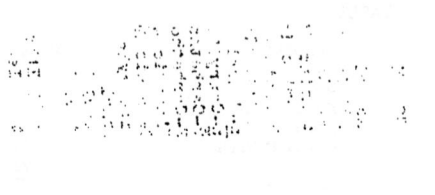

ANTIMALARIALS

An earlier seminar by Wearn in 1941 on antimalarials gave a survey of the subject up to that time. The first part of this seminar deals with the cinchona alkaloid type of antimalarial, the effect of variation in their structure on parasiticidal activity, and recent work on the synthesis of some compounds similar to these cinchona alkaloids.

The cinchona group is composed of about twenty alkaloids extracted from the roots and bark of the cinchona trees. These naturally occurring alkaloids were used as antimalarials as early as 1630, although it was not until 1820 that quinine and cinchonine were isolated.

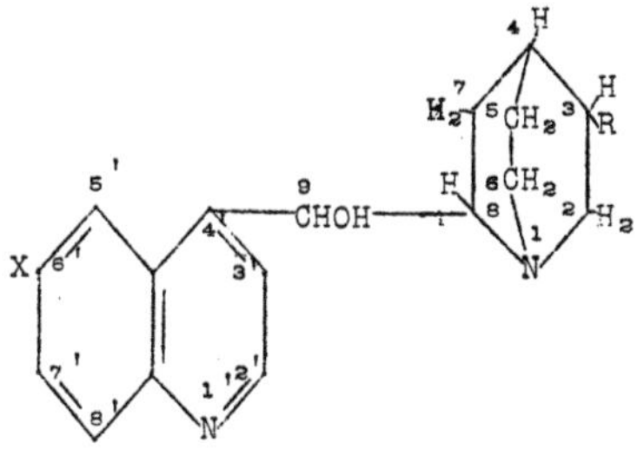
X
CHOH
N
N
R
H
H_2
CH_2

(1) X=H in cinchonine, cinchonidine, dihydrocinchonine, cinchotenidine and dihydrocinchotenidine; in all other cases it is ($-OCH_3$), except in cupreine when it is (-OH), and the cupreine alkyl ethers, quinethyline ($-C_2H_5$), quinpropyline ($-OC_3H_7$) and quinamyline ($-OC_5H_{11}$).

(2) R=($-CH=CH_2$) in cinchonine, cinchonidine, quinine and quinidine and ($-CH_2CH_3$) in their reduced dihydro derivatives.

(3) R=(-COOAlkyl) in the alkyl quitenines, quitenidines, cinchotenines and cinchotenidines.

(4) Carbon atoms 3,4,8 and 9 are all dextrorotatory in cinchone and its derivatives; carbon atoms 3 and 4 are dextrorotatory and 8 and 9 are levo-rotatory in cinchonidine and quinine and their derivatives.

A great deal of study has been made of the relationship between the structure of the cinchona type compounds and their parasiticidal activity.

Hydrogenation of the vinyl group in quinine, giving hydroquinine, produced little change in activity, but hydrogenation of cupreine to hydrocupreine changed an almost inactive compound to one having a good degree of antimalarial activity.

Other changes in the vinyl group were studied. Addition of halogen acids caused only slight loss of activity. In dehydroquinine (>CH-C≡CH), antimalarial action was reduced. Upon oxidation of the vinyl group to carboxyl in quitenine (>CH-CO_2H), antimalarial action disappeared entirely, but reappeared in the esters.

All variations of the (-CHOH) group in quinine and cinchonine and their derivatives were accompanied by loss of activity.

Substitutions in the 5' position in the quinoline nucleus of hydroquinine were without apparent effect.

Azo dyes synthesized from hydroquinine were nearly always inactive; 6-methoxyquinoline-8-azohydrocupreine being an exception.

Whereas in quinine the presence of the (-CHOH-) group appears essential to antimalarial activity, Ainley and King found that reduction to the carbinol of quinicine, shown to be inactive, did not increase its activity.

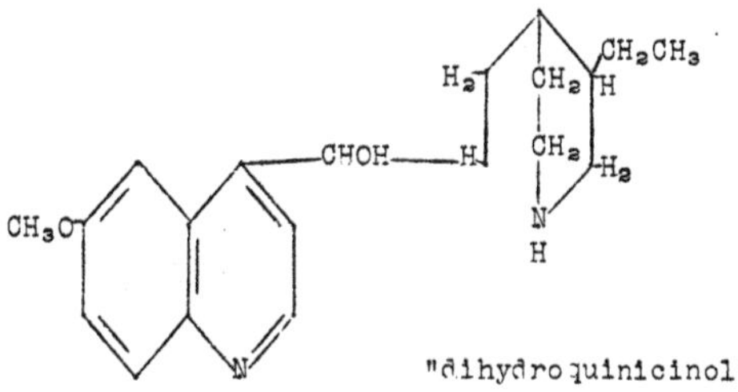
CH₃O
CHOH
H₂
CH₂
CH₂CH₃
H
H
CH₂
H₂
N
H
N

"dihydroquinicinol"

If the -OH group of dihydrocupreine is alkylated with dialkylaminoalkyl sidechains of various types $\left(\begin{matrix}R \\ R'\end{matrix}\right\rangle N-(CH_2)_x-)$, the antimalarial activity of the compound decreases and finally disappears as the chain increases in length.

Hydroquinine was found by Goodson to be the most active of the natural cinchona alkaloids against avian malaria, followed in decreasing effectiveness by quinidine, cinchonidine, and cinchonine.

Solomon, in studying the action of mineral acids upon cinchona alkaloids, has obtained a different series of derivatives in which the quinuclidine ring is ruptured, producing a secondary amine group.

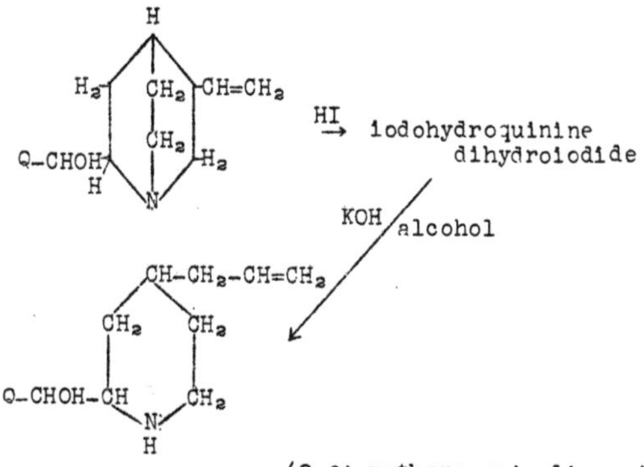
H
H_2
CH_2
$CH=CH_2$
CH_2
Q-CHOH
H
H_2
N
HI
iodohydroquinine
dihydroiodide
KOH
alcohol
$CH-CH_2-CH=CH_2$
CH_2
CH_2
Q-CHOH-CH
CH_2
N
H

(Q=6' methoxy quinoline ring)

Thus quinine is converted to niquine, cinchonine to _-cinchonine, and quinidine to niquidine. These compounds have about the same degree of antimalarial activity as quinine itself.

Nandi, following Kaufman's theory that antimalarial action is largely a property of the grouping

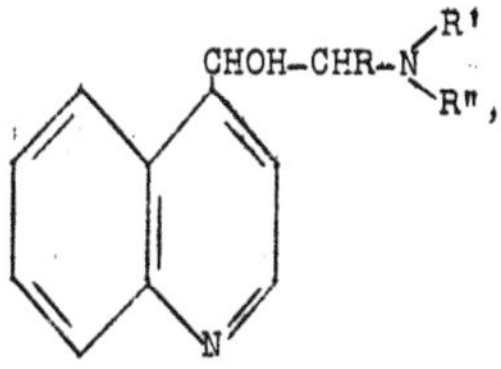
R'
CHOH-CHR-N
R"
N

, in which the alcohol group is not considered essential, or the 4-position in quinoline strictly adhered to, has synthesized a number of compounds related to cinchonine and quinine. The grouping (-CH-OH-CH-R-N) was introduced into the 3-position instead of the 4-position in quinoline, with the ultimate object of testing these compounds for physiological activity.

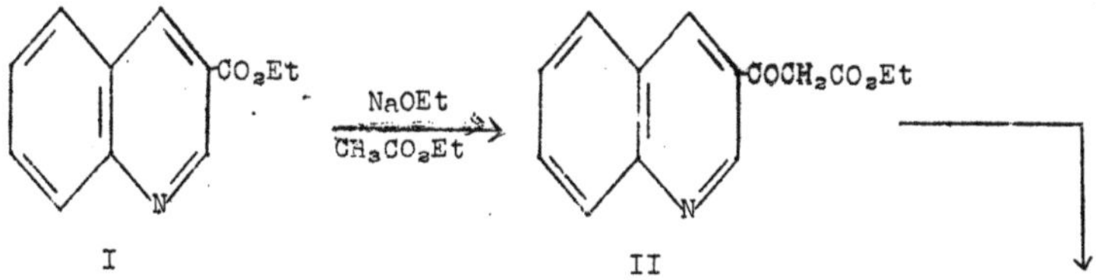
CO_2Et
N
NaOEt
CH_3CO_2Et
$COCH_2CO_2Et$
N

I II

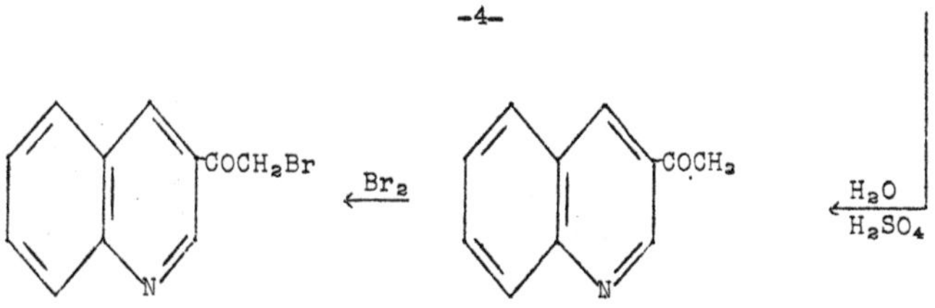
COCH₂Br
Br₂
N
COCH₃
N
H₂O
H₂SO₄

IV III

The 3-quinolyl bromomethyl ketone (IV) was then condensed with different secondary amines such as piperidine, diethylamine, and dimethylamine to prepare the amino ketones. These were then reduced catalytically to the amino alcohols of the general formula:

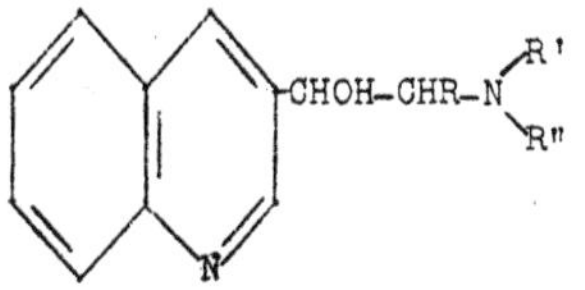
CHOH-CHR-N
R'
R"
N

The ethyl ester of N-benzoyl homocincholoipon was obtained from cinchonine by reduction and subsequent oxidation.

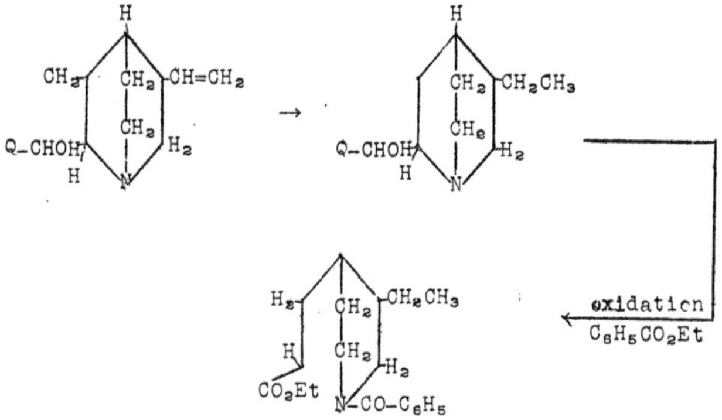
H
CH₂
CH₂
CH=CH₂
CH₂
Q-CHOH
H
H₂
N
H
CH₂
CH₂CH₃
CH₂
Q-CHOH
H
H₂
N
oxidation
C₆H₅CO₂Et
H₂
CH₂
CH₂CH₃
H
CH₂
H₂
CO₂Et
N-CO-C₆H₅

This ethyl ester of β-4-(3-ethylpiperidyl-) propionic acid has also been completely synthesized by Rabe and coworkers, starting with β-collidine. Ethyl quinoline-3-carboxylate is then condensed with

this to give the ethyl ester of α-(3-quinolyl) β-4-(3-ethyl-1-benzoyl-piperidyl) propionic acid.

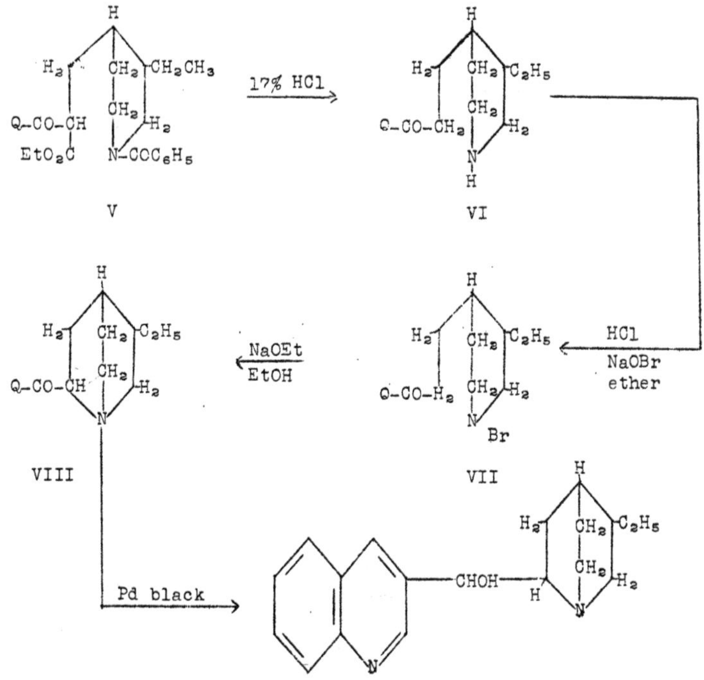
V
17% HCl
VI
HCl
NaOBr
ether
VII
NaOEt
EtOH
VIII
Pd black
CHOH

IX

3'-Quinolyl-8-(3-ethylquinuclidyl) methanol, (IX), is an isomeric analog of dihydrocinchonine.

In a second series of experiments, 2-methoxyquinoline-3-carboxylate was used as starting material and converted as shown above to the bromomethyl ketone.

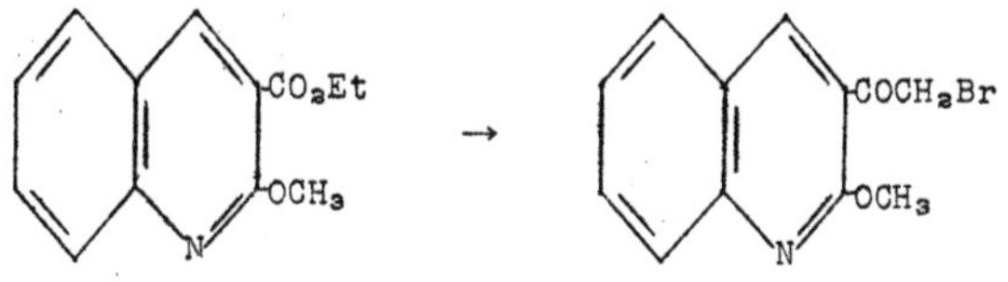

The bromomethyl ketone was then condensed as before with various secondary amines, and these compounds reduced to the hydroxy compounds. To affect a synthesis of the isomeric analog of dihydroquinine, ethyl 2-methoxyquinoline-3-carboxylate was condensed with N-benzoylhomocincholoipon ethyl ester and carried through the same series of reactions as in the synthesis of the isomeric analog of dihydrocinchonine. The final product was 2'-methoxy-3'-quinolyl-8-(3-ethylquinuclidyl)- methanol.

All of the 3'-quinolyl substituted bases of the first series (cinchonise related) were found to be ineffective against avian malaria, but do retain some parasiticidal activity. The second series (quinine related) had not yet been tested.

SYNTHETIC ANTIMALARIALS

Previous to the discovery of plasmochin attempts had been made to synthesize compounds which would be effective against malaria, but these syntheses consisted in variation in the constitution of quinine and little progress in malaria therapy had resulted from them.

The most effective gametocidal drug of all the synthetic antimalarials is plasmochin, 8-(ω-diethylaminoisoamyl)-amino-6-methoxyquinoline. Knunyantz synthesized it as follows:

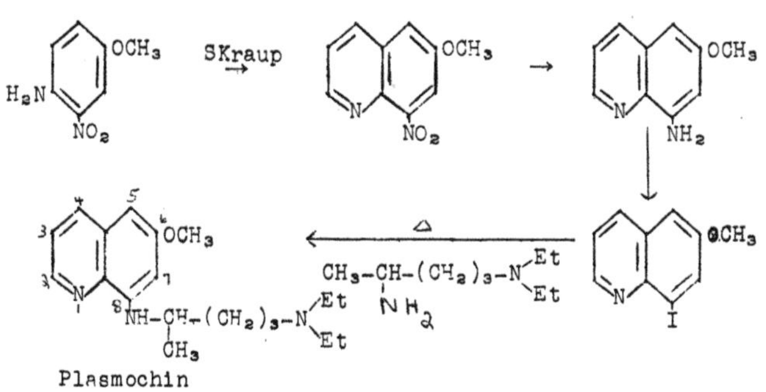

Plasmochin

Since plasmochin is a gametocidal drug, it may be used together with quinine to prevent the spread of malaria. It was selected as the most effective of a large group r. 6-alkoxy-8-alkylaminoquinolines, many of which exhibited activity. Compounds in which the side chain, substituted at position 8 of the quinoline nucleus, was normal showed greater antimalarial activity than those in which it was branched. Though the presence of a methoxy group in position 6 of the quinoline nucleus favored antimalarial activity, it was not indispensable. There were two effective compounds in which the 6-methoxy group was replaced by "H" and "OH". In another compound, replacement of this group by an ethoxy group greatly decreased the activity; replacement by a methyl group was followed by a complete loss of activity.

In studying the influence of the length of the side chain in position 8 on the quinoline nucleus in a series 6-CH_3O-C-8-NH-$(CH_2)_n$-N-Et_2, where C=quinoline nucleus and n=2, 3, 4, or 5, Magidson and Struknow found that the maximum antimalarial activity was reached when n=3. There was another antimalarial maximum at n=9. The wide range of activity was due to the low toxicity. Also it was observed that homologs with an even number of carbon atoms in the side chain at 8 were less active thereapeutically than those containing an uneven number.

Interpolation of an "OH" group in the side chain at position 8 on the quinoline nucleus reduced the chemotherapeutic index, this decreased activity being due mainly to increased toxicity. When an ethyl group was introduced into the amino group at 8, with the formation of a tertiary amine, antimalarial activity disappeared. The replacement of the diethylamino group by a piperidino group lowered the index markedly though the structure of the resulting compound approached that of quinine.

Mauss and Mietzsch reported the first synthetic schizonticidal drug of real importance, "atebrin" (2-methoxy-6-chloro-9-(ω-diethylaminoisoamyl-amino)-acridine). The chlorine atom on position 6 appears to be essential for activity. An increase in the molecular weight of the 2-alkoxy group or the substitution of a methyl group in place of it led to a loss in activity. Magidson and coworkers and Knunyantz synthesized atebrin as follows:

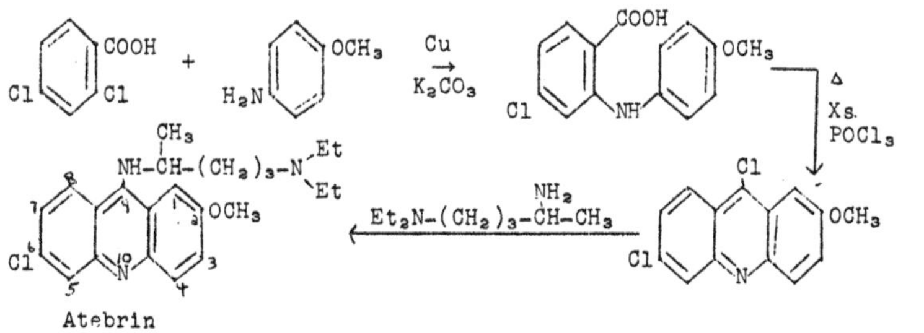
COOH
Cl
Cl
+
H₂N
OCH₃
Cu
K₂CO₃
COOH
OCH₃
Cl
NH
Δ
Xs.
POCl₃
Cl
OCH₃
Cl
N
NH₂
Et₂N-(CH₂)₃-CH-CH₃
CH₃
NH-CH-(CH₂)₃-N
Et
Et
OCH₃
Cl
N
Atebrin

Dr. Adams has worked with two types of compounds in this field. One type (1) is closely related to atebrin. The properly substituted 2-aminobiphenyl is condensed with dichlorobenzoic acid to produce this type, although difficulty has been encountered in effecting this condensation with substituted biphenyls. The other type is the dialkylaminoalkylaminobenzofuroquinoline (2 and 3). 2-Nitro-3-aminodibenzofuran can be treated with an alkylamine and then the nitro group reduced and cyclized by a Skraup reaction. Reversing this procedure produces another isomer. The type (3) is inactive.

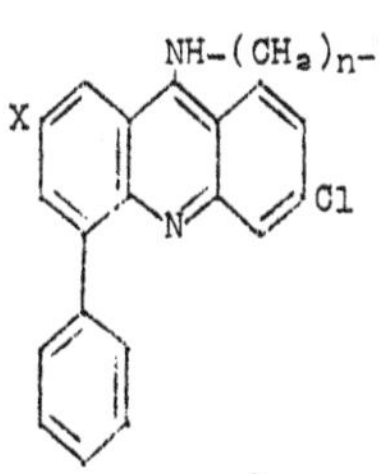
NH-(CH₂)n-N
R
R
X
N
Cl

1

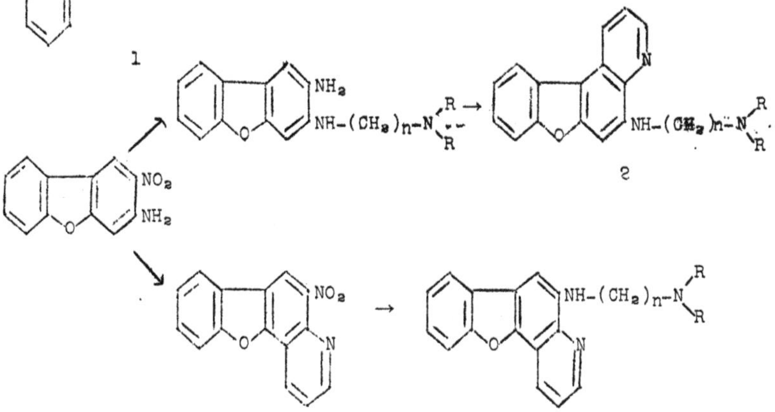
NO₂
NH₂
NH₂
NH-(CH₂)n-N
R
R
N
2
NO₂
N
NH-(CH₂)n-N
3

Laurie and Yorke observed that Synthalin (decamethylenediguanidine) has a definite therapeutic effect on mice infected with various pathogenic trypanosomes. This led to the preparation of a considerable number of guanidines, isothioureas, amidines and amines, with alkyl and alkylene chains. It was found that certain of the diamidines exhibit a powerful trypanocidal action in vitro. With the most active of the series, undecane diamidine, it is possible to produce permanent cures in infected laboratory animals. 1,11-Undecane diamidine is of entirely different chemical constitution from any previously known antimalarial.

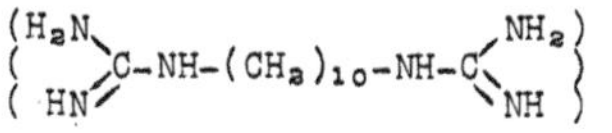
H₂N
C-NH-(CH₂)₁₀-NH-C
NH₂
HN
NH

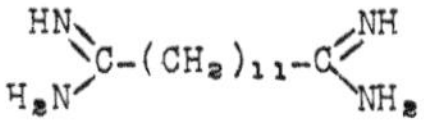
HN
C-(CH₂)₁₁-C
NH
H₂N
NH₂

Other active compounds of this type are:

4,4'-diamidino stilbene 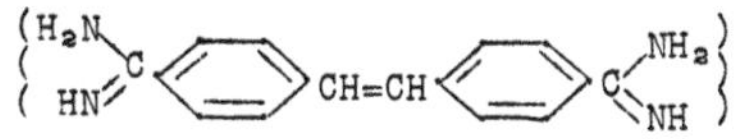

and the members of the series 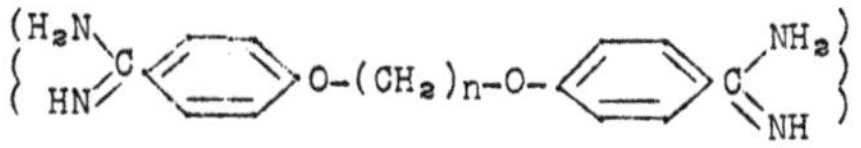

4,4'-diamidino diphenoxypropane, where n=3
4,4'-diamidino diphenoxypentane, where n=5

The activity of 4,4' diamidino compounds was distinctly greater than that of the 3,4' or 3,3' compounds. Relatively slight changes in the radical

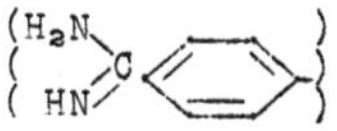 as 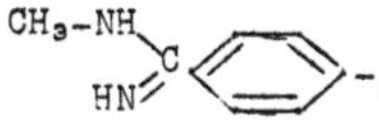,

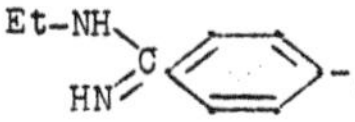, 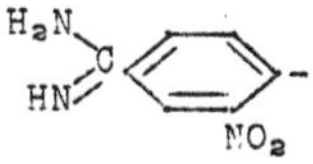, or 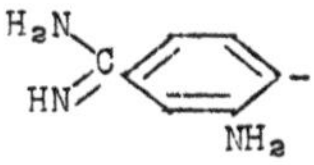

completely destroyed trypanocidal activity.

The method of preparation of the amidines was essentially that of Pinner which involves the conversion of the dinitriles into the imino-ethers and thence into the amidines.

$$RCN + EtOH\ (2.5\text{-}3\ mols) + HCl \rightarrow R\text{-}C(=NH)\text{-}OEt\cdot HCl \xrightarrow{NH_3} R\text{-}C(NH_2)=NH\cdot HCl + EtOH$$

Although successful treatment of malaria with sulphanilamide compounds has been reported by several workers, the results are somewhat contradictory. The protecting action of a single dose appeared to last less than 48 hours. The antimalarial action of compounds of the sulphanilamide group which have been tested up to the present appears to be strongest against monkey malaria, slighter against malaria of man, and on the whole, lacking against avian malaria.

Bibliography

Bishop, Parasitology, 34, 1 (1942).
King and Work, J. Chem. Soc., 1307 (1940).
Nandi, Proc. Ind. Acad. Sci., 12A, 1 (1940).
Prelog, Stern and Seiwerth, Natur., 28, 750 (1940).
Rabe, Huntenburg, Schultze and Volger, Ber., 64B, 2487 (1931).
Solomon, J. Chem. Soc., 77 (1941).
Wearn, Org. Seminars, Univ. of Ill., first semester, 1940-1941.
Ainley and King, Proc. Roy. Soc., (London), 125B, 60 (1938).
Work, J. Chem. Soc., 1315 (1940).
Knunyantz and co-workers, Bull. Acad. Sci. U.S.S.R., 153, 165 (1934); (C.A., 28, 4837 (1934)).
Fourneau and co-workers, Ann. Inst. Pasteur, 46, 514 (1931).
Magidson and Strukow, Archiv. Pharm. Berl, 271, 359, 569 (1933).
Magidson, Grigorovski, Maksimov and Margolina, Khim. Farm. Prom., No. 1, 26 (1935); (C. A., 30, 1516 (1936)).
Bovet, Ann. Inst. Pasteur, 51, 528 (1933).
King, Laurie, and Yorke, Lancet, 233, 1360 (1937).
Glenn, Hughes, Laurie and Yorke, Ann. Trop. Med., 32, 103 (1938).
Yorke, Trans. Roy. Soc. Trop. Med. Hyg., 33, 463 (1940).
Fu and Sah, JACS, 64, 1482 (1942).
Fu, JACS, 64, 1487 (1942).
Ashley, Barber, Ervins, Newberry and Self, J. Chem. Soc., 103 (1942).
Knorr, Ber., 50, 229 (1917).
Sinton, Ann. Trop. Med. Parasit, 33, 37 (1939).

Reported by John M. Stewart and Louis D. Scott
March 10, 1943

THE ADDITION OF ALIPHATIC DIAZO COMPOUNDS TO

α,β-UNSATURATED KETONES

From the work of von Pechmann (15, 16, 17) and von Auwers (12, 13, 14) we find that diazomethane adds to α,β-unsaturated esters to give Δ¹-pyrazolines. In general the nitrogen of the diazomethane is attached to the carbon atom alpha to the ester group. If there is an alpha hydrogen present, the Δ¹-pyrazolines readily isomerize to Δ²-pyrazolines which contain a conjugated system (reaction a). If there is no alpha hydrogen present, the Δ¹-pyrazolines may be isomerized by shaking with concentrated hydrochloric acid (reaction b).

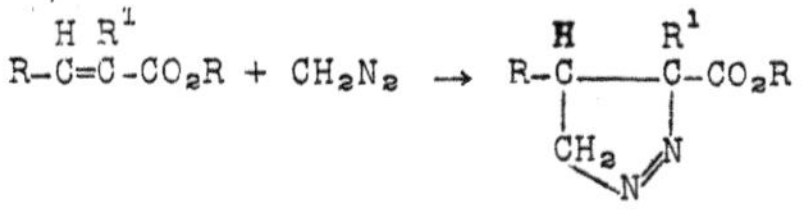

Δ¹-pyrazoline

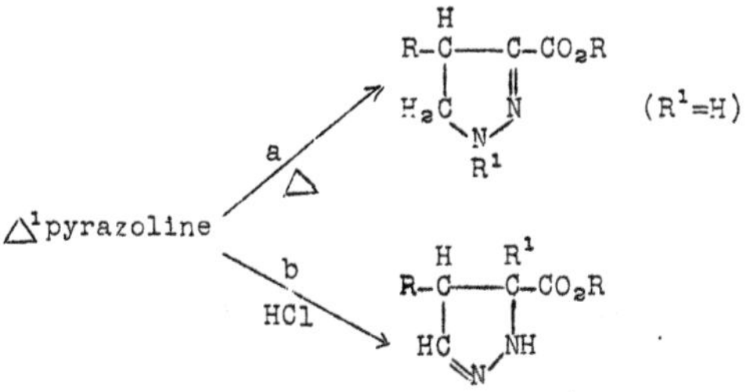

The pyrazoline ring is numbered by starting at the nitrogen atom containing a substituent and proceeding through the next nitrogen atom. In Δ¹-pyrazolines the nitrogen atoms are always numbered 1 and 2. The selection of the nitrogen atom from which the numbering starts depends on the substituents on the ring. Alternate positions may be given by numbers in parentheses. An illustration will show the conventions:

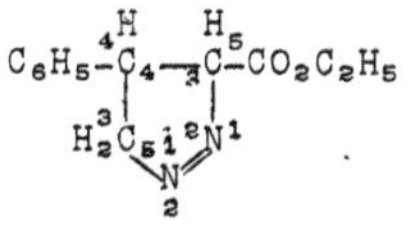

3(5)-carbethoxy-4(4)-phenyl Δ¹-pyrazoline

On pyrolysis the substituted pyrazolines usually evolve nitrogen to form either cyclopropane derivatives or unsaturated compounds. In general the more highly substituted compounds tend to form cyclopropanes, while the simpler ones go almost exclusively to unsaturated esters. Sometimes the compound retains its nitrogen and forms a pyrazole when heated.

Azzarello (1) was the first to add diazomethane to a straight chain unsaturated ketone. He used both benzalacetone and dibenzalacetone. In the first case he was able to identify the product by oxidizing it to the known 3-carboxy-4-phenylpyrazole (I).

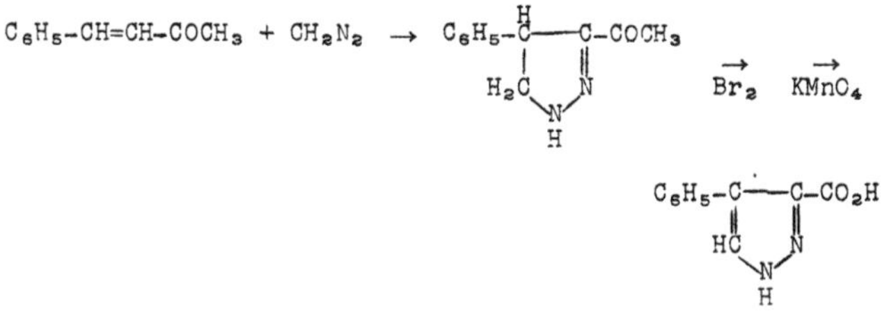

I

Smith and Howard (10) repeated this work with similar results. They were able to obtain β-methylbenzalacetone from the pyrolysis of the Δ^2-pyrazoline. This shows that diazomethane adds to unsaturated ketones in the same way it does to unsaturated esters (i.e. the nitrogen atom goes to the alpha carbon). Apparently the Δ^1-pyrazoline first formed is immediately converted into Δ^2-pyrazoline.

When Smith and Pings (11) treated benzalacetophenone with diazomethane, they obtained two products. One of these was converted into the other on gentle heating. By analogy with the ester addition compounds they assumed these to be Δ^1- and Δ^2-pyrazolines respectively. Either of these on pyrolysis gave dypnone (II) and 3-benzoyl-4-phenyl pyrazole III.

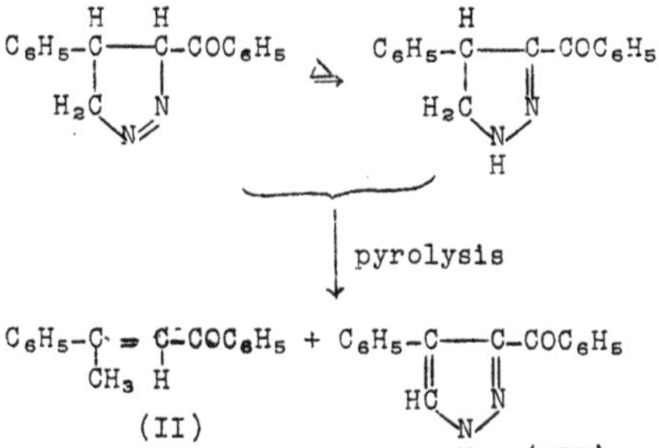

(II) (III)

Kohler and Steele (7) added diazoacetic ester to benzalacetophenone in an attempt to prepare a cyclopropane. Instead of the desired product, they obtained chiefly a pyrone (IV) when the pyrazoline was thermally decomposed. However, the yield of the cyclopropane (V) jumped from less than 1% to 40% when the pyrolysis was carried out in the presence of polished platinum.

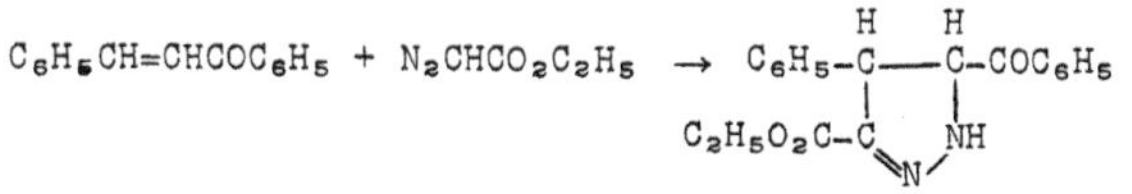

$$C_6H_5CH=CHCOC_6H_5 + N_2CHCO_2C_2H_5 \rightarrow \underset{\underset{\overset{|}{N}\diagdown_N NH}{C_2H_5O_2C-C}}{C_6H_5-\overset{H}{\underset{|}{C}}-\overset{H}{\underset{|}{C}}-COC_6H_5}$$

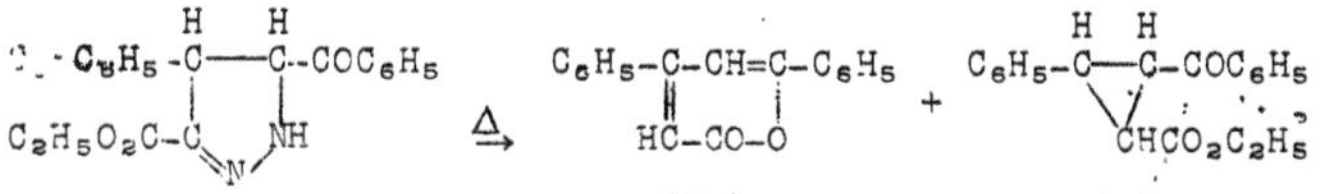

$$\underset{\underset{\overset{|}{N}\diagdown_N NH}{C_2H_5O_2C-C}}{\cdot \cdot C_6H_5-\overset{H}{\underset{|}{C}}-\overset{H}{\underset{|}{C}}-COC_6H_5} \quad \overset{\Delta}{\rightarrow} \quad \underset{HC-CO-O}{C_6H_5-C-CH=C-C_6H_5} \quad + \quad \underset{CHCO_2C_2H_5}{C_6H_5-\overset{H}{\underset{|}{C}}-\overset{H}{\underset{|}{C}}-COC_6H_5}$$

(IV)   (V)

Smith and Howard (9) have recently reported work on the addition of aliphatic diazo compounds to cis- and trans-dibenzoylethylene. They observed that the addition takes place more readily with the trans-diketone. However, the addition products in both cases are identical. When diazomethane was used, a Δ^1-pyrazoline (VI) was obtained which changed to the Δ^2-pyrazoline (VII) on gentle heating. These were distinguished by the fact that (VII) gave a urea (VIII) with phenylisocyanate, while (VI) gave no reaction. No cyclopropane derivative was isolated when the pyrazolines were decomposed by heating. Some dibenzoylpyrazole was obtained from the pyrolysis and from the treatment of (VII) with bromine.

(VI)

No reaction

(VII)

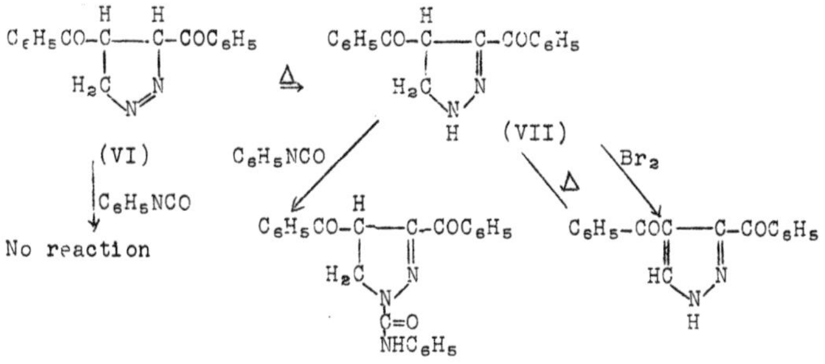

(VIII)   (IX)

Diphenyldiazomethane reacted with trans-dibenzoylethylene to give two products. It was shown that one was a pyrazoline (X) and the other was a highly substituted cyclopropane (XI). This is the first example of direct cyclopropane formation in this type of reaction. The structure of (XI) was proven by comparing it with a sample made by another procedure.

Phenyl diazomethane gave only oils when added to dibenzoylethylene.

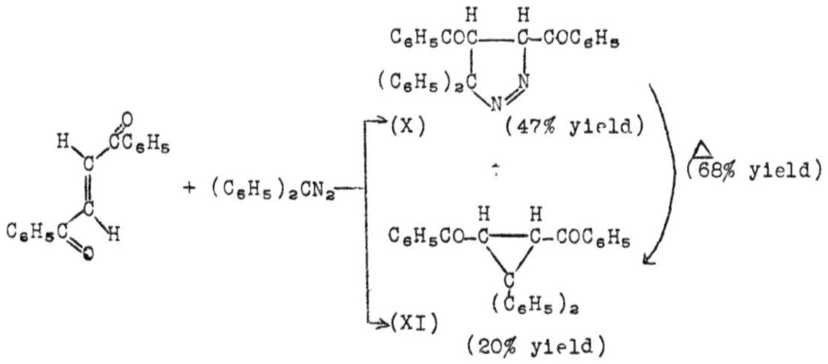
H
CC_6H_5
O
C
C
C_6H_5C
H
O
+ $(C_6H_5)_2CN_2$
(X)
H H
C_6H_5COC
$C-COC_6H_5$
$(C_6H_5)_2C$
N
N
(47% yield)
(68% yield)
(XI)
H H
C_6H_5CO-C
$C-COC_6H_5$
C
$(C_6H_5)_2$
(20% yield)

Several investigators (2, 3, 4, 5, 17) have reported the addition of diazo compounds to quinones. In general, these form the same kind of pyrazoline derivatives that are formed with straight chain unsaturated ketones. If the quinone is partially alkylated, the pyrazolines form with difficulty and usually lose nitrogen to methylate the ring.

Bibliography

Azzarello, Gazz. chim. ital., 36, II, 50 (1906). (1)
Bergmann and Bergmann, J. Org. Chem., 3, 125 (1938). (2)
Fieser and Hartwell, J. Am. Chem. Soc., 57, 1479 (1935). (3)
Fieser and Seligman, ibid., 56, 2691 (1934). (4)
Fieser and Peters, ibid., 53, 4080 (1931). (5)
Howard, Seminar Univ. of Minn., 4/16 (1941). (6)
Kohler and Steele, J. Am. Chem. Soc., 41, 1093 (1919). (7)
Smith, Chem. Rev., 23, 193 (1938). (8)
Smith and Howard, J. Am. Chem. Soc., 65, 159 (1943). (9)
Smith and Howard, ibid., 65, 165 (1943). (10)
Smith and Pings, J. Org. Chem., 2, 23 (1937). (11)
Von Auwers, Ber., 66, 1198 (1933). (12)
Von Auwers and Cauer, Ann., 470, 284 (1929). (13)
Von Auwers and Konig, Ann., 496, 27, 252 (1932). (14)
Von Pechmann, Ber., 27, 1890 (1894). (15)
Von Pechmann and Burkard, Ber., 33, 3591 (1900). (16)
Von Pechmann and Seel, Ber., 32, 2292 (1899). (17)

Reported by Stanley Parmerter
March 17, 1943

RECENT INVESTIGATIONS ON ANTI-HEMORRHAGICS

In 1929 Dam first noticed that chicks on an ether-extracted diet suffered from intramuscular hemorrhages. This fact was confirmed by McFarlane and coworkers in 1931, and interest in an antihemorrhagic factor rapidly grew as it was shown that none of the known vitamins cured this condition, apparently due to some vitamin deficiency. Almquist worked out a diet of ether-extracted fish meal, ether-extracted Brewer's yeast, ground polished rice, cod liver oil and a mixture of $CuSO_4$, NaCl and $FeSO_4$, which was found to be very satisfactory as a standard diet for producing a hemorrhagic condition in the chick. The usual duration of the diet was two to three weeks, after which the chick was acceptable if the clotting time was over one hour. Potential anti-hemorrhagic concentrates were then administered and the clotting time taken after 18 hours. In 1939 Doisy et al. isolated the factor, vitamin K_1, which was apparently responsible for preventing the hemorrhage. Strenuous work to prove the structure of the new compound followed, and the race to synthesize vitamin K_1 or 2-methyl-3-phytyl-1,4-naphthoquinone was climaxed by the almost simultaneous success of Fieser and Doisy.

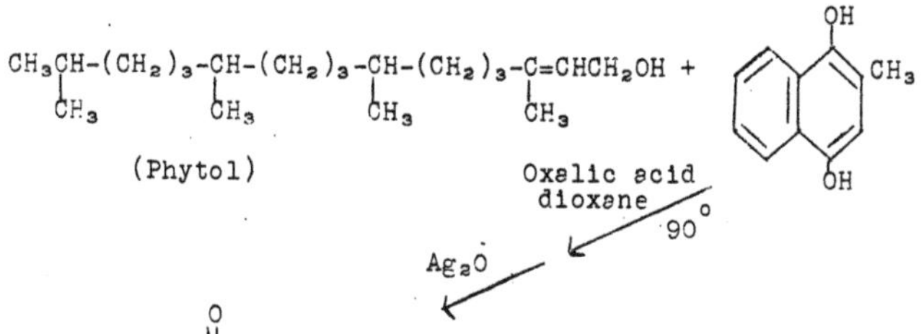

CH3CH-(CH2)3-CH-(CH2)3-CH-(CH2)3-C=CHCH2OH +
CH3
(Phytol)
OH
CH3
Oxalic acid
dioxane
90°
Ag2O

K_1

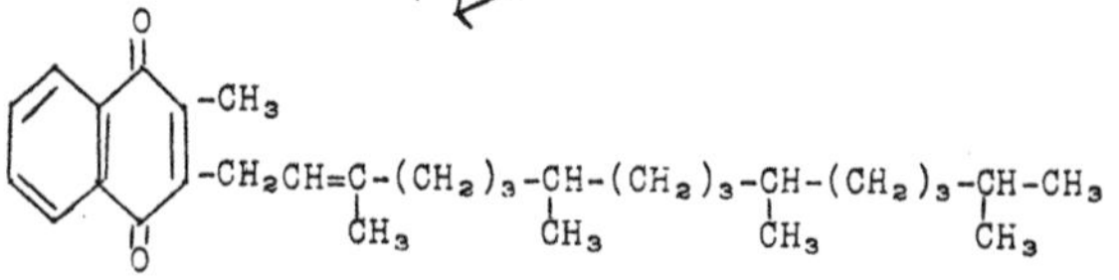

O
-CH3
-CH2CH=C-(CH2)3-CH-(CH2)3-CH-(CH2)3-CH-CH3
CH3

K_2

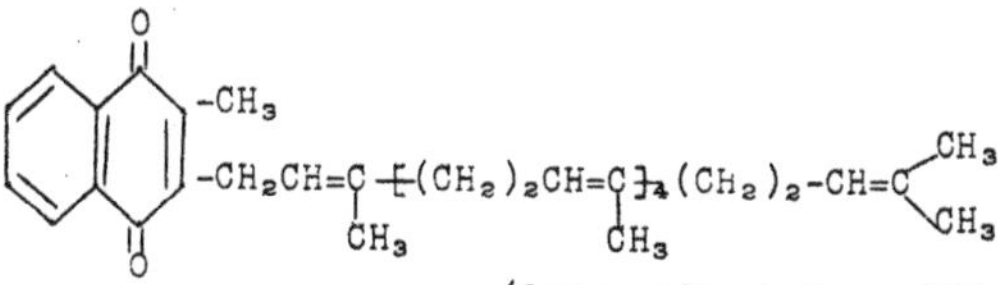

O
-CH3
-CH2CH=C-[(CH2)2CH=C]4(CH2)2-CH=C
CH3

(Long side chain = difarnesyl)

K_5

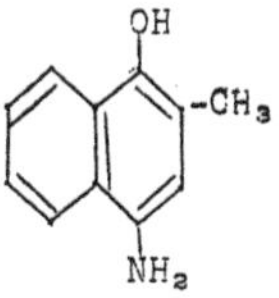

OH
-CH3
NH2

[illegible] INVESTIGATIONS ON [illegible]

In 1929 Dam first noticed that [illegible] fed on [illegible] diet [illegible] from intramuscular [illegible] and coworkers in 1935, [illegible] rapidly grew as it was shown that [illegible] vitamin [illegible] this condition, apparently due to [illegible] Almquist worked out a diet [illegible] showed [illegible] ground putrefied fish [illegible] 1935 and 1939, [illegible] was found to [illegible] satisfactory [illegible] isolated [illegible] from alfalfa [illegible] the major function of the [illegible] in the clotting [illegible] hemorrhagic concentrates [illegible] factor, vitamin K, which was [illegible] structure of [illegible] followed, and the [illegible] synthetic vitamin K, or 2-methyl-1,4-naphthoquinone [illegible] direct [illegible] Fieser and Doisy [illegible]

It is the purpose of this paper to briefly review the investigations that have gone on since the synthesis and proof of structure of vitamin K_1. The investigation of vitamin K activity followed several paths - the naming and proof of structure of vitamin K_2 and vitamin K_5, investigation of the vitamin K activity of related compounds and possible correlation of structure to activity.

Perhaps the least of all has been accomplished on postulation of a theory of mechanism. Weir mentions that the action of injected vitamin K_1 is so rapid that enzymatic activity is suggested. It has been postulated that K is not a part of the prothrombin molecule but stimulates the cells to produce prothrombin. Engelkes says that vitamin K factors have an activating effect on the thrombocytes and endothelial cells. McCawley has measured the oxidation-reduction potential of the 1,4-quinoid part of the K type molecule and suggests this potential influences the action of cathepsin to synthesize prothrombin. Up until the present, however, postulation of mechanism has almost been pure guess work, and no theory has any pertinent substantial evidence to back it up.

Administration of K_1 to infants has been found to prevent the normal decline of the prothrombin level which occurs 3-5 days after birth, and it is now standard procedure to give some K concentrate with bile salts in oil orally to mothers in labor, a procedure which keeps the prothrombin level up in the infant after birth. Vitamin K_5 concentrate has been used successfully even in cases of obstructional jaundice, since the K_5 salt in water solution can be given intravenously. However, in spite of occasional reports to the contrary, vitamin K concentrate up to the present has been found helpful only in hemorrhagic diseases which involve the actual deficiency of vitamin K in the body.

There naturally has been a good deal of interest in the toxic effects of the vitamin K factors. It is known that an average oral dose of .2 gm/kg. body weight of phthiocol or .5 gm/kg. of 2-methyl-1,4-naphthoquinone will cause death in mice, due to vascular congestion and depression of respiration. Naphthoquinone in vitro is said to convert hemoglobin into methemoglobin. However, 180-200 mgs. of 2-methyl-1,4-naphthoquinone have no noticeable toxic effect on the human, and vitamin K_1 has been found to be a great deal less toxic than 2-methyl-1,4-naphthoquinone.

The table on page 5 summarizes the physiological data for many of the compounds which have been tested. In general, most have been found to have the 1,4-naphthoquinone structure, usually with a 2-methyl group and some 3-substituted long side chain, or must be readily convertible to such a structure. From the table it is apparent that the 2-methyl group is very essential for high potency, and that either removing this group or enlargening it, even if only by one carbon atom, decreases the potency completely out of proportion to the change in weight. Fieser explains this great activity of 2-methyl-1,4-naphthoquinone by postulating that if it is given orally, it combines with a reactive alcohol furnished in foodstuffs or is parenterally supplied with a side chain by a liver process. Through such a process, one part of 2-methyl-1,4-naphthoquinone would give 2.6 parts of K_1 and 3.4 parts of K_2. This compares favorably with Dam's report that the 2-methyl-1,4-naphthoquinone is 2.11 times as active as

vitamin K_1 and 3.1 times as active as K_2. There is no direct evidence, however, to prove that vitamin K_1 and K_2 are not instead broken down to 2-methyl-1,4-naphthoquinone in the body, but there is no laboratory evidence for this degradation. On the contrary, vitamin K_1 forms phythiocol under very mild conditions, and the proportionate activities of the phytyl, geranyl and farnesyl derivatives of 2-methyl-1,4-naphthoquinone do not indicate that they derive their K activity through ability to break down to 2-methyl-1,4-naphthoquinone in the body.

It has been found that, on conversion to the oxides by treatment with sodium peroxide, the potency of the quinones is lowered only slightly. This can be explained by the ease of reduction back to the naphthohydroquinone. This is easily done in the laboratory with sodium hydrosulfite.

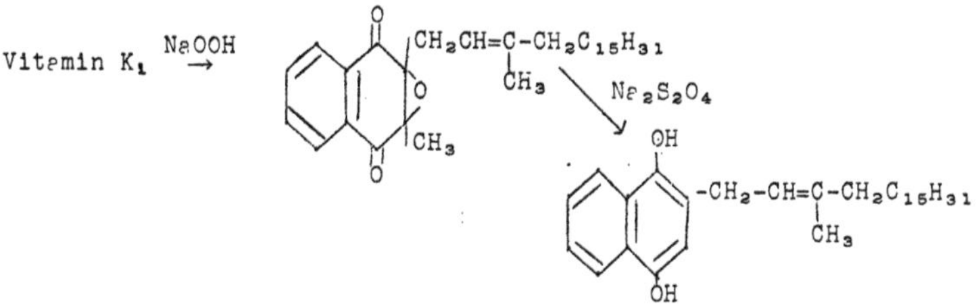
Vitamin K1
NaOOH
CH2CH=C-CH2C15H31
CH3
O
CH3
Na2S2O4
OH
-CH2-CH=C-CH2C15H31
CH3
OH

That the oxides are not active in the same proportion as the reduction products might be explained by differences in absorption. Inefficient conversion has been given by Fieser as a reason for 2-methyl-1,4-naphthoquinone being 17 times as active as its oxide. This is substantiated by the fact that the oxides of mono-alkylated quinones can undergo a second type reaction; ie: 2-methyl-1,4-naphthoquinone oxide forms phthiocol on treatment with sulfuric acid at low temperature. The fact that the 2,3-di-methyl-1,4-naphthoquinone oxide is twice as active as its quinone is inexplicable and inconsistent with the idea of the oxides being precursors to the quinones.

The esters of the various hydroquinones are usually of about the same activity as the free hydroquinones. The fact that the diacetate, dipropionate, and the dibenzoate esters of 2-methyl-1,4-naphthohydroquinone have the same activity would seem to indicate saponification taking place in the body. This is well supported by the low activity of the highly hindered and hence difficultly saponifed dimesitoyl derivative. Small differences may be explained by a difference in the rate of absorption, or it might mean that these derivatives act as a whole.

Ansbacher was unable to explain the high activity of the dimethyl ether of 2-methyl-1,4-naphthohydroquinone, since he believed it very difficult to split to the quinone <u>in vitro.</u> Fieser, however, has been able to split this ether very smoothly in the laboratory at 60° with chromic acid. Ethers then can be split to the quinone either through hydrolysis or oxidation.

The Diels-Alder synthesis has proven very useful in the synthesis of many of the substances listed in the table. Two examples are the preparation of 5,8-dihydro-vitamin K_1 (IV) and 2-(δ-methyl- -pentenyl)-1,4-dihydro-anthraquinone (V).

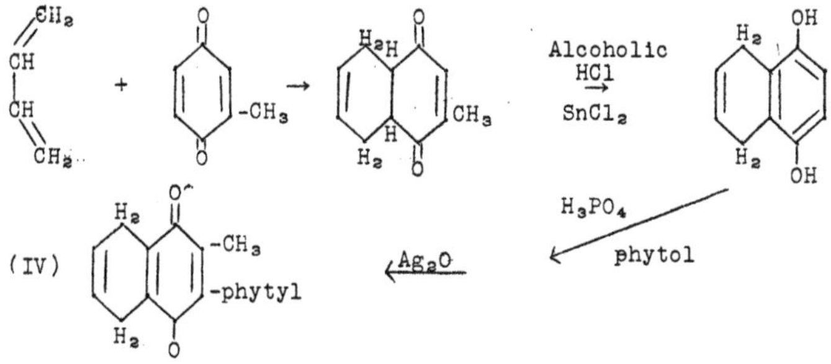

Alcoholic
HCl
SnCl₂
H₃PO₄
phytol
Ag₂O
(IV)
-CH₃
-phytyl

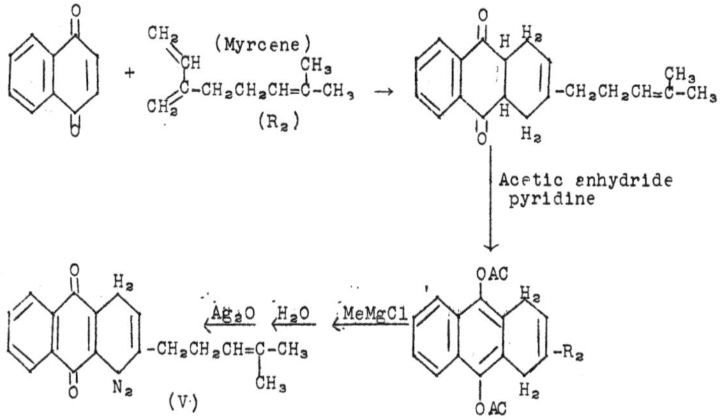

(Myrcene)
(R₂)
Acetic anhydride
pyridine
MeMgCl
H₂O
Ag₂O
-R₂
(V)

The striking potency of the 2- and 3-methyl-1-naphthols and the inactivity of the other isomers seem to point again to potency depending on ability to be converted to the 2-methyl-1,4-naphthoquinone. This must take place in the body for the 2- and 3-methyl-1-naphthols by hydroxylation para to the hydroxyl group.

2-Methyl-1-naphthyl amine is 1/5 as potent as its corresponding naphthol, probably because conversion involves deamination or para-hydroxylation followed by oxidation with unavoidable loss. The p-hydroxy-amino compound is nearly as potent as methyl-naphthoquinone, whereas methyl-naphthyl amine is only 1/17 as effective. This possibly can be explained by the fact that the p-amino-naphthol requires no deamination, this occurring spontaneously after oxidation to a quinone-imine.

Compound	Effective dose (γ)
2-methyl-3-phytyl-1,4-naphthoquinone (Vitamin K_1)	1
2-methyl-3-farnesyl-1,4-naphthoquinone	5
2-methyl-3-β,γ-dihydrophytyl-1,4-naphthoquinone	8
2-methyl-3-geranyl-1,4-naphthoquinone	25
2-methyl-3-cinnamyl-1,4-naphthoquinone	25
2-methyl-3-(β,γ,γ-trimethyl-allyl-1,4-naphthoquinone)	50
2,3-dimethyl-1,4-naphthoquinone	50
2-methyl-3-benzyl-1,4-naphthoquinone	200
2-methyl-3-hydrocinnamyl-1,4-naphthoquinone	300
2-methyl-1,4-naphthoquinone	0.3
2-phytyl-1,4-naphthoquinone	50
2-farnesyl-1,4-naphthoquinone	500
2-β,γ-dihydrophytyl-1,4-naphthoquinone	600
2-allyl-1,4-naphthoquinone	800
2-geranyl-1,4-naphthoquinone	1000-slight
2-ethyl-1,4-naphthoquinone	1000-inactive
2-n-propyl-1,4-naphthoquinone	1000-inactive
2-ethyl-3-phytyl-1,4-naphthoquinone	1000
2,3-diallyl-1,4-naphthoquinone	1000
1,1-dimethyl-3-t-butyl-1,4-dihydro-anthraquinone	1000-inactive
2-(δ-methyl-γ-pentenyl)-1,4-dihydro-anthraquinone	1000-inactive
2,6-dimethyl-3-phytyl-1,4-naphthoquinone	1000-inactive
2,5-dimethyl-3-phytyl-1,4-naphthoquinone	500-slight
2,7-dimethyl-3-phytyl-1,4-naphthoquinone	1000
2,8-dimethyl-3-phytyl-1,4-naphthoquinone	500-slight
6,7-dimethyl-3-phytyl-1,4-naphthoquinone	1000-inactive
Plumbagin (2-methyl-5-hydroxy-1,4-naphthoquinone)	400
Phthiocol (2-methyl-3-hydroxy-1,4-naphthoquinone)	500
Juglone (5-hydroxy-1,4-naphthoquinone)	1000-inactive
Lawsone (2-hydroxy-1,4-naphthoquinone)	1000-inactive
Lapachol	1000-inactive
2-β-heptenyl-3-hydroxy-1,4-naphthoquinone	1000-inactive
Crude 2-farnesyl-3-hydroxy-1,4-naphthoquinone	1000-inactive
5,8-dihydro-vitamin K_1	4
β,γ,5,6,7,8-hexahydro K_1	1000-very slight
2-methyl-5,8-dihydro-1,4-naphthoquinone	6
2-methyl-5,8,9,10-tetrahydro-1,4-naphthoquinone	8
2-methyl-5,6,7,8-tetrahydro-1,4-naphthoquinone	500
2-methyl-1,4-naphthohydroquinone	0.5
2-methyl-1-naphthol	1
3-methyl-1-naphthol	0.6
4-methyl-1-naphthol	1000-inactive
1-methyl-2-naphthol	1000-inactive
3-methyl-2-naphthol	1000-inactive
1-naphthol	1000-slight
2-methyl-1-naphthylamine	5

BIBLIOGRAPHY

Fieser, J. Am. Chem. Soc., 61, 3467 (1939)
Ansbacher, Fernholz and Dolliver, J. Am. Chem. Soc., 62, 155 (1940)
Dam, Glavind and Karrer, Helv. Chim. Acta, 23, 224 (1940)
Tischler, Sampson and Fieser, J. Am. Chem. Soc., 62, 1881 (1940)
Fieser, Tishler, Sampson and Woodford, J. Biol. Chem., 137, 659 (1940)
Doisy, Binkley and Thayer, Chem. Rev., 28, 477 (1941)
Binkley, McKee, Thayer and Doisy, J. Biol. Chem., 133, 721 (1940)
Shimkin, J. Pharmacol., 71, 210 (1941)
Binkley, Cheney, Holcomb, McKee, Thayer, MacCorquodale, Doisy, J. Am. Chem. Soc., 61, 2559 (1939)
Fieser, Campell, Fry and Gates, J. Am. Chem. Soc., 61, 2559 (1939)
Fieser, Tishler and Wendler, J. Am. Chem. Soc., 62, 2861 (1940)
Anderson, Karabin, Udesky and Seed, Arch. Surg., 41, 1244 (1940)-- C.A., 35, 518 (1941)
Kugelmass, Arch. Disease Childhood, 15, 97 (1940) -- C.A. 34, 6676 (1940)
Bollman, Butt and Snell, J. Am. Med. Assoc., 115, 1087 (1940) -- C.A., 34, 8006 (1940)
Hellman, Moore and Shettles, Bull. Johns Hopkins Hosp., 66, 379 (1940) -- C.A., 34, 6679 (1940)
Koller, Schweiz. med. Wochschr., 69, 1159 (1939) -- C.A., 34, 5934 (1940)
Molitor and Robinson, Proc. Soc. Exptl. Biol. Med., 43, 125 (1940) -- C.A., 34, 2066 (1940)
McCawley and Gurchot, Univ. Calif. Pub. Pharmacol, 1, 325 (1940) -- C.A., 35, 1466 (1941)
Dam, Franklin Inst., 231, 300 (1941) -- C.A., 35, 3299 (1941)
Engelkes, Nederland. Tijdschr. Geneeskunde., 85, 1985 (1941) -- C.A., 35, 8026 (1941)
Brinkhaus, Medicine, 19, 329 (1940) -- C.A., 35, 2201 (1941)

Reported by H. J. Sampson Jr.
March 17, 1943

CINCHONA ALKALOIDS

Rabe, Prelog

In 1904, Rabe began the study of the cinchona alkaloids. In a series of thirty-three papers since then, exhaustive studies have been made, especially with the alkaloids quinine, cinchonine, hydroquinine, hydrocinchonine, their optical isomers, and certain ruban derivatives. This seminar will attempt to summarize the more important synthetic work.

In order to follow the nomenclature, the following scheme will be referred to

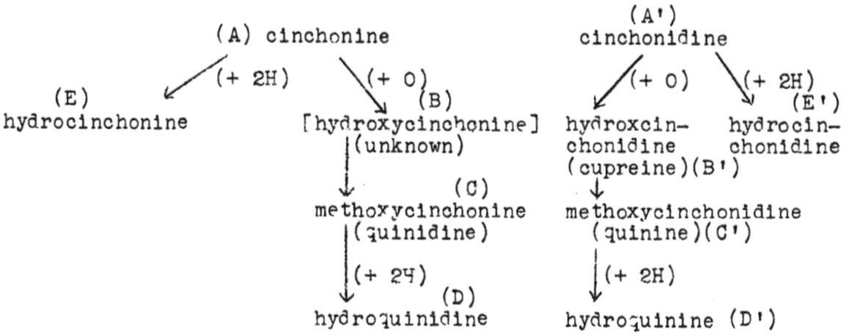

(A) cinchonine
(+ 2H)
(+ O)
(E) hydrocinchonine
(B) [hydroxycinchonine] (unknown)
(C) methoxycinchonine (quinidine)
(+ 2H)
(D) hydroquinidine
(A') cinchonidine
(+ O)
(+ 2H)
hydroxcinchonidine (cupreine)(B')
(E') hydrocinchonidine
methoxycinchonidine (quinine)(C')
(+ 2H)
hydroquinine (D')

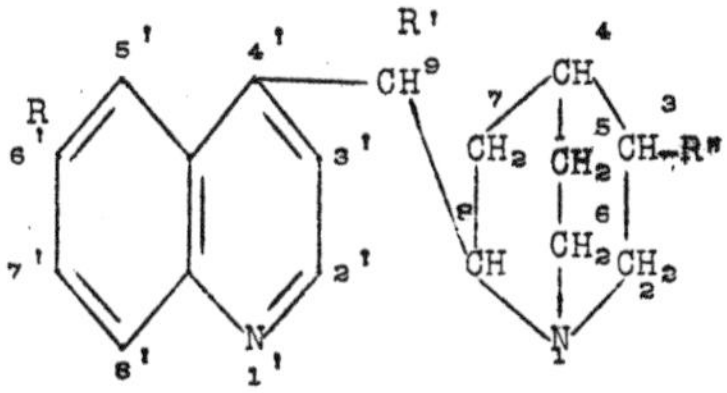

R
R'
CH⁹
CH
CH₂
CH₂
CH-R"
CH₂
CH₂
CH
N
N

(A) R=H, R'=OH, R"=CH=CH_2
C_8+ C_9+

(B) R=OH, R'=OH, R"=CH=CH_2

(C) R=OCH_3, R'=OH, R"=CH=CH_2
C_8+ C_9+

(D) R=OCH_3, R'=OH, R"=-CH_2-CH_3
C_8+ C_9+

(E) R=H, R'=OH, R"=-CH_2-CH_3
C_8+ C_9+

The table is the same for the (A'-E') series except that the rotations of C_8 and C_9 are both (-). If an epi appears before the name in the cinchonine series, C_8 is (+) rotation and C_9 is (-) and visa-versa

in the cinchonidine series. If R'=(H), the compounds are called desoxyhydroquinidine, etc. In all of the compounds to be discussed the configuration of carbons 3 and 4 are constant (two exceptions) and the net effect is always dextro-rotatory. If there is no vinyl or ethyl group at R", they are named as ruban derivatives.

In 1922,[4] Rabe proposed the name ruban for 2-quinuclidyl-4-quinolylmethane (R=R'=R"=H) the mother substance of the cinchona alkaloids and the name rubatoxan for the mother substance of the quinotoxins or quinicines resulting from the rupture of the 1,8 bond in ruban. Hence, quinine would be 6'-methoxy-3-vinyl-9-rubanol.

The general method used by Rabe[6] for his recent work may be illustrated by his first total synthesis of two cinchona alkaloids, hydroquinine and hydroquinidine.

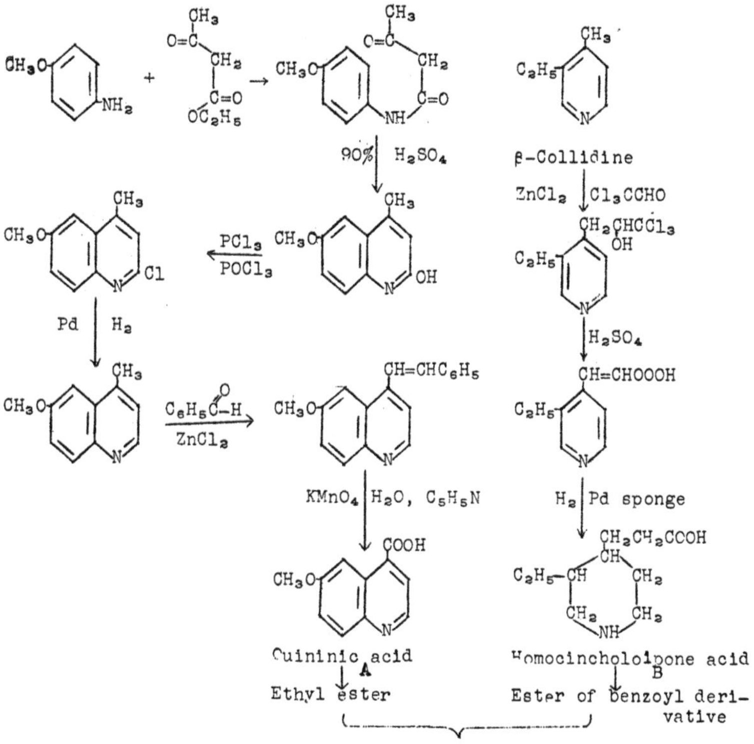

CH₃O
NH₂
CH₃
O=C
CH₂
C=O
OC₂H₅
90% H₂SO₄
PCl₃
POCl₃
OH
Cl
Pd H₂
C₆H₅C-H
ZnCl₂
CH=CHC₆H₅
KMnO₄ H₂O, C₅H₅N
COOH
Quininic acid A
Ethyl ester
C₂H₅
β-Collidine
ZnCl₂ Cl₃CCHO
CH₂CHCCl₃
OH
H₂SO₄
CH=CHOOOH
H₂ Pd sponge
CH₂CH₂CCOH
NH
Homocincholoipone acid B
Ester of benzoyl derivative

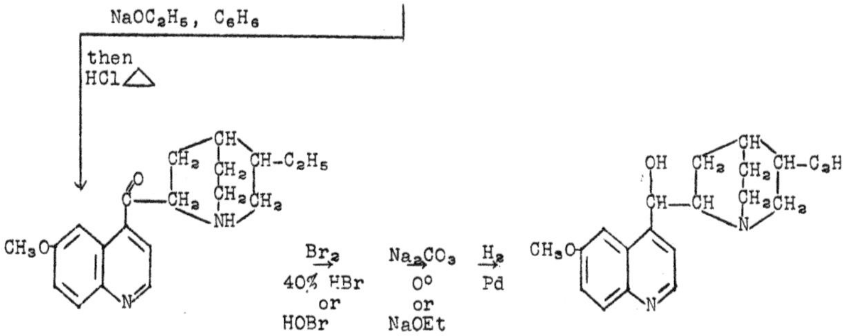
NaOC₂H₅, C₆H₆
then
HCl △
Br₂
40% HBr
or
HOBr
Na₂CO₃
0°
or
NaOEt
H₂
Pd

This beautiful piece of synthetic work is all the more remarkable because the homocincholoipone acid (B) is one of four isomers of ethyl piperidylpropionic acid and must be isolated. The enantiomorphs of hydroquinine and of hydroquinidine have also been synthesized[10] using (-) homocincholoipone acid as a starting material.

In like manner,[9] the four stereoisomeric 9-rubanols have been synthesized (see table). By conversion to the chloride followed by reduction, the (++) (refers to carbons 8 and 9, respectively) and (+-) chlorides are reduced to the (+) ruban while the (-+) and (--) chlorides gave the (-) ruban.

By treatment of quinine with potassium hydroxide in amyl alcohol Rabe[11,12] has succeeded in isolating epiquinine (A) and epiquinidine (B) by a very neat procedure - precipitation as the double sulfate $A,B \cdot H_2SO_4 \cdot 6H_2O$ and subsequent separation. This confirmed their isolation by Thron[26] and Dirscherl in 1935. The quinidine was separated as the tartrate, epiquinine and epiquinidine precipitated as the double sulfate and epiquinidine isolated as the thiocyanate and epiquinine as the hydrobromide in 66% and 77% yield, respectively.

By 1934,[13] the 6'-methoxy-9-hydroxyrubans had been synthesized but due to negative chemotherapeutic tests, the results were not published. The hope was that by comparison of the (--)-6'-methoxy-9-rubanol with the corresponding cinchona alkaloids, it might be possible to determine whether it is the configuration of the vinyl groups which is responsible for the physiological properties of the latter. The synthesis and characterization of the four 6'-methoxy-9-rubanols have now been completed and the procedures improved.

The two racemates were easily separated but resolution of the racemates was a difficult task and was effected through the dianisoyltartrates. Comparison of the chemical behavior of the 4-members of each series (vinyl and vinyl-free) unexpectedly shows the same difference in basicity. The (--)-isomer does not show quinine action against bird malaria and the (++) has a slight action.

Prelog[15] has synthesized a racemic 6'-methoxy-9-rubanol and improved the method of Rabe by using powdered sodium in the condensation to form the rubatoxanone. However, results of physiological tests with the dihydrochloride of his compound are in disagreement with Rabe since Prelog found great activity for his racemic dihydrochloride. There has been no explanation for this.

Prelog[14] has used a novel method in synthesizing rubatoxan.

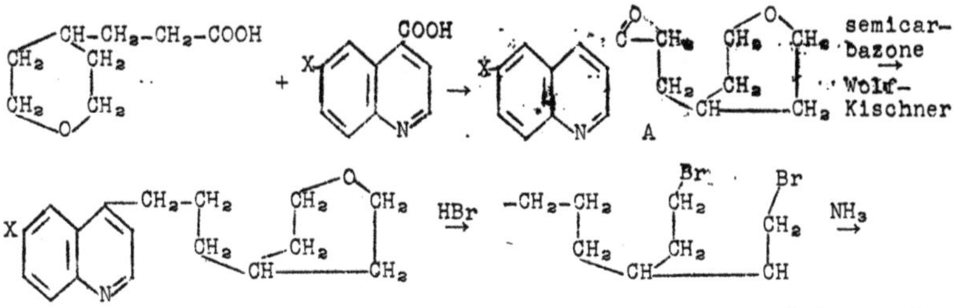

CH-CH2-CH2-COOH
COOH
X
A
semicar-bazone
Wolff-Kischner
HBr
NH3

α(tetrahydro-4-pyranyl)-
4-quinolyl propane

rubatoxan if
X=H

He has also synthesized aminoalcohols of the following type:

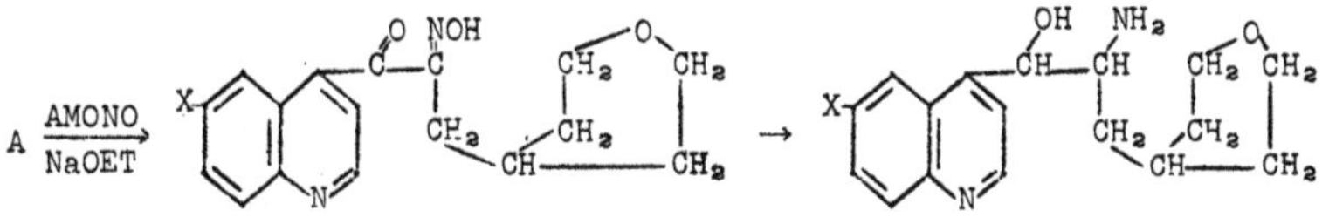

A
AMONO
NaOET
NOH
OH
NH2

Prelog[16] has also recently published a smooth synthesis of collidine to supplement that of Rabe[3] and Jantzen and Chichibabin and Oparina.[19]

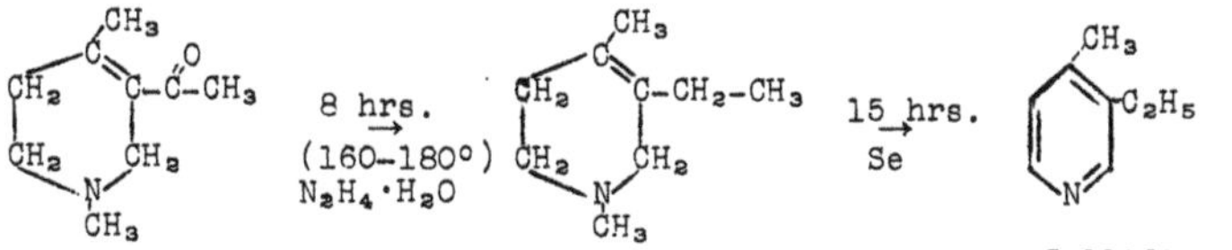

8 hrs.
(160-180°)
N2H4·H2O
15 hrs.
Se

1,4-dimethyl-
3-acetyl-1,2,5,6-
tetrahydropipridine

3-ethyl-1,2,5,6-
tetrahydropyridine

β-Collidine

The β-collidine can be obtained in a very pure form by this method.

Although no structural isomers of quinine have been mentioned in this seminar, Rabe[18] has recently reported the isolation of heteroquinine from purified precipitated quinine having the following formula:

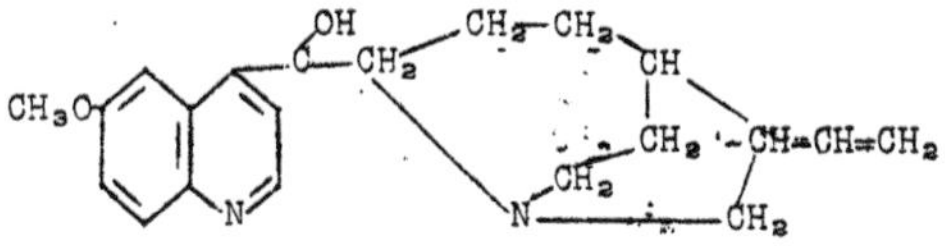

OH
C
CH₂
CH₂–CH₂
CH
CH₃O
N
CH₂
CH₂
N
CH₂
CH–CH=CH₂

I

As quinine is a derivative of quinuclidine so I is a derivative of 1-azabicyclo(3,3,2) nonane which is designated as homoquinuclidine.

Rabe has characterized all of the stereoisomeric quinines, hydroquinines, cinchonines, hydrocinchonines, desoxyquinines, desoxyhydroquinines, desoxycinchonines, desoxyhydrocinchonines, rubans, 9-rubanols, and the 6'-methoxyrubanols. He has extended his total synthesis of hydroquinine to all the compounds in the table except quinine, its three stereoisomers, and desoxyquinine and its isomer and completed partial synthesis of these.

TABLE

The net of carbons 3 and 4 are dextrorotatory in all cases except the two exceptions mentioned.

Compound	C_8	C_9	M.P. °C.	$[\alpha]_D^T$	M.P. of derivative
Quinine	–	–	172.8	-158.2^{15}	Q·HCl·$2H_2O$ 158-160
Epiquinine	–	+	011	43.3^{22}	di-HCl, 196
Quinidine	+	+	171.5	243.5^{15}	Q·HCl·H_2O, 258-259
Epiquinidine	+	–	113	102.4^{19}	di-HCl, 195-196
(+) Hydroquinine	+	+	171.5	143.5^{16}	H·HCl·$2H_2O$, 206-208
(–) Hydroquinine	–	–	169	140.4^{18}	

Compound	C_8	C_9	M.P. °C.	$[\alpha]_D^T$	M.P. of derivative
(+) Hydroquinidine	+	+	168-169	230.6^{18}	H·HCl, 223-274
(-) Hydroquinidine	-	-	171	-237.7^{18}	
Epihydroquinine	+	-	Oil	32.5^{22}	Chloride, (123)
Epihydroquinidine	+	+	122	73.7^{16}	Chloride, (oil)
Desoxydihydroquinine	-	0	70	-77.3^{20}	
Desoxyquinine	-	0	48	-97.7^{20}	
Desoxyhydroquinidine	+	0	85-87	167.7^{20}	
Desoxyquinidine	+	0	80-82	211^{20}	
Cinchonine	+	+	264	133^{25} $CHCl_3$	$C\cdot HCl\cdot 2H_2O$, 217-218
Epicinchonine	+	-	82-83	120.3^{20}	
Cinchonidine	-	-	204.5	-178^{15} 0.1 N H_2SO_4	$HCl\cdot H_2O$ 242
Epicinchonidine	-	+	62.8	$103\text{-}104^{20}$	
Hydrocinchonine	+	+		225.8^{15} ·1 N H_2SO_4	H HCl $2H_2O$ 220-221
Epihydrocinchonine	+	1	88.4	126^{20}	
Hydrocinchonidine	-	-	232	-95.8^{26}	H HCl $2H_2O$ 202-203
Epihydrocinchonidine	-	+	106	48.3^{20}	
Desoxyhydrocinchonine	+	0	72	143^{20}	
Desoxyhydrocinchonidine	-	0	52	-21.2^{20}	
Desoxycinchonine	+	0	91	179.3^{20}	
Desoxycinchonidine	-	0	60-62	-29.9^{20}	

Compound	C_8	C_9	M.P. °C.	$[\alpha]_D^T$	M.P. of derivative
9-Rubanols	+	+	229.5-30	132.5^{15}	Chloride 135-7.5
	+	-	118-119	14.3^{16}	140-142
	-	-	228.5-30.5	-131.8^{15}	136-138
	-	+	117-118	-14.9^{18}	141-142
ruban	+		Oil	80.5^{16}	
ruban	-		Oil	-78.4^{16}	
6'-Methoxy-9-rubanols	+	+	187	173.8^{17}	mono HCl salt 221 (decomp)
	-	-	187	-173.5^{17}	mono HCl salt 219 decomp.
	-	+	Oil-solidifying to glass	-23.25^{20}	mono HCl salt -223.3
	+	-	Oil-solidifying to glass	23.5^{20}	mono HCl salt -221.3 (decomp.)

Bibliography

1. Rabe and Kinder, Chem. Staatstab-Hamburg, Ber., 51, 1360 (1918).
2. Rabe and Kinger, Univ. Hamburg, Ber., 52B, 1842 (1919).
3. Rabe and Jantzen, Univ. Hamburg, Ber., 54B, 924 (1921).
4. Rabe, Ber., 55B, 522 (1922).
5. Rabe, Kindler and Wagner Ber., 55B, 532 (1922).
6. Rabe, Hunterburg, Schultz and Volger, Ber., 64B, 2487 (1931).
7. Rabe, Irshick, Suszka, Muller, Nielsen, Kolbe,
8. Von Riegen, Hochstatter, Ann. 492, 151 (1932).
9. Rabe, and Riza, Ann, 493, 151 (1932).
10. Rabe and Schultz, Ber., 66B, 120 (1933).
11. Rabe and Kindler, Ber., 72B, 263-264 (1939).
12. Rabe and Hater, J. Prakt. Chem., 154, 66 (1939).
13. Rabe and Hogen, Ber., 74B, 636 (1941); see C.A., 35, 7867 (1941).
14. Prelog, Seiwerth, Hahn and Cerkovnikov, Ber., 72B, 1325 (1939).
15. Prelog, Seiwerth, Heimbach and Stern, Ber., 74B, 647 (1941); see C.A. 35, 7970 (1941).

16. Prelog and Komzak, Ber., 74B, 1705 (1941); see C.A., 37, 133 (1943).
17. Chichibabin and Oparin, Ber., 60B, 1877-1879 (1927).
18. Rabe, Ber., 74B, 647 (1941); see C.A., 35, 7969 (1941).
19. Rabe, Univ. of Jena, Ann., 373, 85.
20. Thron and Dirscherl, Ann., 521, 48 (1935).
21. Koenigs and Husmann, Ber., 29, 2185 (1896).

Reported by C. G. Overberger
March 24, 1943

THE PTERINS

A Summary of the Structure - Proof by Wieland, _et al_, of Some Pigments from the Wings of the _Pieridae_.

The first work with the wing pigments of butterflies was done by Hopkins, who in 1889 extracted the white wing scales with boiling water. The pigments were given the generic name of lepidopterins, or simply pterins. The first crystalline form of any of these was the barium salt obtained by Schöpf and Wieland in 1925. Analytical values for this compound checked well with the corresponding values for xanthine; it was named xanthopterin. The deep-yellow compound gave the murexide test and yielded uric acid on stringent hydrolysis. It was thought, therefore, to be a derivative of uric acid.

A second pigment investigated by these workers had also been extracted from the wings of the _pieridae_ or cabbage butterflies by Hopkins, who used a dilute ammoniacal extractive and precipitated the product with acid. Schöpf found that it gave the murexide test only after repeated evaporations with nitric acid, and that also in other respects it was materially different from uric acid.

In general the pterins are insoluble in neutral solvents, permitting no recrystallizations, chromatographic separation or molecular weight determinations therein. They occur naturally as mixtures, difficult to purify either for analysis or characterization; no melting point can be obtained for them.

The synthesis of leucopterin (I) was accomplished by Purrmann in 1940 by heating 2,4,5-triamino-6-hydroxypyrimidine with oxalic acid until the excess oxalic acid had decomposed. A successful molecular weight determination in phenol with the trimethyl derivative confirmed this structure a year later.

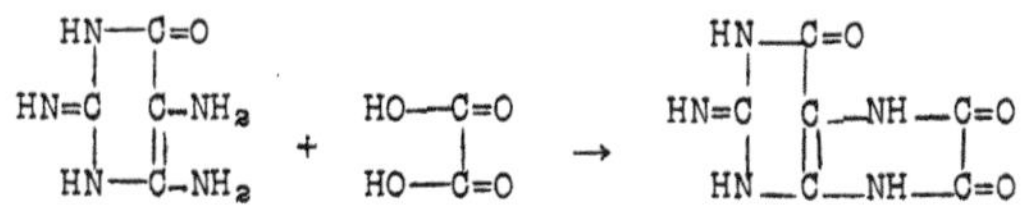

I

In the latter year Purrmann also synthesized xanthopterin (II) by reacting 2,4,5-triamino-6-hydroxypyrimidine with dichloroacetic acid to form the amide and subsequently closing the ring with silver acetate or carbonate:

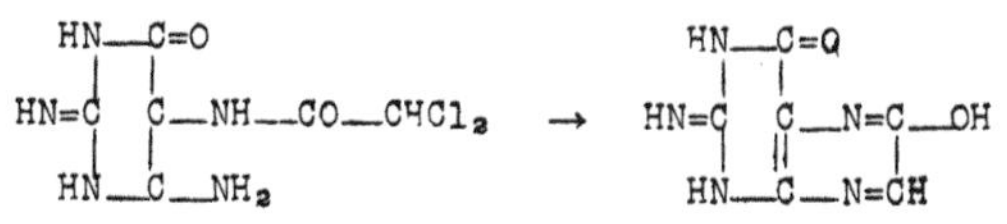

II

Some other pterins and purines isolated from the wings of these butterflies are desoxy-leucopterin (III), xanthine (IV) and guanopterin (isoguanine) V. Though isoguanine had been observed in plants, this was its first occurrence in the animal kingdom.

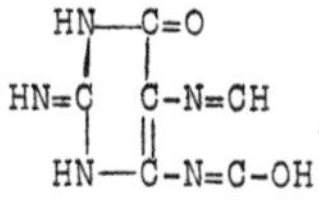

III

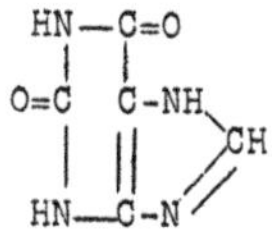

IV

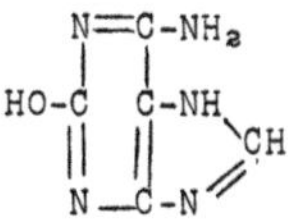

V

The relationship of the pterins or azine-purines, to purines and the flavins may be illustrated by a typical member of each family:

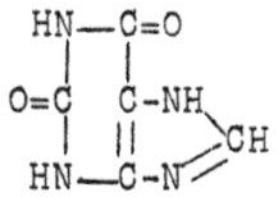

Xanthine
a purine

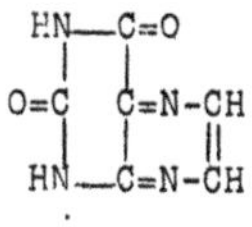

Lumazine
a pterin

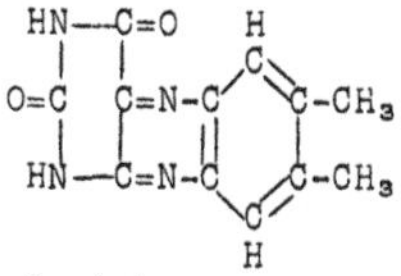

Lumichrome
a flavin

The principle reactions which led finally to the synthesis of leucopterin and xanthopterin may be mentioned, interpreted here in the light of the more recent work on these pigments.

Nitrous acid caused one-fifth of the nitrogen present in leucopterin to be lost, replacing this with oxygen and yielding desiminoleucopterin. This reaction supplements the fact that on hydrolysis uric acid gives urea while guanidine was obtained from the pterin.

The products of hydrolytic cleavage have given the most useful indications as to this structure. Alloxan or allantoin are obtained on hydrolysis of uric acid; these were not formed from leucopterin. In hydrochloric acid at 170° the reaction proceeded with leucopterin and uric acid each in much the same manner:

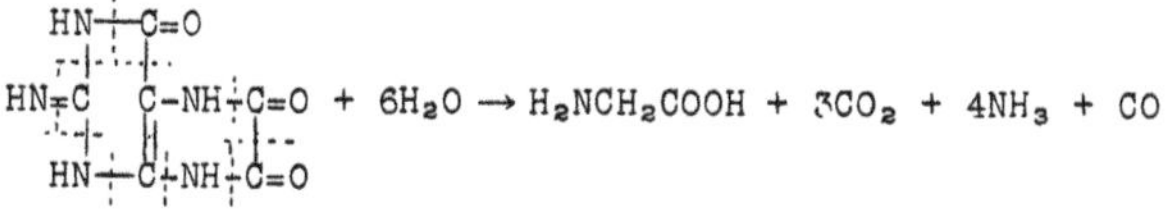

The action of chlorine on leucopterin in acetic acid or in methyl alcohol formed respectively either the glycol or glycol dimethyl ether. Hydrochloric acid split leucopterin glycol under much milder conditions, permitting the isolation of guanidine, oxalic acid and glyoxalic acid (resin). Since similar treatment of uric-acid glycol furnishes no oxalic acid, this then did not come from carbons 4 and 5; furthermore its presence explained the carbon oxides appearing in the previous reaction under conditions at which oxalic acid is unstable.

[illegible] have been observed [illegible], [illegible] factor [illegible].

III IV V

The [illegible] of the [illegible] or [illegible] and the [illegible] may be illustrated by [illegible].

[illegible]

Nitrous acid [illegible] of the [illegible] nitrosamine [illegible] with [illegible] yielding [illegible] from the [illegible].

[illegible]

[illegible]

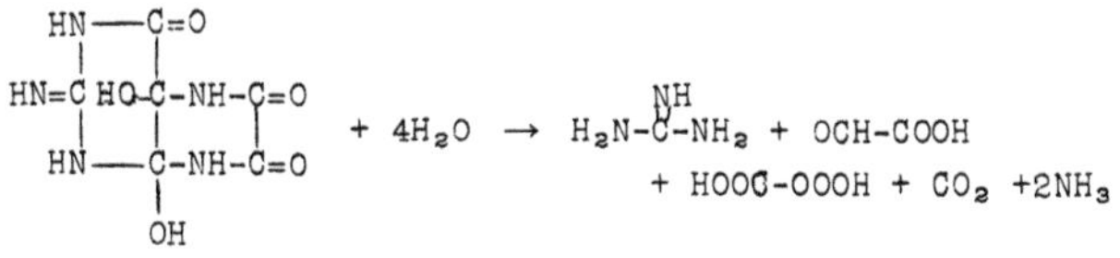

$$\begin{matrix} HN\!-\!\!-C\!=\!O \\ | \quad | \\ HN\!=\!C \; HO\!-\!C\!-\!NH\!-\!C\!=\!O \\ | \qquad | \\ HN\!-\!\!-\!\!-C\!-\!NH\!-\!C\!=\!O \\ | \\ OH \end{matrix} + 4H_2O \;\rightarrow\; H_2N\!-\!\overset{\overset{\displaystyle NH}{\|}}{C}\!-\!NH_2 + OCH\!-\!COOH$$
$$+ \; HOOC\!-\!COOH + CO_2 + 2NH_3$$

Leucopterin-glycol on careful oxidation gave 2-imino-hydantoin oxamide (VI) in much the same manner as that in which allantoin is had from uric-acid glycol. This suggested VII as the correct structure of leucopterin glycol.

$$\begin{matrix} HN\!-\!\!-C\!=\!O \\ | \quad | \\ HN\!=\!C \qquad | \\ | \qquad\; O\;\;O \\ HN\!-\!\!-C\!-\!NH\!-\!\overset{\|}{C}\!-\!\overset{\|}{C}\!-\!NH_2 \end{matrix}$$

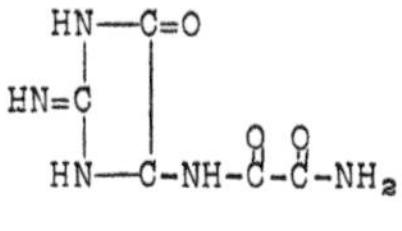

VI

$$\begin{matrix} HN\!-\!\!-C\!=\!O \\ | \qquad\; OH \\ HN\!=\!C \quad C{<} \\ | \qquad \backslash NH\!-\!\overset{\overset{O}{\|}}{C}\!-\!\overset{\overset{O}{\|}}{C}\!-\!NH_2 \\ | \\ HN\!-\!\!-C\!=\!O \end{matrix}$$

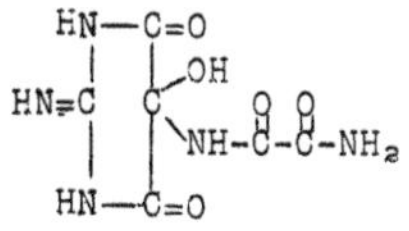

VII

The dimethyl ether of leucopterin glycol lost methyl alcohol on warming, and the monomethyl ether so formed when warmed to 65° in aqueous medium readily lost carbon dioxide and guanidine.

An acid (VIII), isolated from this decomposition as the methyl ester, was judged to be the same as that found by Biltz in a similar reaction.

$$\begin{matrix} HN\!-\!\!-\!-C\!=\!O \\ | \qquad\qquad | \\ HN\!=\!C \; CH_3O\!-\!C\!-\!NH\!-\!C\!=\!O \\ | \qquad\qquad | \\ HN\!-\!\!-\!-C\!-\!NH\!-\!C\!=\!O \\ | \\ OH \end{matrix} \;\rightarrow\; H_2N\!-\!\overset{\overset{\displaystyle NH}{\|}}{C}\!-\!NH_2 + CO_2 + H_2N\!-\!\overset{\overset{O}{\|}}{C}\!-\!\overset{\overset{H}{|}}{\underset{\underset{OCH_3}{|}}{C}}\!-\!NH\!-\!\overset{\overset{O}{\|}}{C}\!-\!\overset{\overset{O}{\|}}{C}\!-\!OH$$

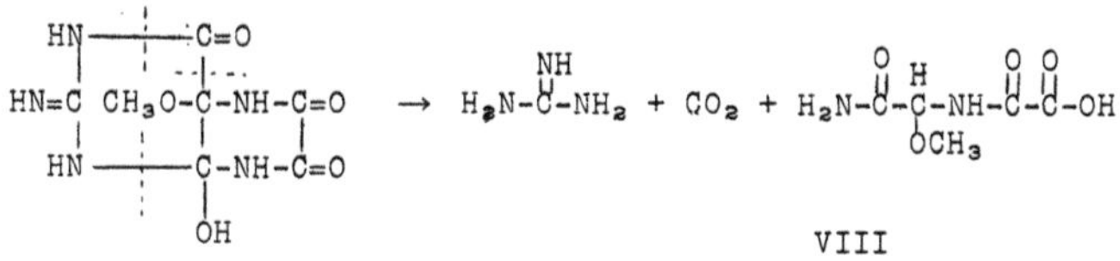

VIII

Hydrolysis in hydrochloric acid of the dimethyl ether of desoxy-leucopterin glycol gave alloxan (IX), oxalic acid, ammonia and two moles of methyl alcohol; no guanidine was derived from desoxy-leucopterin. 5-Methoxyuramil oxalic acid (X) was isolated as the methyl ester.

$$\begin{matrix} HN\!-\!\!-\!-C\!=\!O \\ | \qquad\qquad | \\ O\!=\!C \; CH_3O\!-\!C\!-\!NH\!-\!C\!=\!O \\ | \qquad\qquad | \\ HN\!-\!\!-\!-C\!-\!NH\!-\!C\!=\!O \\ | \\ OCH_3 \end{matrix} \;\rightarrow\; \begin{matrix} HN\!-\!\!-C\!=\!O \\ | \qquad | \;\; OH \\ O\!=\!C \quad C{<} \\ | \qquad\;\; \backslash OH \\ HN\!-\!\!-C\!=\!O \end{matrix} + (COOH)_2 + 2\,CH_3OH +$$

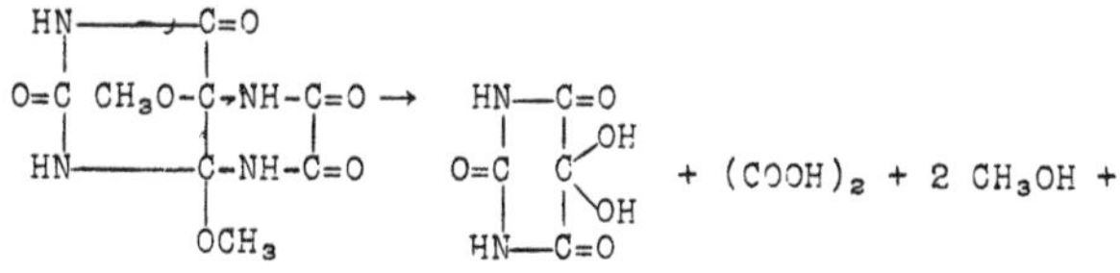

IX

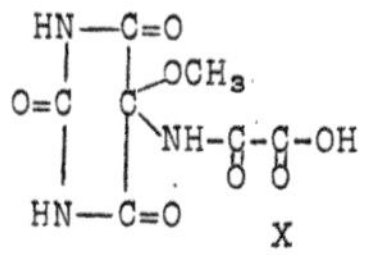

X

The relationship between leucopterin and xanthopterin was not known until 1939. In that year Purrmann accidentally discovered this relationship, before the structure of either had been determined, thus materially simplifying the problem. This worker was studying the decolorization that accompanied the reversible addition of hydrogen peroxide and sulfurous acid to xanthopterin, when he found that cold 30% hydrogen peroxide reacted irreversibly to give a colorless compound. This compound was shown by comparison of analyses, crystal form, physical constants, Debye-Scherrer x-ray pattern and van Slyke nitrogen determination to be identical with leucopterin. The transformation can be accomplished in better than 60% yields by auto-oxidation of eanthopterin in the presence of platinum and acid.

BIBLIOGRAPHY

Hopkins, Trans. Roy. Soc. London 186B, 661 (1895).
Schöpf and Wieland, Ber. 58, 2178 (1925).
Schöpf and Wieland, Ber. 59, 2067 (1926).
Wieland, Metzger, Schöpf and Bülow, Ann. 507, 226 (1933).
Wieland and Kotzschmar, Ann. 530, 152 (1937).
Wieland and Purrmann, Ann. 539, 179 (1939).
Wieland and Tartter, Ann. 543, 287 (1940).
Wieland and Purrmann, Ann. 544, 163 (1940).
Purrmann, Ann. 544, 182 (1940).
Wieland, Tartter and Purrman, Ann. 545, 209 (1940).
Purrmann, Ann. 546, 98 (1941).
Wieland and Decker, Ann. 547, 180 (1941); C.A. 35, 5901[8].
Wieland and Tartter, Ann. 545, 197 (1940)

Reported by George Mueller
March 24, 1943

A NEW SYNTHESIS OF HETEROCYCLIC COMPOUNDS

Amines of the type $ClCH_2CH_2NR_2$ and $(ClCH_2CH_2)_2NR$ are remarkably stable toward sodium amide. Below 100° they do not react to any appreciable extent with either sodium amide or with added or liberated ammonia. If however, a substance containing hydrogen replaceable by sodium is added to the mixture of haloethylamine and sodium amide, it can form the sodium derivative which then reacts with the halo alkyl amine.

Most of the compounds to be described can also be obtained by using sodium phenolate instead of sodium amide, or with the pure sodium derivative of the compound to be alkylated, but the latter are not always obtainable and the sodium phenolate requires a greater expenditure of material.

Compounds of the type $X(CH_2CH_2Cl)_2$ where X can be O, S, NR, NSO_2Ph condense with benzyl cyanide under the influence of sodium amide to give the heterocyclic compounds.

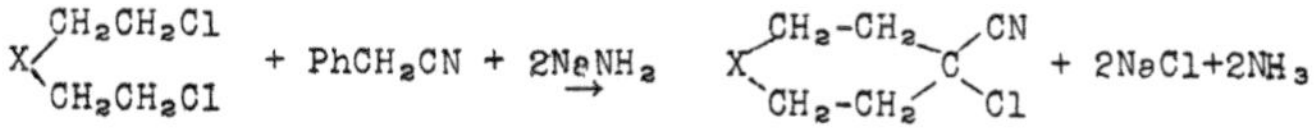

This reaction is of especially great importance where X=NR, as it makes possible the preparation of piperidine derivatives for which no other method was before feasible. The reaction is not limited to benzyl cyanide. The active methylene groups of fluorene, 1-methyloxindole, $MeSO_2NEt_2$, $PhCH_2SO_2NMe_2$, $MeSO_2Ph$, and $PhCH_2SO_2Ph$ condense to yield substituted piperidines as well.

Many of the compounds prepared in this way have noticeably good spasmolytic or local anesthetic action. The hydrochloride of Ethyl-1-methyl-4-phenylpiperidine-4-carboxylate has been introduced into therapeutics under the name of Dolatin:

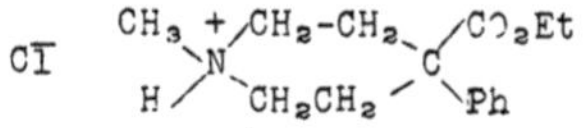

Piperidine derivatives have been obtained from N-methyl-2,2'-dichlorodiethylamine and the N-benzyl analogue.

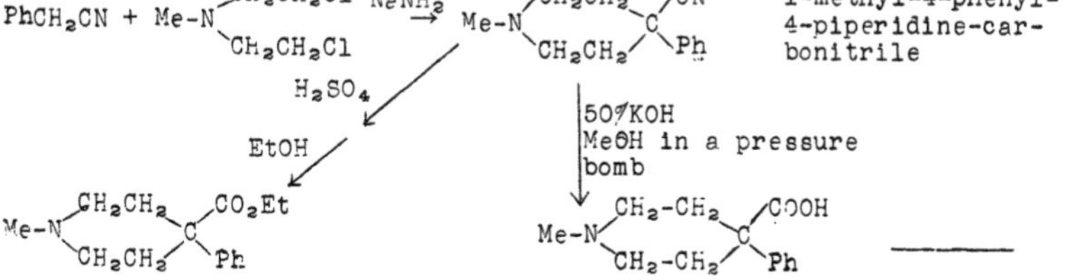

HCl salt is Dolatin. loses CO_2 on melting

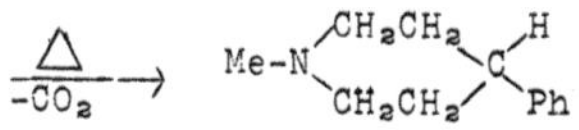

$$\xrightarrow[-CO_2]{\triangle} \quad Me-N\begin{array}{c} CH_2CH_2 \\ \diagdown \\ CH_2CH_2 \end{array}C\begin{array}{c} H \\ \diagup \\ \diagdown Ph \end{array}$$

1-methyl-4-phenyl-piperidine.

The N_1N'-Bis(2-hydroxyethyl)-p-toluenesulfonamide is prepared in the following way.

$$H-N\begin{array}{c} CH_2-CH_2OH \\ \diagup \\ \diagdown \\ CH_2CH_2OH \end{array} + Me-\bigcirc-SO_2Cl \rightarrow CH_3-\bigcirc-SO_2-N\begin{array}{c} CH_2CH_2OH \\ \diagup \\ \diagdown \\ CH_2CH_2OH \end{array}$$

$$\Big\downarrow SOCl_2$$

$$CH_3-\bigcirc-SO_2-N\begin{array}{c} CH_2CH_2 \\ \diagup \\ \diagdown \\ CH_2CH_2 \end{array}C\begin{array}{c} CN \\ \diagup \\ \diagdown Ph \end{array}\xleftarrow[PhCH_2CN]{NaNH_2} \quad CH_3-\bigcirc-SO_2-N\begin{array}{c} CH_2CH_2Cl \\ \diagup \\ \diagdown \\ CH_2CH_2Cl \end{array}$$

$$H_2SO_4 \Big\downarrow$$

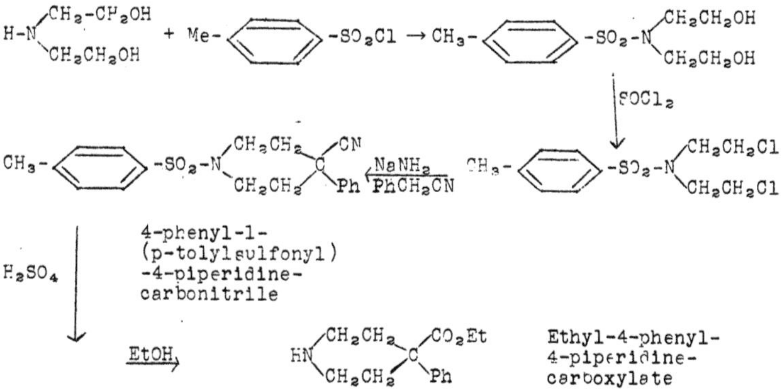

4-phenyl-1-
(p-tolylsulfonyl)
-4-piperidine-
carbonitrile

$$\xrightarrow{EtOH} \quad HN\begin{array}{c} CH_2CH_2 \\ \diagup \\ \diagdown \\ CH_2CH_2 \end{array}C\begin{array}{c} CO_2Et \\ \diagup \\ \diagdown Ph \end{array}$$

Ethyl-4-phenyl-
4-piperidine-
carboxylate

The active hydrogens in phenylmethylsulfone undergo similar condensations.

$$CH_3SO_2Ph + MeN\begin{array}{c} CH_2CH_2-Cl \\ \diagup \\ \diagdown \\ CH_2CH_2-Cl \end{array}\xrightarrow{NaNH_2} \quad MeN\begin{array}{c} CH_2CH_2 \\ \diagup \\ \diagdown \\ CH_2CH_2 \end{array}C\begin{array}{c} SO_2Ph \\ \diagup \\ \diagdown h \end{array}$$

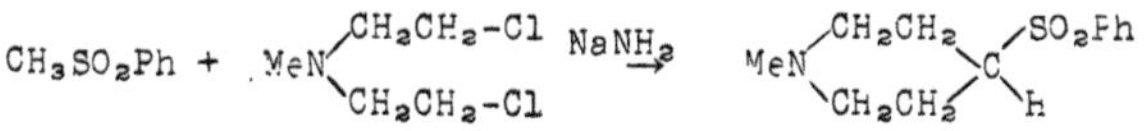

1-methyl-4-piperidyl
phenylsulfone.

Alkylations of the following general type are summarized in the table.

$$\begin{matrix} G_1 \\ \end{matrix}\!\!\!\searrow CH_2 \;(NH)\! \quad G_2\nearrow \quad + \quad Et_2NCH_2CH_2Cl \xrightarrow{NaNH_2} (G_1)(G_2)CHCH_2CH_2NEt_2 \;/\; (G_1)(G_2)N\text{-}CH_2CH_2NEt_2$$

Reactant	Product, name and formula		Remarks
$PhCH_2COPh$	$PhCH(COPh)CH_2CH_2NEt_2$,	3-Diethylamino-1-phenylpropyl phenyl ketone	HCl salt has noteworthy spasmolytic action.
$PhCH_2SO_2Ph$	$PhCH(SO_2Ph)CH_2CH_2NEt_2$,	3-Diethylamino-1-phenylpropyl phenyl sulfone	HCl salt is neutral to litmus.
$PhCH_2Ph$	$Ph_2CHCH_2CH_2NEt_2$,	N,N-Diethyl-3,3-diphenyl propylamine.	HCl salt soluble in water. Local anesthetic, spasmolytic.
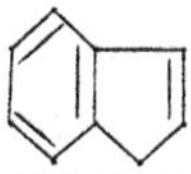	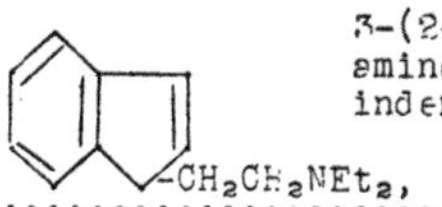	3-(2-Diethylamino ethyl) indene	HCl salt a local anesthetic.
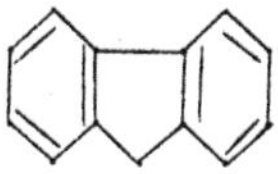	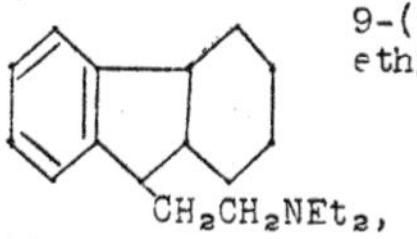	9-(2-Diethylamino ethyl) fluorene	HSO_4^- salt is a local anesthetic
Ph_2NH	$Cl_2NCH_2CH_2NEt_2$,	1-Diethylamino-2-diphenylamino-ethane.	Mono HCl salt as well as the mono etho bromide have weak local anesthetic action.
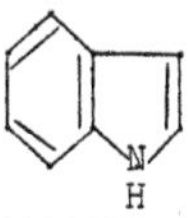	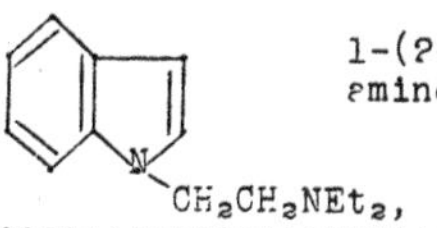	1-(2-diethylamino) indole	Mono HCl salt is indifferent pharmoologically.
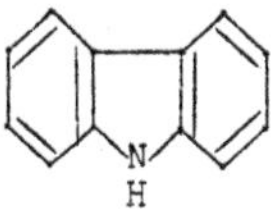	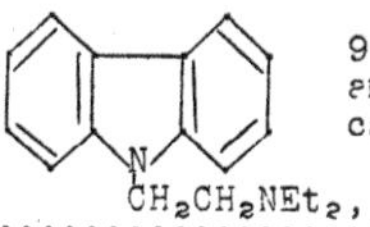	9-(2-diethylamino-ethyl carbazole.	Magnificent blue fluorescent substance. Phosphate is a local anesthetic.

BIBLIOGRAPHY

Eisleb, O., Ber., 74B, 1433 (1941). C.A. 36, 5465 (1941).
Meyer, V., Ber., 21, 1344 (1888).

Reported by J. W. Macorney - March 31, 1943

ABIETIC ACID

Abietic acid or sylvic acid is a diterpene obtained from rosin and is probably the most abundantly available and the cheapest of the organic acids. The name abietic (ăb´ĭ ĕt´ ik) was first applied by Baup in 1826 to a product from Pinus abies. Rosin or colophony, which is the commercial source of the diterpenes, is the acidic residue remaining after steam distillation of the oil of turpentine from the exudation of American pine trees. Annual production of rosin in the United States has been valued at forty million dollars.

Essentially there are two methods for the conversion of rosin to abietic acid. In the laboratory, yields of 80-90% can be obtained by vacuum distillation, the acid boiling at 200-210°/(<1 mm.). Steele's method of conversion consists of boiling rosin in 98% acetic acid for two hours; the acid crystallizes out on cooling and is obtained in 40% yield on recrystallization from acetic acid, m.p. 159-161°, $[\alpha]_D$ - 77°.

The structure I for abietic acid is now fairly well established. One of the principal difficulties has been the determination of the position of the two double bonds. This was complicated by the ease of isomerization and disproportionation. The following facts, however, were known. Abietic acid forms a (maleic) anhydride adduct which indicates a conjugated system. However, this adduct, which forms at 100°, was shown to be the same as that formed from levopimaric acid, a structural isomer of abietic acid. This has been shown to be the result of heat isomerization of abietic acid to pimaric acid. Fieser has shown that abietic acid forms a crystalline derivative with diazotized p-nitroaniline, indicating a conjugated system, and also, on the basis of absorption spectra, that the double bonds are in two different rings. Oxidation of abietic acid with ozone or permanganate consistently yielded isobutyric acid as one of the products, which indicates that C_7 must be involved in the conjugated system.

The structure of the oxidation product (II) had been proved conclusively. This established the structure of (III) which in turn proved that one ethylene link was between C_8 and C_{14}. Careful oxidation of (I) with permanganate yielded a dihydroxyabietic acid (IV) which on oxidation with permonophthalic acid formed an oxide (VII). Hydration gave the same tetrahydroxyabietic acid (V) that was obtained directly from abietic acid with permanganate. Dehydrogenation of (IV) with selenium gave 7-hydroxy-1-methylphenanthrene and retene. Oxidation of (IV) with lead tetra-acetate gave the unsaturated ketoaldehyde. These reactions indicate an α-glycol grouping for the dihydroxy acid (IV) including C_7.

In the formation of the tetrahydroxyabietic acid (V) from (I) the intermediate product on treatment with hydrochloric acid yields a chlorotrihydroxy acid. This chloro-acid on treatment with alkali forms an oxidodihydroxy acid (VIIa) stereoisomeric with (VII), which, on hydrolysis, also forms the tetrahydroxyabietic acid (V). Oxidation with lead tetra-acetate of these two isomeric oxido compounds (VIIa) and (VII) gave the same oxidoketoaldehyde.

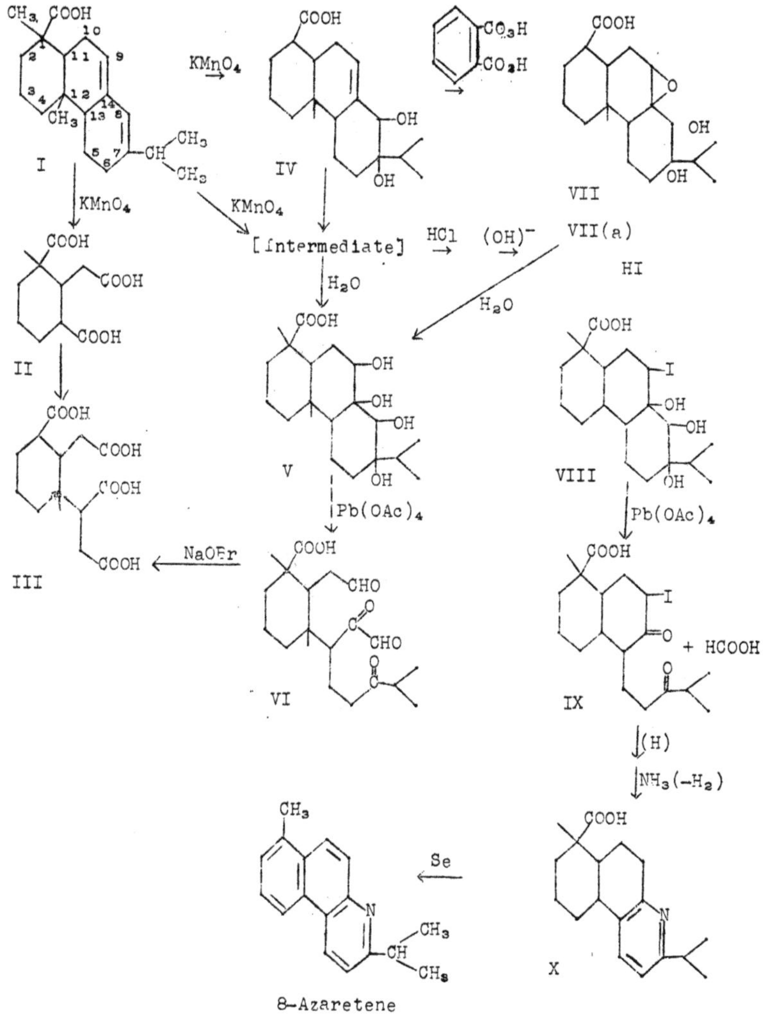
KMnO4
KMnO4
KMnO4
[Intermediate]
HCl
(OH)⁻
VII(a)
HI
H2O
H2O
Pb(OAc)4
Pb(OAc)4
NaOBr
+ HCOOH
(H)
NH3(-H2)
Se
I
II
III
IV
V
VI
VII
VIII
IX
X
8-Azaretene

The oxido acid (VIIa) has also been converted to a compound believed to be 8-azaretene. The formation of the diketo-acid (IX) indicates a glycerol grouping in the iodotrihydroxy-acid (VIII) and shows that the four hydroxyls in V are attached to adjacent carbons. This is the final proof that structure I correctly represents abietic acid. An independent synthesis of 8-azaretene will be unequivocal evidence for that structure.

A great deal of confusion and ambiguity has existed in the literature concerning the identity of the acids originally present in rosin and the various reactions of these acids. The following relationships have been shown to exist.

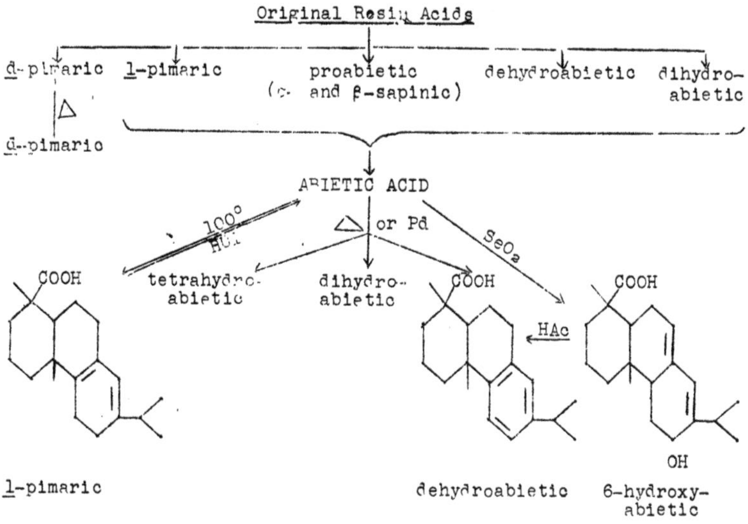
Original Resin Acids
d-pimaric
l-pimaric
proabietic
(α- and β-sapinic)
dehydroabietic
dihydro-abietic
Δ
d-pimaric
ABIETIC ACID
100°
HCl
Δ or Pd
SeO₂
COOH
tetrahydro-abietic
dihydro-abietic
COOH
HAc
COOH
OH
l-pimaric
dehydroabietic
6-hydroxy-abietic

The price and availability of abietic acid together with its hydrophenanthrene type of nucleus has suggested the possibility of using the acid in the synthesis of various physiologically active compounds. The following reactions have been a successful start in this direction.

Methyl dehydroabietate $\xrightarrow[AlCl_3]{CH_3COCl}$ Methyl 6-acetyldehydro abietate $\xrightarrow{NH_2OH}$ oxime

$\downarrow$ HCl

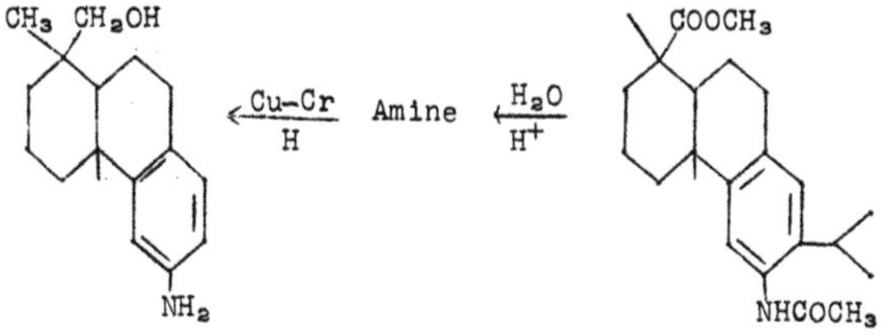

CH_3 CH_2OH
Cu-Cr
H
Amine
H_2O
H^+
$COOCH_3$
NH_2
$NHCOCH_3$

6-Aminodehydro-abietinol

Methyl 6-acetaminodehydro-abietate

Nitrosyl-sulfuric pyridine → Diazo

CH_3I, Ag_2O → Methyl ether of 6-dimethyl aminodehydroabietinol hydrochloride

(similar to morphine in structure -showed no activity)

Diazo → H_2O → 6-Hydroxydehydro-abietinol

(5γ doses showed high oestrogenic acitivity)

The liquid esters of abietic acid are useful plasticizers and softeners in the manufacture of lacquers and inks. The manganese and cobalt resinates are important driers for paint oils and varnishes. The sodium salt is used in paper sizing and the making of hard water and laundry soaps. However, the most important commercial property of abietic acid is its ability to form simple and mixed polymers which are the basis for the "four hour enamels" and spar varnishes. Besides abietic acid the hydrogenated or stable dihydroabietic acid, "stabelite", is also used in polymer formation as well as a flux, dental cement, pressure and laminating adhesive and for the above mentioned uses of abietic acid and its salts.

Abietic acid polymers, "Polypale" or "Nuroz", are used in ester gum, gloss oils, and limed resins. Abietic acid-maleic anhydride-glycerol polymers, "Amberol" or "Beckacite" maleics, are used in fast drying varnishes, spar varnishes, baking enamels, primers and surfacers, printing ink vehicles, and plywood bonding. Abietic acid-phenol-formaldehyde-glycerol polymers, "Beckacite" or "Amberol" phenolics, and abietic acid-pentaerythritol polymers, known as "Pentalyn", have the same uses.

References

Review, Chem. Soc. - Annual Reports, 38 (1941).
Bain, J. Am. Chem. Soc., 64, 871 (1942).
Fieser, Chem. of Natural Products, Second Ed.; J. Am. Chem. Soc., 60, 160, 2631 (1938); 61, 2528 (1939).
Hasselstrom, ibid., 57, 2118 (1935); 60, 2340 (1938); 63, 421 (1941).
Palkins, ibid., 60, 921, 1419 (1938); 61, 247, 1230, 3197 (1939); J. Chem. Ed., 12, 35 (1935).
Ruzicka, Helv. Chim. Acta, 23, 333, 341, 355 (1940); 24, 223, 504 (1941); Chem. and Ind., 55, 546 (1936).
Steele, J. Am. Chem. Soc., 44, 1333 (1922).
Chem. and Eng. News, 20, 538 (1942).

Reported by Sidney Melamed
March 31, 1943

THE OXIDATION OF OLEFINIC DOUBLE BONDS TO GLYCOLS

The formation of glycols by oxidation of an olefinic double bonds can procede in two ways, i.e.

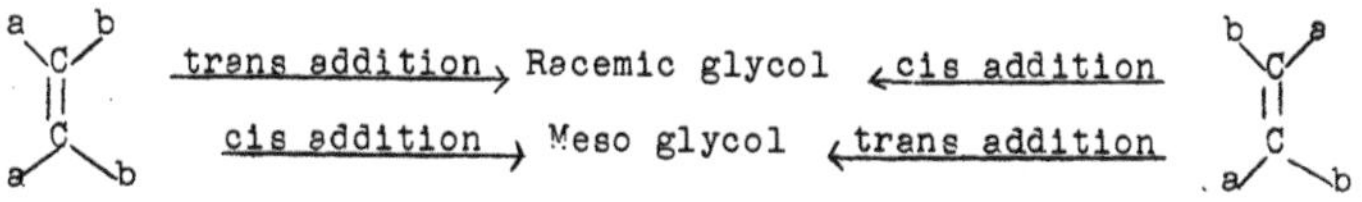
trans addition
Racemic glycol
cis addition
cis addition
Meso glycol
trans addition

The work of Boeseken (1) and Kuhn (2) demonstrated that oxidizing agents like OsO_4 and $KMnO_4$ gave glycols derived from the hydrolysis of a cis addition product and that others like alkaline peroxide and perbenzoic acid gave the corresponding steriosomers.

The use of alkaline permanganate as a preparative method for cis glycols was developed by Wagner (3) in 1888. A 1% solution of $KMnO_4$ is added to an aqueous solution or suspension of the olefin at room temperature or lower. The use of higher concentration of permanganate or more vigorous conditions results in further oxidation of the glycol and cleavage of the molecule. The oxidation of highly insoluble compounds is facilitated by the use of ethyl alcohol as a solvent but the reaction must be run at -47° (4). By this method cis glycols derived from olefinic hydrocarbons have been prepared in yields of 50% or higher, but the high oxidation potential of alkaline permanganate has prevented its use on olefins with functional groups that are sensitive to oxidation.

Osmium tetroxide is a highly specific oxidizing agent and affords a means of hydroxylating olefinic compounds under less vigorous conditions that those required for permanganate oxidations. As with $KMnO_4$, only the glycol produced by the hydrolysis of the cis addition product is formed. Criegee (5), in 1936, succeeded in isolationg several addition products of the olefin and the OsO_4.

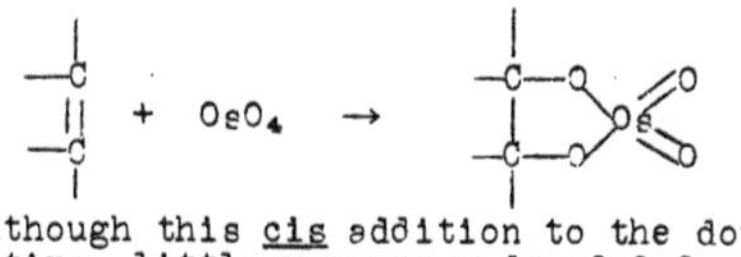

Although this cis addition to the double bond is practically quantitative, little use was made of OsO_4 as a hydroxylating agent because of the extreme stability of the osmium ester towards hydrolysis. Apparently the equilibrium

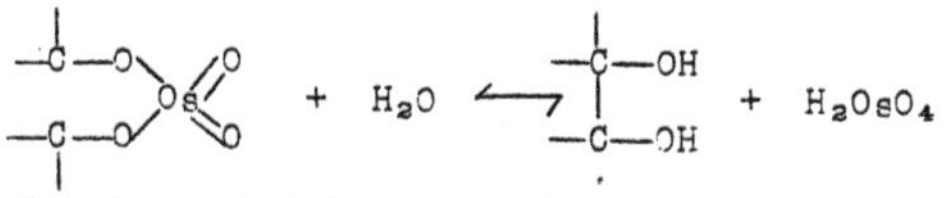

is almost completely on the left even in water solution. The equilibrium may be shifted to the right by removal of the acid either by oxidation to OsO_4, or by reduction to lower oxides which are insoluble in water.

The oxidative shift of the equilibrium was affected by Hoffman (6) with $NaClO_3$. Since the osmium tetroxide is constantly regenerated only catalytic amounts are required. The main side reaction is the

addition to the double bond of hypochlorous acid which is formed in small amounts in the reduction of the chlorate. Braun (7) found that the amount of chlorine containing by-product was greatly reduced if silver chlorate was used instead of $NaClO_3$, probably because the silver hypochlorite decomposes spontaneously

$$3\ AgOCl \rightarrow AgClO_3 + \underline{2AgCl}$$

Hoffman investigated the stability of functional groups that are readily oxidized by alkaline permanganate toward OsO_4-$NaClO_3$. It was found that aliphatic alcohols, aldehydes, ketones, and acids, and most aromatic hydrocarbons are unaffected by this reagent at the temperatures usually employed for the oxidation of the double bond. Anthracene, phenolic compounds, and aromatic amines are oxidized to an appreciable extent.

Miles (8) developed a smooth synthesis of cis glycols using H_2O_2-t-butyl alcohol - OsO_4. A 5% solution of peroxide in t-butyl alcohol is added slowly to a mixture of the unsaturated compound and traces of OsO_4 in t-butyl alcohol. The reaction is complete in 3 to 47 hours at room temperature, and the yields of cis glycols reported are superior to those obtained by the Hoffman method.

Presumably the peroxide plays the same role as the $NaClO_3$ in the cleavage of the osmium ester. This is substantiated by the fact that hydroxylation of double bonds with alkaline peroxide or H_2O_2 in acetic acid gives the trans isomer. Miles (9) found that the photochemical hydroxylation of olefins with hydrogen peroxide yielded cis glycols by the simultaneous addition of two hydroxyl radicals to the double bond, so the possibility that the function of the osmium tetroxide is to catalyze the decomposition

$$H_2O_2 \xrightarrow{OsO_4} 2HO.$$

cannot be ruled out.

Criegee found that the hydrolysis of the osmium ester formed by the cis addition of OsO_4 to a double bond could be affected by reducing agents like sodium sulfite, formaldehyde, and ascorbic acid. Since the osmium tetroxide is removed from the reaction mixture by precipitation in the form of a complex of lower oxide, mole for mole amounts of the reagent must be used. This method has not been used much for the preparation of cis glycols because osimum tetroxide is both poisonous and expensive. In the hydroxylation of certain sterols, the absence of side reactions and the high yields obtained compensate for the above disadvantages, and this method has been used extensively. The reaction is run at room temperature in absolute ether or dioxane and requires 3 to 5 days to go to completion. The osmic ester is isolated and hydrolized by refluxing for several hours with sodium bisulfite in dilute alcohol or by treatment with alkaline formaldehyde at room temperature.

This method has been used frequently in the preparation of cis glycols of sensitive compounds since the highly specific action of osmium tetroxide is not nullified by the presence of another oxidizing agent as is the case when oxidative cleavage is used. Serini and his coworkers (10) have used reductive cleavage of the osmic ester to pre-

pare glycols in the allopregnane series

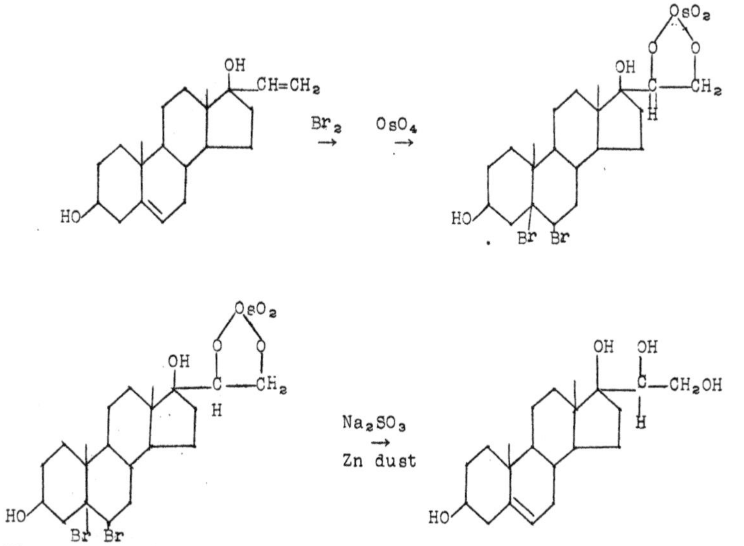

OH
CH=CH2
Br2
OsO4
HO
OsO2
O
O
OH
C
CH2
H
HO
Br Br
OsO2
O
O
OH
C
CH2
H
HO
Br Br
Na2SO3
Zn dust
OH OH
C—CH2OH
H
HO

It was later found that it was not necessary to protect the cyclic double bond by bromination if the hydroxyl group were acetylated (11). H. Fischer (12) has reported yields of 20% in the cis hydroxylation of porphyrin- like compounds.

TRANS ADDITION

Peracids

Nikolaus Prileschajew in 1909 first published work showing that perbenzoic acid would convert olefins into epoxy compounds (13). He used a neutral solvent such as ether or chloroform to which the perbenzoic acid was added and the solution cooled to zero. The olefin was then added in quantities almost equivalent to the perbenzoic acid and allowed to stand for some time at room temperature. The benzoic acid produced was extracted, and the solvent and oxide separated by distillation. In this way Prileschajew made oxides of such compounds as allyl alcohol, decylene, diisobutylene, citral, and limonene, and found they could be easily hydrated to glycols.

The reaction appears to be a general one for peracids, and epoxy compounds have been made by peracetic, percamphoric, perfuroic, and monoperphthalic acids (14). Perbenzoic and peracetic acid have been the only ones to be used or studied in any detail. However,

monoperphthalic acid is gaining more favor as it is more easily and cheaply prepared and is more stable towards alkali and water than is perbenzoic. The phthalic acid formed from it is insoluble in chloroform and so may be removed easily in many cases where the use of water must be avoided in working up the reaction products (15). Directions for preparing perbenzoic and monoperphthalic acids are given in "Organic Syntheses" (16). Peracetic acid is made from acetic acid and hydrogen peroxide with a trace of sulfuric acid used as a catalyst.

The reactivity of the carbon-carbon double bonds towards the peracids is profoundly influenced by the substituents attached to the unsaturated carbon atoms, and Prileschajew noted in his first studies that conjugated double bonds were less reactive than isolated double bonds. Böeseken found that a double bond that was alpha-beta to a carbonyl group has lessened reactivity with peracids, regardless of whether the carbonyl group was in an aldehyde, ketone, acid, or ester (17). Maleic and fumaric acids are not attacked at all. Cinnamyl alcohol is readily converted to the epoxy derivative, but cinnamaldehyde is attacked slowly, and cinnamic acid and its esters give a barely visible reaction with perbenzoic acid (18).

In studies of the action of perbenzoic acid on conjugated double bonds of natural products, it has been found that frequently some of the double bonds are left unattacked so that the use of perbenzoic to determine the number of double bonds in a compound has been disappointing. Geraniol will give either the monoxide or dioxide as desired, but lycopene will show only twelve out of thirteen double bonds affected by perbenzoic, and xanthophyl, and carotene show only eight out of eleven (19).

In studies of reaction rates of peracetic acid in acetic acid by Sturrman, (20) it has been found that the reaction is bimolecular. He found the following velocity constants with the time being in minutes, the concentration in moles per liter, and the temperature used being 25.8°.

HYDROCARBON	$K \times 10^3$
Ethylene	0.19
Propylene	4.2
1-Pentene	4.3
2-Butene	93
2-Pentene	95
$Me_2C:CH_2$	92
$Me_2C:CHMe$	1240
Cyclopentene	195
1-Methylcyclopentene	2220
Cyclohexene	129
Styrene	11.2
Allylbenzene	1.9
Alpha-phenylpropene	46
Stilbene	6.7
Iso-stilbene	11.1
$Ph_2C:CH_2$	48

The results of this work and similar studies by Böeseken show clearly that the reactivity of the double bond towards the per acids is increased as the degree of substitution on the unsaturated carbon atoms increases.

Epoxy compounds are readily hydrated to glycols by water having a trace of mineral acid as a catalyst. Van Loon, in his thesis at Delft in 1918, first showed that an epoxy compound was converted into a glycol by trans addition of water. Since then Böeseken, Kuhn and Ebel, (2) and Wilson and Lucas (21) have presented enough examples to show that this hydration in a trans manner is general, and in few cases is any of the cis addition compound formed.

Böeseken, in working with peracetic acid in acetic acid, found that the monoacetate of the glycol or the diacetate was formed instead of an epoxy compound (22). The glycols were of the trans addition form, but up until 1930 it was believed that peracetic acid acted in an entirely different manner upon olefins than did perbenzoic acid, and it was thought that peracetic acid was incapable of forming epoxy compounds from an olefin. Then Arbusow and Michailow (23) showed by using chloroform as the solvent instead of acetic acid that epoxy derivatives of olefins are readily obtainable by the use of peracetic acid, and that these oxides were converted into monoacetates and diacetates upon standing with acetic acid.

Hydrogen Peroxide and Acetic Acid

Hilditch (24) reported that acetic acid and hydrogen peroxide would convert an olefin into the corresponding glycol by allowing a reaction mixture of these components to stand for several days. Scanlon and Swern (25) improved this proceedure by first heating a mixture of 30 per cent hydrogen peroxide and acetic acid together for one hour at 85° and then cooling to 25°. The heat evolved on addition of the olefin is about sufficient to heat the mixture up to 75° and the mixture is kept at that temperature for about four to five hours to complete the reaction. None of these workers mention that what they are probably working with is peracetic acid. Hydrogen peroxide and acetic acid will react with each other to give peracetic acid. The reaction is a reversible one and equilibrium at room temperatures with 1 per cent sulfuric acid as the catalyst is obtained only after 56 hours. The equilibrium point lies well over on the peracetic acid side. However, equilibrium is quickly obtained at a temperature of 70-80° as Smit has shown (26).

Sodium Hydrogen Peroxide

Weitz and Scheffer (27) showed in 1921 that alkaline hydrogen peroxide was an excellent agent to convert a double bond to the oxide if it was alpha-beta to a carbonyl group. Such compounds as mesityl oxide, benzalanthrone, and benzylacetone are readily converted into the epoxy derivatives by this reagent. The reaction has been improved by Kohler (28) and Fieser (29), and for sensitive compounds hydrogen peroxide in alcohol or dioxane along with a little sodium bicarbonate is used and gives practically quantitative yields. In this manner Fieser synthesized 2-methyl-1,4-naphthoquinone oxide. A simpler methyl quinone, 2,6-dimethylquinone is not known to react with perbenzoic acid, and in general the peracids react very slowly if at all with carbob-carbon double bonds alpha-beta to carbonyl groups of

any sort. Alkaline hydrogen peroxide has little if any effect on the straight olefins which react with perbenzoic acid, so the uses of these reagents do not overlap in the formation of epoxy compounds.

The Prevost Reaction

In 1933 Charles Prevost reported that the following reaction would take place with benzene as the solvent to give good yields:

$$2C_6H_5CO_2Ag + I_2 + R\text{-}CH{=}CH_2 \rightarrow 2AgI + R\text{-}CH(OCOC_6H_5)\text{-}CH_2OCOC_6H_5.$$

The reactants are used in the molecular proportions indicated in the above equation, and about 300 cc. of anhydrous benzene is used as a solvent for a tenth of a mole of the olefin. In some cases the reaction proceeds almost instantly in the cold, and in others the mixture may be refluxed for fifty hours to complete the reaction.

The scope of the reaction is quite wide, and all varieties of olefins appear to react. Chlorine and bromine may be used in place of the iodine, (32) and it appears that almost any silver salt of a carboxylic acid can be used. Solvents such as dry benzene, chloroform, carbontetrachloride and so on have been used. Yields seem to be the best if silver benzoate, iodine and the olefin are used in a benzene solution. The following yields were reported by Prevost (31): $R\text{-}CH{=}CH_2$, 90% or above; $R\text{-}CH{=}CH_2\text{-}OCOC_6H_5$, 70% and up; $R\text{-}CH{=}CHCO_2C_2H_5$, 35% and better. Compounds such as $R\text{-}CH{=}CHBr$, $Ph_2C{=}CH_2$, and $R\text{-}CH{=}CH\text{-}CH{=}CH\text{-}R$ will also react.

The mechanism of the reaction has been nicely substantiated by the isolation of intermediates and appears to go in three stages according to the following equations: (33)

1. $2R\text{-}CO_2Ag + X_2 \rightarrow AgX + Ag(R\text{-}CO_2)_2^- + X^+$

2. $Ag(R\text{-}CO_2)_2X + R'\text{-}CH{=}CH\text{-}R' \rightarrow R\text{-}CO_2Ag + R'\text{-}CH(OCOR)\text{-}CH(X)R'$

3. $R\text{-}CO_2Ag + R'CH(OCOR)\text{-}CH(X)R' \rightarrow AgX + R'\text{-}CH(OCOR)\text{-}CH(OCOR)R'$

The final result is <u>trans</u> addition to the double bond, and the product is preponderantly or exclusively the <u>trans</u> form. Of all the ethylenes studied, ethylene itself seems to react the slowest. Biallyl gives the tetrabenzoate. With asymmetrical ethylenes the halogen is found to add to the carbon atom with the least substitution; that is the more negative carbon atom attracts the positive halogen indicated in the first equation. Thus styrene forms $C_6H_5\text{-}CH(OCOC_6H_5)\text{-}CH_2I$. The silver iodine benzoate complex is not believed to add in a 1,4 manner to butadiene since the unsaturated 1,2-dibenzoate may be readily obtained.

Applications of this reaction have not been too numerous. Some aldehydes have been prepared from the glycols formed from the hydrolysis of the dibenzoate by splitting the glycols with lead tetraacetate (34). Compounds of the general formula RCHOHCHOHCHOHR' may be made by the action of a Grignard on the ester formed from an α,β unsaturated acid (35).

BIBLIOGRAPHY

(1) Boeseken - Univ. of Delft- Rec. trav. chim. 47, 683, (1928)
(2) Kuhn and Ebel, Ber., 58, 921, (1925).
(3) Wagner, Ber., 21, 1230, (1888).
(4) Maan, Rec. trav. chim., 48, 332, (1929),
(5) Criegee,-Univ. of Marburg, Ger. - Ann., 522, 75, (1936).
(6) Hoffman, Ber., 45, 3329, (1912); Ber., 46, 1657, (1913).
(7) Braun, J.A.C.S., 51, 228, (1929).
(8) Miles and Sussman - M.I.T.-, J.A.C.S., 58, 1302, (1936). ibid, 59, 2342, (1937).
(9) Miles, Kurz, and Anslow, J.A.C.S., 59, 543, (1937).
(10) Serini and Logemann, Ber., 71, 1362, (1938).
Serini, Logemann, and Hildebrand, ibid,, 72, 391, (1939) .
(11) Reich, Sutter, and Reichstein, Helv. Chim Acta., 23, 170
Butenandt, Schmidt-Thome, and Paul, Ber., 72, 1112, (1939).
(12) H. Fischer and Eckoldt, Ann., 544, 138, (1940).
(13) Prileschajew, Ber., 42, 4811 (1909).
(14) Miles and Cliff, J.A.C.S., 55, 349, 352 (1933).
(15) Böhme, Ber., 70; 379 (1937).
(16) "Organic Syntheses", Col. Vol,, 2nd Ed., 431 (1941).
(17) Boeseken, Rec. trav. chim., 45, 838 (1926).
(18) Bodedorf, Arciv. Pharm., 268, 491 (1930).
(19) Pumerer, Ber., 62, 1411 (1929).
(20) Sturrman, Proc. Acad. Sci. Amsterdam, 38, 450 (1935); see C.A. 29, 4657[8] (1935).
(21) Wilson and Lucas, J.A.C.S., 58, 2396 (1936).
(22) Boeseken, Rec. trav. chim., 47, 683 (1928).
(23) Arbusow and Michailow, J. Prakt. Chem., 127, 1-15, 92-102 (1930).
(24) Hilditch, J. Chem. Soc., 1828 (1926).
(25) Scanlon and Swern, J.A.C.S., 62, 2305 (1940).
(26) Smit, Rec. trav. chim., 49, 675 (1930).
(27) Weitz and Scheffer, Ber., 54, 2329 (1921).
(28) Kohler, Richtmeyer, and Hester, J.A.C.S., 53, 213 (1931).
(29) Fieser, Campbell, Fry and Gates, ibid., 61, 3216 (1939).
(30) Prevost, Compt. rend., 196, 1129 (1933).
(31) Prevost, ibid., 197, 1661 (1933).
(32) Prevost and Wieman, ibid, 204, 989 (1937).
(33) Prevost, Atti X° Congr. Intern. Chim., 3, 318, (1939).
(34) Hershberg, Helv. Chim. Acta., 17, 351 (1934).
(35) Prevost and Losson, Compt. rend., 198, 659 (1934).

Reported by Jack Mills and John S. Meek
April 7, 1943

OXIDATIONS WITH LEAD TETRACETATE

In 1920 O. Dimroth demonstrated the possibility of using lead tetraacetate in place of lead dioxide for the oxidation of certain hydroquinones to quinones. Unlike lead dioxide, lead tetraacetate can be prepared easily in a pure state and can therefore be used in calculated amounts. More important, however, is that it is soluble in many organic solvents (acetic acid, benzene, chloroform, tetrachloroethane, nitrobenzene); thus through its use difficulties due to the uncertainty of the particle size and surface character of the oxixizing agent are obviated. Lead tetraacetate has made possible many oxidations which run either very poorly or not at all with other oxidizing agents.

In an oxidation reaction lead tetraacetate may be considered as splitting off two acetoxy groups. These may (1) add to a carbon to carbon double bond, (2) replace an activated hydrogen atom, or (3) unite with two hydrogen atoms causing dehydrogenation. The dehydrogenation may or may not be accompanied by the cleavage of a carbon to carbon bond; the glycol-cleavage is a significant example of dehydrogenation with cleavage.

<u>Dehydrogenations</u>

Because of its extremely high oxidation potential lead tetraacetate can convert all known hydroquinones to the corresponding quinones and all known leuco forms of dyes to the colored forms. It has made possible the preparation with excellent yields of the otherwise difficult to prepare di- and triquinones (the triquinones only in solution, however) of the anthraquinone and naphthazarinquinone series.

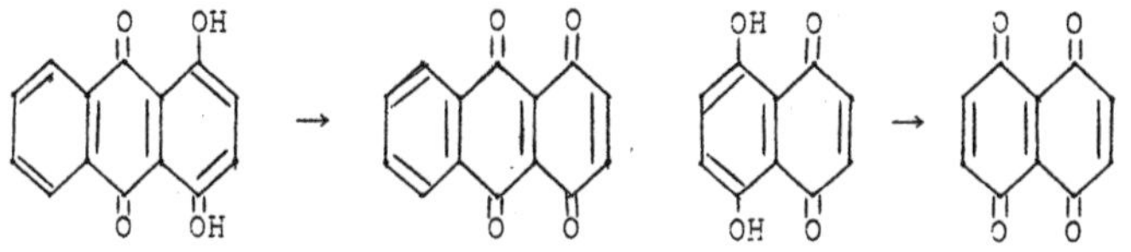

The dehydrogenation of hydroaromatic compounds with lead tetraacetate is not always smooth. Δ^{2}-Dihydronaphthalene can be converted to naphthalene with a 70% yield, but with the Δ^{1}-isomer addition of acetoxy groups to the double bond rather than dehydrogenation occurs. Tetralin, cyclohexene, and cyclohexadiene cannot be oxidized to naphthalene and benzene respectively. The dehydrogenation of 9, 10-dihydroanthracene gives only a 25% yield of anthracene, because anthracene itself undergoes substitution.

In general dehydrogenations of hetrocyclic compounds seem to proceed well. The product obtained, however, may depend upon the solvent used. The dihydroacridine derivative (I) gives (II) in acetic acid and (III) in benzene.

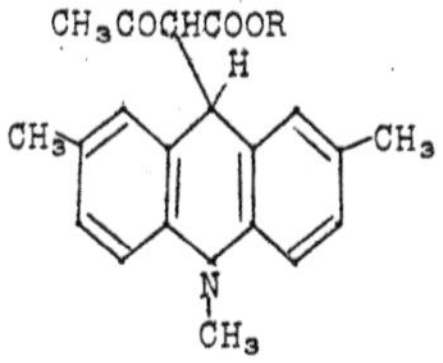
CH3COCHCOOR
H
CH3
CH3
N
CH3

I

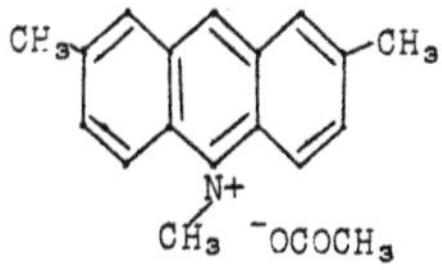
CH3
CH3
N+
CH3
-OCOCH3

II

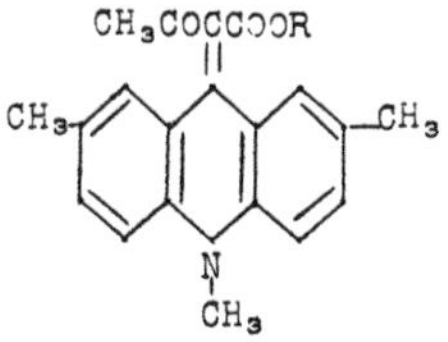
CH3COCCOOR
CH3
CH3
N
CH3

III

The dehydrogenation of alcohols to aldehydes and ketones can usually be carried out in benzene solution only. The presence of pyridine to combine with the acetic acid formed by the reaction seems to be favorable. Since the reaction has a velocity of a much smaller order of magnitude than that of the dehydrogenative cleavage of glycols, it is not a competing reaction in the glycol-cleavage.

Substitution of -H by $-OCOCH_3$

With lead tetraacetate the direct substitution of an activated hydrogen atom by a hydroxyl group can often be effected. Since the hydroxyl group is acetylated it is protected against further oxidation. Naphthalene is only slowly attacked, but anthracene yields anthranol acetate (IV) and oxanthrol acetate (V). 1,2-Benzanthracene reacts at 100° in acetic acid to give a 52% yield of (VI). 1,2,5,6-Dibenzanthracene (VII), however, is stable to oxidation perhaps due to steric hindrance; by virtue of this fact it can be separated from chrysene (VIII) which is more easily oxidized. 3,4-Benzpyrene in benzene solution at 20° gives an 80% yield of (IX) in 30 minutes.

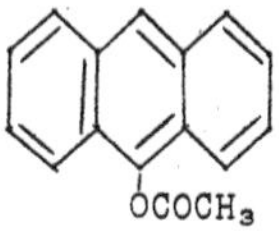
OCOCH3

IV

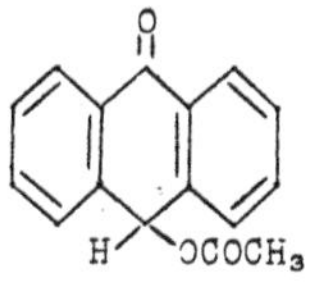
O
H
OCOCH3

V

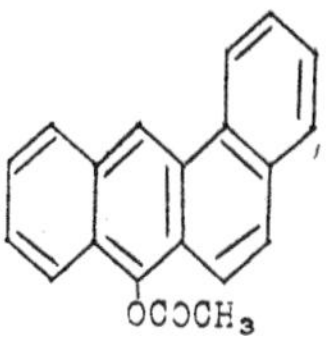
OCOCH3

VI

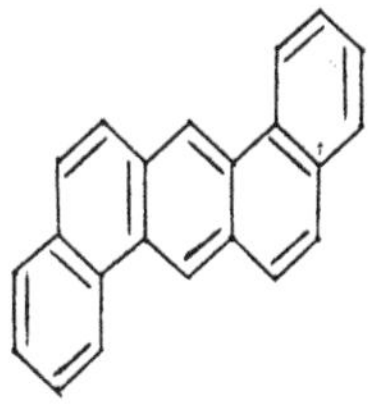

VII

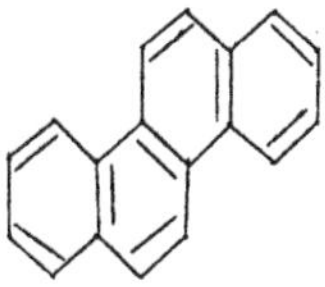

VIII

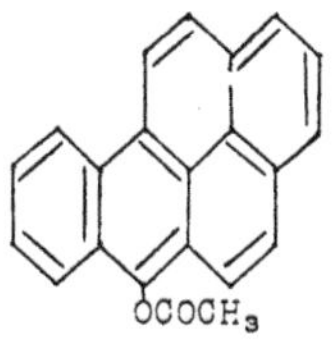
OCOCH3

IX

That substituents in an aromatic nucleus influence the orientation of the acetoxy groups is shown by the synthesis of the naphthoquinone derivative (XI) from (X). A systematic investigation of the influence of substituents does not appear to have been made.

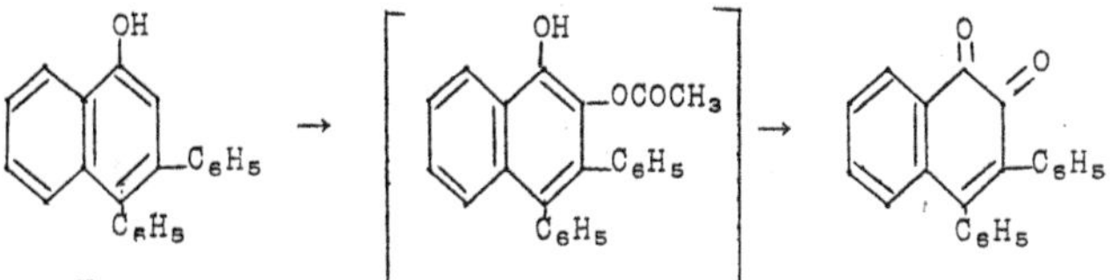

X XI

Saturated side chains are attacked in the α-position even more readily than nuclear hydrogen atoms. Thus toluene, diphenyl methane, and triphenyl methane are converted with increasing ease into the acetates of benzyl alcohol, benzhydrol, and triphenyl carbinol, respectively. Tetralin and acenaphthene give products substituted in the saturated ring only:

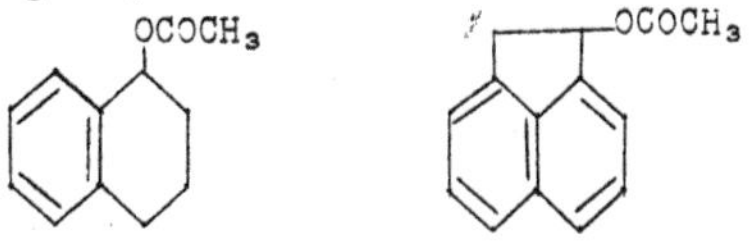

Methylene groups adjacent to a double bond can also be substituted. In this case, however, addition to the double bond occurs at the same time. Cyclopentene, cyclohexene, and indene may be converted to the esters of the corresponding unsaturated alcohols with yields of about 20%. In contrast with these results camphene reacts smoothly to give a 75% yield of the enol acetate. Here unlike all other olefins thus far investigated a hydrogen on a carbon of the double bond is substituted.

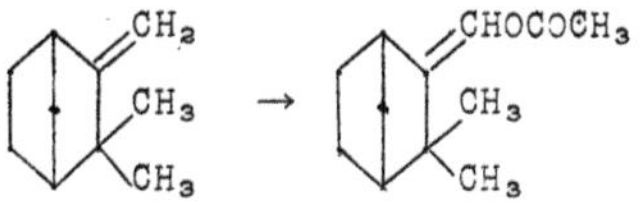

Many ketones, α-ketoacid esters, and α-dicarboxylic acid esters may be substituted in an active methyl or methylene group. Acetone yields the acetic acid esters of hydroxy and dihydroxy acetone; acetophenone gives the ester of benzoyl carbinol. Although the yields are not very good, this reaction has been used in work on the sex hormones to proceed from a $-COCH_3$ to a $-COCH_2OH$. Simple aliphatic acids, nitriles, and nitro compounds are not attacked.

<u>Addition of Acetoxy Groups to Double Bonds</u>

The action of lead tetraacetate on simple aliphatic olefins does not appear to have been investigated. Cycloolefins, as mentioned previously, undergo substitution in the α-position and addition

to the double bond simultaneously. Cyclopentene when oxidized at 50° gives an 18% yield of the diacetates of cis and trans cyclopentane 1,2-diol. Cyclohexene correspondingly gives a mixture of the cyclohexane 1,2-diol acetates. The pure trans diols were isolated from the oxidations of indene (18%) and Δ^1-dihydronaphthalene (45%), but the absence of the cis forms was not proved.

Addition of acetoxy groups to a double bond proceeds much more easily and smoothly when positive substituents, especially the methoxy group, are present. For example, anethole and isoeugenol methyl ether react in acetic acid solution at room temperature to give mixtures of the cis and trans diacetoxy adducts. Unsaturated compounds with negative substituents, on the other hand, are attacked only with difficulty. For example, crotonaldehyde reacts noticeably only at 80° to give a mixture of products.

The behavior of compounds with conjugated double bonds has been investigated rather thoroughly. The acetoxyl groups add predominantly 1,2 to dimethyl butadiene and cyclohexadiene, while with cyclopentadiene the addition is both 1,2 and 1,4; each product consists of a cis and trans form. The oxidation of cyclopentadiene is further complicated by the oxidation of an acetoxy group. The main products of the oxidation are the mixed esters (XII) and (XIII). Quite analogous reaction products result with lead tetra-salts of propionic and butyric acid, while lead tetrabenzoate gives with good yield the trans form of (XIV).

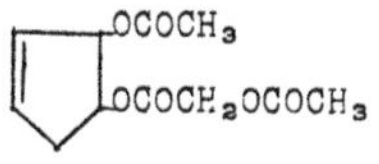

$OCOCH_3$
$OCOCH_2OCOCH_3$

XII

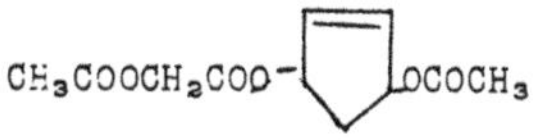

CH_3COOCH_2COO
$OCOCH_3$

XIII

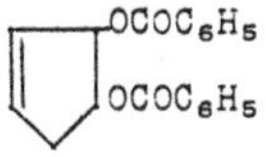

$OCOC_6H_5$
$OCOC_6H_5$

XIV

Glycol Cleavage

The most important reaction of lead tetraacetate is that by which a carbon chain is broken between two atoms each of which carries a free hydroxyl group. The reaction proceeds at room temperature in benzene, dilute acetic acid, or water to yield aldehydes and ketones.

$$\begin{matrix} >C-OH \\ | \\ >C-OH \end{matrix} + Pb(OCOCH_3)_4 \rightarrow \begin{matrix} >C=O \\ \\ >C=O \end{matrix} + Pb(OCOCH_3)_2 + 2CH_3COOH$$

Thus one can obtain glyoxalic acid ester from tartaric acid ester. Dihydroxy stearic acid can serve as the starting material for pelargonic aldehyde and the half aldehyde of azelaic acid. In many cases it is possible to proceed from the olefin by adding two hydroxyl groups to the double bond by known methods and then oxidizing with lead tetraacetate. In this manner the same cleavage of a double bond as one would obtain from an ozonization is achieved, and the yields are usually much better.

The glycol cleavage has special meaning the the preparation of rare sugars and sugar derivatives. For example by this method it

is very simple to prepare d-glyceraldehyde from mannose.

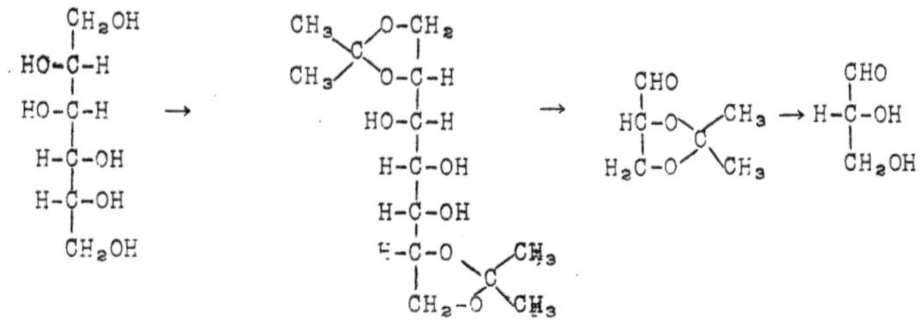

If a cyclic glycol is subjected to oxidation with lead tetraacetate a dialdehyde, a diketone, or a ketoaldehyde is obtained. This method has been used with considerable success in the preparation of many new compounds. The yields vary considerably, because of the instability of the products. In some cases a dialdehyde cannot be isolated, because it is readily oxidized to the dicarboxylic acid.

The oxidative cleavage is not confined to 1,2 glycols. α-Hydroxy acids react just as smoothly to give an aldehyde or ketone and carbon dioxide. The cleavage of α-hydroxy ketones does not proceed so well. 1,2-Diamines, 1,2-aminoalcohols, and α-aminoacids have all been cleaved with lead tetraacetate. The imine which would be expected as the product, however, is usually dehydrogenated to a nitrile.

$$C_6H_5CH(NH_2)COOH \rightarrow C_6H_5CH{=}NH \rightarrow C_6H_5C{\equiv}N$$

BIBLIOGRAPHY

Criegee, Angew. Chem., 53, 321 (1940)

Reported by E. W. Meynert
April 14, 1943

SUBSTITUTION REACTIONS AT BRIDGEHEAD CARBON ATOMS

Bartlett and Coworkers, Harvard University

The mechanism of substitution reactions at a saturated carbon atom, according to the Hughes-Ingold Theory, has been adequately discussed in several previous seminars. A bridgehead carbon atom, that is, a carbon atom occurring at the junction of three rings, presents a unique type of saturated carbon atom. It is structurally incapable of undergoing a Walden inversion, hence could not be party to a bimolecular replacement of a substituent group. According to Bredt's Rule, a bridgehead carbon atom cannot form a double bond; only one apparent exception to this rule has been found. A recent confirmation of the hypothesis is the case of bicyclo-(2,2,2)-octanedione-2,6 (I). This diketone, having a hydrogen atom between two carbonyls, gives no color with ferric chloride, no copper salt, is no more soluble in aqueous alkali than in pure water, and in the Zerewitinoff determination consumes two moles of methylmagnesium iodide, but liberates only 0.15 mole of hydrogen.

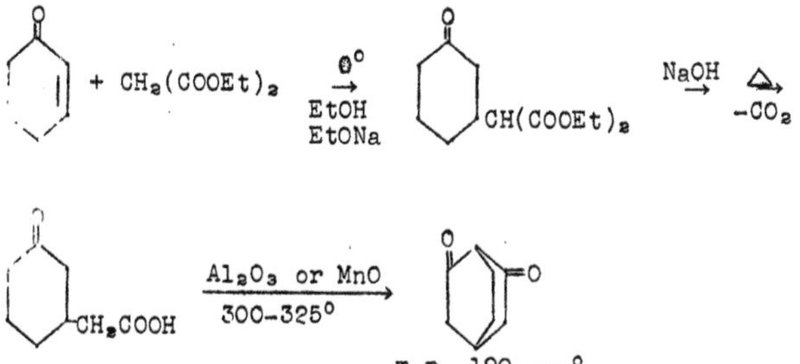
+ $CH_2(COOEt)_2$
0°
EtOH
EtONa
$CH(COOEt)_2$
NaOH
$-CO_2$
CH_2COOH
Al_2O_3 or MnO
300-325°

m.p. 190-191°

I

A great deal of evidence has been presented by Bartlett to show that these bridgehead carbon atoms are also incapable of forming carbonium ions. This evidence is the behavior of a number of compounds with negative substituents at the bridgehead toward conditions under which an analogous acylic compound would immediately react. The compounds were of types II, III, and IV, and the results are summarized in the table.

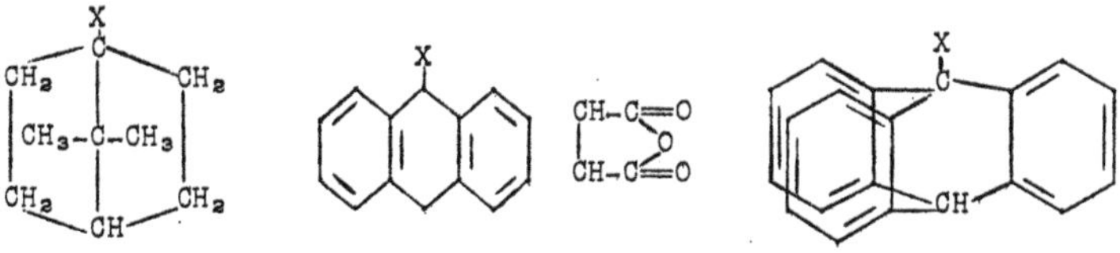
X
CH_2
CH_3-C-CH_3
CH

II III IV

Compound	Treatment	Result, Remarks	Ref.
II, X = $-CONH_2$	Hoffman amide degradation, hydrolysis.	II, X = NH_2 Hydrolysis of methyl carbamate difficult	5
II, X = $-NH_2$	H_2SO_4 + $NaNO_2$	II, X = OH (Proof of structure indirect)	5
	Nitrosyl chloride	II, X = Cl	5
II, X = -OH	PCl_5 or HBr	gives only an oxonium salt.	5
	$SOCl_2$	sulfite only.	5
	Lithium iodide on p-toluenesulfonic ester in acetone.	No reaction.	5
II, X = -Cl	30% KOH in 80% EtOH 24 hrs.	No reaction.	5
	$AgNO_3$ in alcohol, 48 hrs.	No opalescence.	5
	Mg + Et_2O	No reaction.	5
III, X = Br	KOH in EtOH, reflux 15-18 hrs.	No ionized Br., anhydride ring opened, cis-acid → trans-acid.	1,3
X = $-OOCCH_3$	KOH in EtOH, boil 30 min.	Anthrone + maleic anhydride	1
X= $-NHCOCH_3$	35% KOH, Boil 16 hrs.	N acetylated free (trans) acid	3
X = -OH	Alkali, heat	Anthraquinone or dihydrodianthrone	3
X = $-NHC(=O)OCH_3$	20% NaOH, basic hydrolysis	X = $-NH_2$	
III, X = NH_2	$NaNO_2$ + H_2SO_4	III, X = -OH(?)	
III, X = Br	Na metal + alcohol	About 10% of reduced product, X = H.	3
	Ag powder, boiled 16 days in acetone or xylene.	No reaction.	3
	$AgNO_3$ in alcohol	Silver salt of bromo-acid.	3

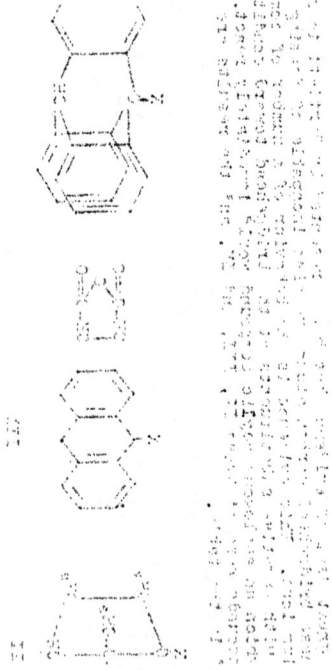

Compound	Treatment	Result, Remarks	Ref.
II, X = $-CONH_2$	Hoffman amide degradation, hydrolysis.	II, X = NH_2 Hydrolysis of methyl carbamate difficult	5
II, X = $-NH_2$	H_2SO_4 + $NaNO_2$	II, X = OH (Proof of structure indirect)	5
	Nitrosyl chloride	II, X = Cl	5
II, X = -OH	PCl_5 <u>or</u> HBr	gives only an oxonium salt.	5
	$SOCl_2$	sulfite only.	5
	Lithium iodide on <u>p</u>-toluenesulfonic ester in acetone.	No reaction.	5
II, X = -Cl	30% KOH in 80% EtOH 24 hrs.	No reaction.	5
	$AgNO_3$ in alcohol, 48 hrs.	No opalescence.	5
	Mg + Et_2O	No reaction.	5
III, X = Br	KOH in EtOH, reflux 15-18 hrs.	No ionized Br., anhydride ring opened, <u>cis</u>-acid → <u>trans</u>-acid.	1,3
X = $-OOCCH_3$	KOH in EtOH, boil 30 min.	Anthrone + maleic anhydride	1
X = $-NHCOCH_3$	35% KOH, Boil 16 hrs.	N acetylated free (<u>trans</u>) acid	3
X = -OH	Alkali, heat	Anthraquinone or dihydrodianthrone	3
X = $-NHC(=O)OCH_3$	20% NaOH, basic hydrolysis	X = $-NH_2$	
III, X = NH_2	$NaNO_2$ + H_2SO_4	III, X = -OH(?)	
III, X = Br	Na metal + alcohol	About 10% of reduced product, X = H.	3
	Ag powder, boiled 16 days in acetone <u>or</u> xylene.	No reaction.	3
	$AgNO_3$ in alcohol	Silver salt of bromo-acid.	3

Compound	Treatment	Result, Remarks	Ref.
IV, X = H	Phenyltriisopropyl potassium in ether.	No reaction.	7
	Chlorination, bromination.	No reaction	7
	Oxidation with CrO_3 in ACOH.	Anthraquinone + CO_2. No intermediate isolated .	7
	Maleic anhydride.	No reaction.	7

With regard to those replacemtns which will occur, it appears justifiable to conclude that they must take place by means of an intramoleculer process. The case of the replacements which do not occur is somewhat less certain. Two things are definitely known: elimination of water or HCl is impossible (Bredt's Rule) and the compounds are incapable of undergoing a Wagner-Meerwein type of rearrangement to give a structure capable of such elimination, since for this reaction to occur, inversion at the carbon atom holding the functional group is necessary.[2]

Two possible explanations present themselves. One is that, in order for replacement to occur, a Walden inversion must take place. This though possibly correct, is not an attractive explanation, since it invalidates some very satisfactory features of the theory of dual mechanism of solvolytic reactions.

If, however, it is assumed that in the unimolecular process, the rate determining step is the "pull" of the solvent molecules on the substituent group, rather than a contribution (to the transition energy) of a donor molecule, then it is only necessary to explain why the bridgehead carbon atom cannot form a positive ion, while the diethyl-_t_-butyl carbinyl cation, for example, may do so. A considerable amount of evidence indicates that the carbonium ion is either permanently planar, or oscillates rapidly between the _d_- and _d_-tetrahedral forms; it has been shown that trimethyl boron, which is analogous to the carbonium ion with a sextet of electrons, actually has such a planar configuration.[8] If then, in the case of the 1-apocamphane derivatives, a planar configuration is necessary for the formation of the ion, then the formation of this ion will be attended by a strain amounting to at least 22.5 kcal, which is sufficient to account for the experimental observations.

An even more convincing case is that of compounds of types III and IV, which are (structurally) triphenyl methane derivatives. In these cases, the stabilizing effect of the resonance forms of the triphenylmethyl cations (and free radicals) is unattainable, since there is no conceivable way in which the phenyl group could become co-planar with the central carbon atom, hence a marked increase in the energy barrier to the formation of the ion is to be expected.

Flow Sheets for Syntheses of Compounds Discussed

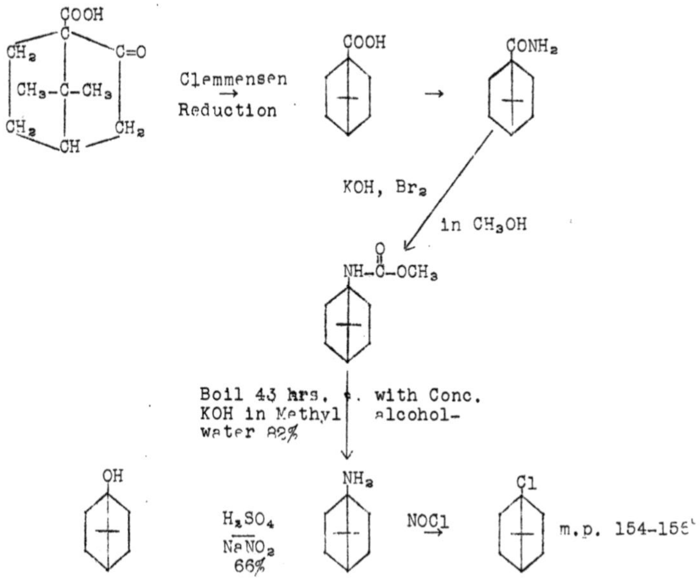
COOH
C
CH2
C=O
CH3-C-CH3
CH2
CH2
CH
Clemmensen
Reduction
COOH
CONH2
KOH, Br2
in CH3OH
NH-C-OCH3
Boil 43 hrs. with Conc.
KOH in Methyl alcohol-
water 82%
OH
H2SO4
NaNO2
66%
NH2
NOCl
Cl
m.p. 154-155°

1-apocamphanol

1-apocamphylchloride

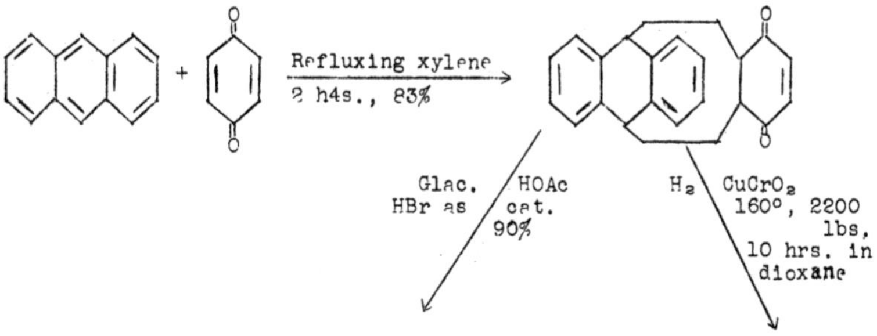
Refluxing xylene
2 h4s., 83%
Glac. HOAc
HBr as cat.
90%
H2 CuCrO2
160°, 2200
lbs.
10 hrs. in
dioxane

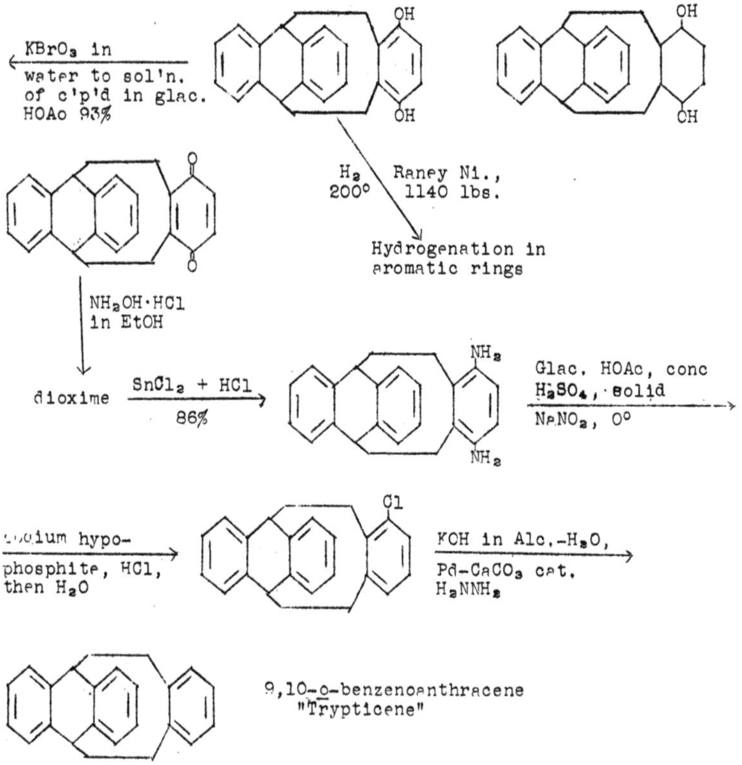
OH
OH
OH
OH
KBrO3 in
water to sol'n.
of c'p'd in glac.
HOAc 93%
O
O
H2 200°
Raney Ni., 1140 lbs.
Hydrogenation in
aromatic rings
NH2OH·HCl
in EtOH
dioxime
SnCl2 + HCl
86%
NH2
NH2
Glac. HOAc, conc
H2SO4, solid
NaNO2, 0°
sodium hypo-
phosphite, HCl,
then H2O
Cl
KOH in Alc.-H2O,
Pd-CaCO3 cat.
H2NNH2
9,10-o-benzenoanthracene
"Trypticene"

Bibliography

(1) Barnett, Goodway, Higgins, and Lawrence, J. Chem. Soc., 1224 (1934).
(2) Bartlett and Bowley, J. Am. Chem. Soc., 60, 2146 (1938).
Bartlett and Pöckel, ibid., 59, 820 (1937).
(3) Bartlett and Cohen, ibid., 62, 1183 (1940).
(4) Bartlett and Jones, ibid., 64, 1837 (1942).
(5) Bartlett and Knox, ibid., 61, 3184 (1939).
(6) Bartlett, Ryan, and Cohen, ibid., 64, 2649 (1942).
(7) Bartlett and Woods, ibid., 62, 2933 (1940).
(8) Bredt, Ann., 437, 1 (1924).
(9) Levy and Brockway, J. Am. Chem. Soc., 59, 2085 (1937).

Presented by John A. McBride, April 14, 1943

THE DIELS-ALDER DIENE SYNTHESIS[1]

Although isolated examples of the diene synthesis appear in the literature previous to 1925-1928, it was not until this time that extensive generalization of the reaction was undertaken. It has since been shown to be very wide in scope, and has come to include reactions of ethylenic compounds with enynes, dienynes, etc. The greater majority of such reactions may be represented by the following types:

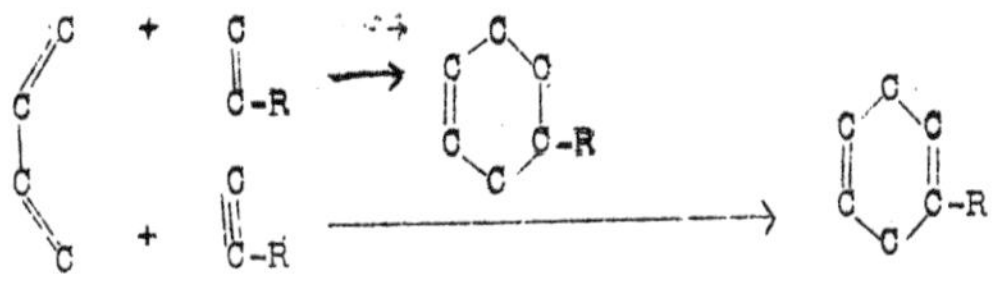

diene   dienophile   adduct

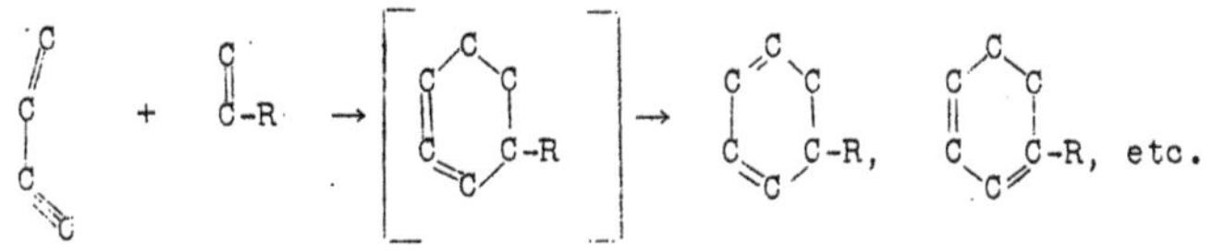

enyne   dienophile   product

R in the general equation is usually of the type $-\underset{\text{O}}{\text{C}}-\text{R}'$, since this group tends to activate the olefinic bond. Other groups may be employed however, and Joshel and Butz (2) have recently shown that ethylene itself will react. Usually other groups decrease the activity of the dienophile, requiring higher temperatures for reaction to occur. The extent of suppression of activity depends upon the type of dienophile.

A. Alkyl and aryl derivatives of acrolein, maleic anhydride, etc. decrease the activity of the dienophile only slightly.
B. In fused ring derivatives of acrolein, maleic anhydride, etc. increased unsaturation of the fused ring other than in the C=C-C=O (carbonylenic group) results in diminished activity. Thus phthalic acid has never been observed to add dienes.
C. Alkyl or aryl groups substituted in quinones reduce the activity of the C=C-C=O group to which they are attached, but have little effect on the other C=C-C=O group (IV, V, VI).
D. Quinones with fused rings resemble Class B. Anthraquinone adds no dienes, but naphthacenediquinone (I) does, since the net effect is the substitution of carbonyl groups on the C=C of the C=C-C=O group.

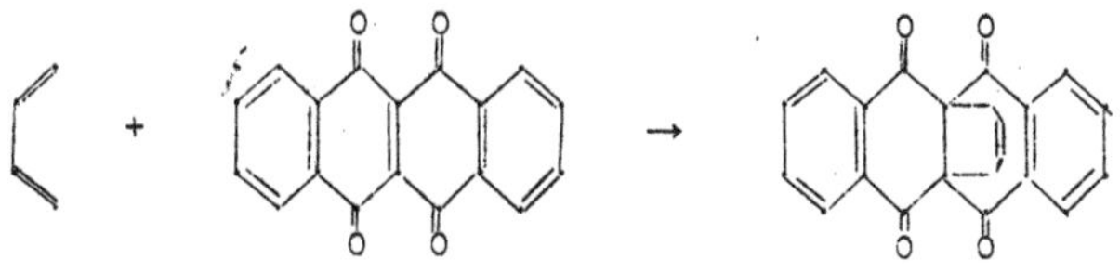

I

In general, dienes and dienophiles will react to form products possessing no angular, geminal, or spirane groupings on the carbon atoms which originated in the dienophilic double bond.

Procedures for the reaction vary widely as to solvent and temperature, depending upon the generators employed. Catalysts have been reported (3) but are seldom used.

<u>Non-carbonylenic Dienophiles</u>

This type of dienophile usually reacts less readily, requiring higher temperatures and sometimes an antioxidant to prevent polymerization of the diene. The substituents on the -C=C- group may be varied widely, however. For example, dihydrothiophene dioxide and vinyl-p-tolylsulfone (II) have been found to react with butadiene, as do 1-nitro-1-butene, vinyl formate, trichloroethylene, vinyl-p-tolylsulfide, crotyl alcohol, allyl iodide, and vinyl chloride with cyclopentadiene (III), to mention only a few.

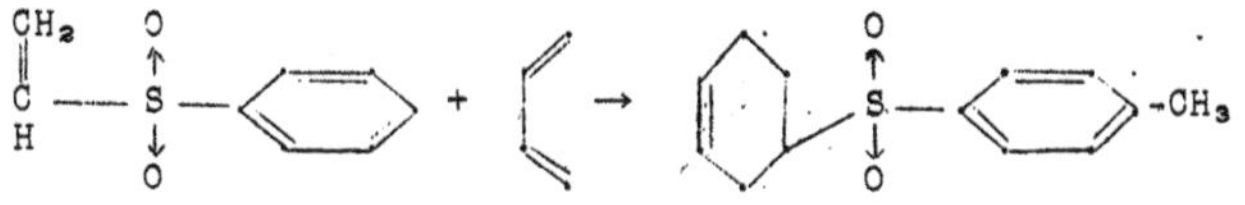

II

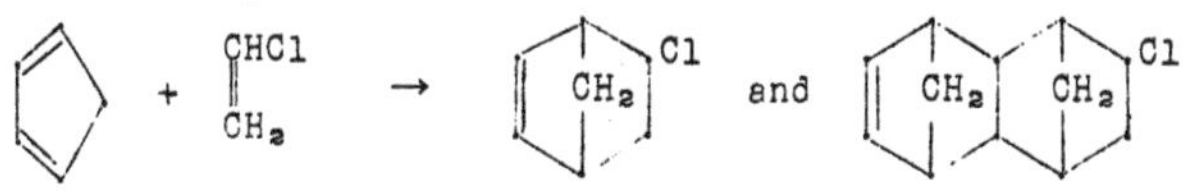

III

Secondary addition often occurs with cyclopentadiene, since it is quite reactive and the primary adduct, having the same type of structure as the dienophile, is nearly as active as the latter.

<u>Quinones as Dienophiles</u>

p-Quinones, especially those derived from or homologous with p-benzoquinone have been studied extensively. If either the diene or the dienophile is unsymmetrical, one or both of two products may be formed. p-Toluquinone (IV) adds one molecule of butadiene easily to form V, rather than the possible product possessing an angular methyl group. Furthermore, the addition of a second molecule of

butadiene to produce VI is more difficult, since both the fused ring and methyl side chain, which produces an angular methyl, deactivate the dienophile. The general tendency to avoid formation of an angular group is useful in predicting the products of a reaction.

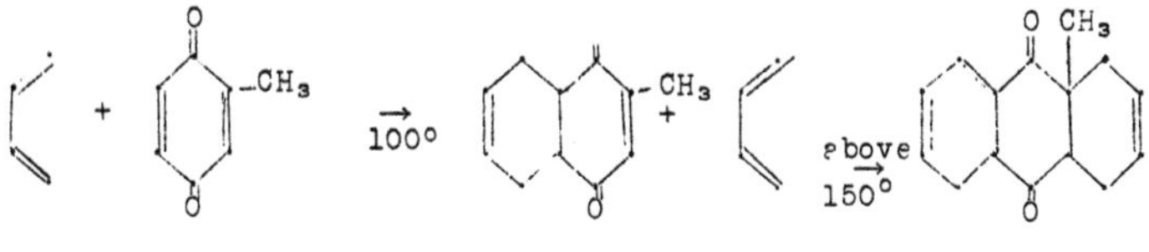

O
CH_3
100°
CH_3
+
above
150°
O CH_3

IV V VI

Fusion of a saturated ring to p-quinone, as in α-naphthoquinone, completely suppresses addition of a second molecule of diene. Further decrease in dienophilic activity is often noted when the fused benzene ring carries substituents such as hydroxyl or acetoxyl. Thus heating 5,7,8-triacetoxy-1,4-naphthoquinone with 2,3-dimethyl butadiene for twenty-seven hours is required to produce the same yield of adduct as is obtained in three hours from the 5,8-diacetoxy compound.

Bergmann, Haskelberg, and Bergmann (4) have found that quinone-type dienophiles often lead to a completely aromatic product (VII) if nitrobenzene is used as solvent. This is especially valuable where the adduct tends to decompose to its generators.

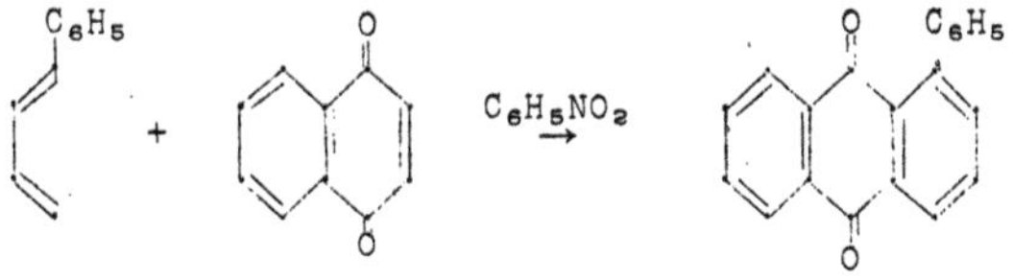

C_6H_5
+
O
$C_6H_5NO_2$
O
C_6H_5
O

VII

The tautomeric form (IX) of 2,4-dinitrobenzene-azo-p-phenol (VIII) is a nitrogen analog of p-quinone and adds dienes in a like manner.

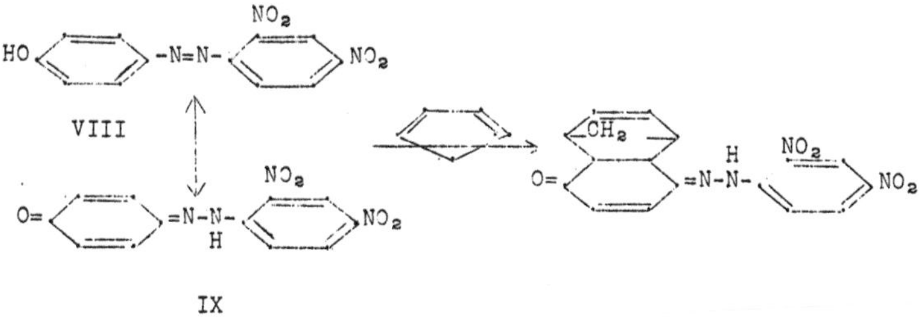

NO_2
HO
-N=N-
NO_2
NO_2
O=
=N-N-
H
NO_2
CH_2
O=
H
=N-N-
NO_2
NO_2

VIII

IX

o-Quinones have been less studied than p-quinones, since the simpler orthoquinones are easily decomposed by heat. 3,4- and 3-substituted 1,2-naphthoquinones (X) have been shown to add dienes slowly to give reduced 9,10-phenanthrenequinone derivatives (XI), the substituents in most cases being halogen, which is quite easily removed as hydrogen halide (XII).

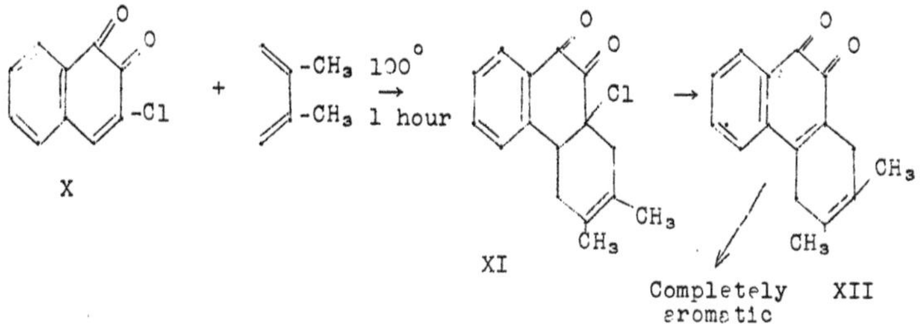
O
O
-Cl
X
+
$-CH_3$
$-CH_3$
100°
1 hour
O
O
Cl
CH_3
CH_3
XI
O
O
CH_3
CH_3
Completely aromatic
XII

In the phenanthrene quinone series, the 1,2- and 3,4- quinones add dienes, while the 9,10- do not.

Dienes.

The forms which the diene may take vary from the simplest, butadiene, to heterocyclics such as furan, and polynuclear systems such as 1,2,5,6-dibenzanthracene. Benzene and thiophene do not add dienophiles, but in compounds such as α-vinyl naphthalene one of the double bonds is a Kekulé double bond. Nitrogen heterocycles such as pyrrole react abnormally. Otherwise, the diene synthesis usually leads to the expected products. It should also be noted that it must be possible for the butadienoid system to exist in the cis-form if reaction is to occur.

The butadienes which have been most employed in diene syntheses are butadiene, piperylene, isoprene, 1,4-dimethylbutadiene, 2,3-dimethylbutadiene, and 1,4-diphenylbutadiene. Other aryl, alkyl, and alkoxy substituted butadienes have been used to a lesser extent. The ability of halogenated butadienoid systems to react in the diene synthesis is dependent upon the position of the halogen atoms. Dienes of the structure Cl-C=C-C=C or C=C(Cl)-C(Cl)=C do not add dienophiles, while dienes of the structure C=C(Cl)-C=C usually add dienophiles in the normal fashion. The use of hexatrienes and dienynes in the synthesis of condensed ring compounds, especially the total synthesis of the steroid nucleus has been discussed by J. W. Mecorney in a recent seminar (Jan. 20, 1943).

Homocyclic Dienes

The general reaction may be represented by the form

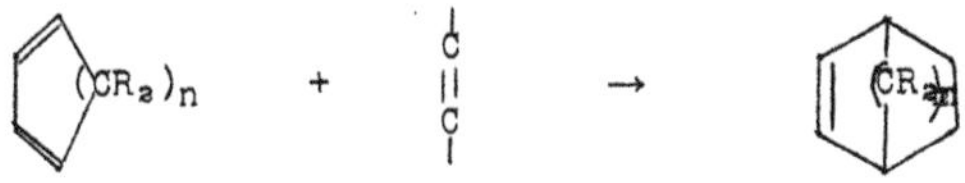
(CR₂)n
+
C
C
→
CR₂

where R may be hydrogen, oxygen, or an aryl or alkyl radical. The highest value of n studies is 3, in 1,3-cycloheptadiene.

The adducts of cyclopentadiene and 1,3-cyclopeptadiene distil at atmospheric pressure unchanged or else decompose into their generators,

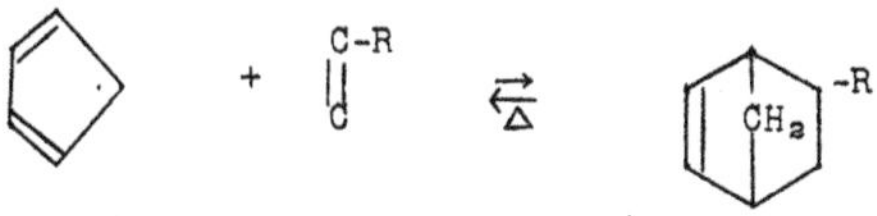
+
C-R
C
Δ
-R
CH₂

while those of cyclopentadienone usually lose CO to form a cyclohexadiene, which may be easily oxidized to the benzenoid form.

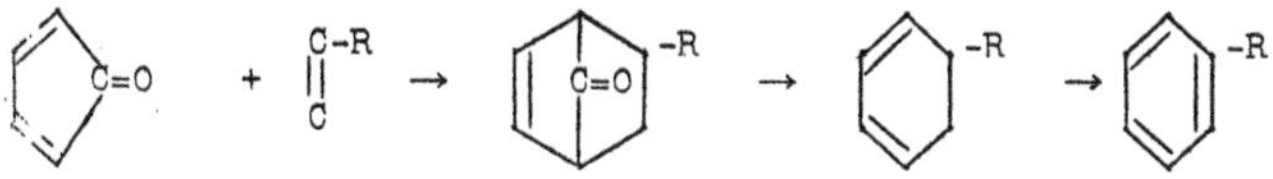
C=O
+
C-R
C
→
-R
C=O
→
-R
→
-R

Fusion of 1,3-cyclohexadiene-quinone adducts in air results in oxidation and loss of olefin to yield an aromatic quinone. Distillation of the adduct (XIV) from a 1,3-cyclohexadiene (XIII) and an olefinic dienophile usually produces a benzene derivative (XV) and the reduced dienophile (XVI).

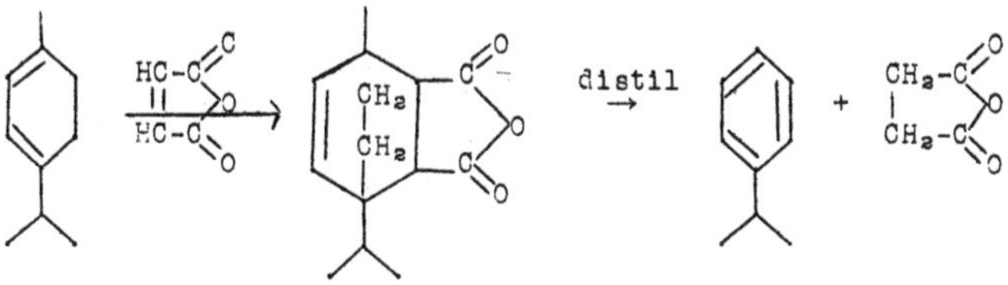
HC-C
HC-C
O
CH₂
CH₂
C
O
distil
→
+
CH₂-C
CH₂-C
O

XIII XIV XV XVI

The phenylated cyclopentadienes are highly reactive, adding inert dienophiles fairly readily as indicated by the reaction of tetracyclone (XVII) with benzonitrile, a compound not ordinarily thought of as a dienophile, to form pentaphenylpyridine (XVIII).

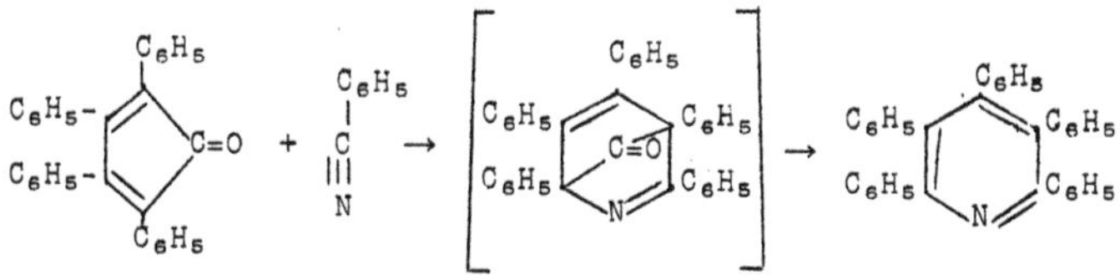

XVII XVIII + CO

Semicyclic and Dicyclic Dienes

Semicyclic and dicyclic dienes are quite active, and react to produce the expected products, even when this contains an angular methyl group. 6-Methoxy-1-vinyl-3,4-dihydronaphthalene (XIX) has been studied extensively because the products formed are important in sterol syntheses.

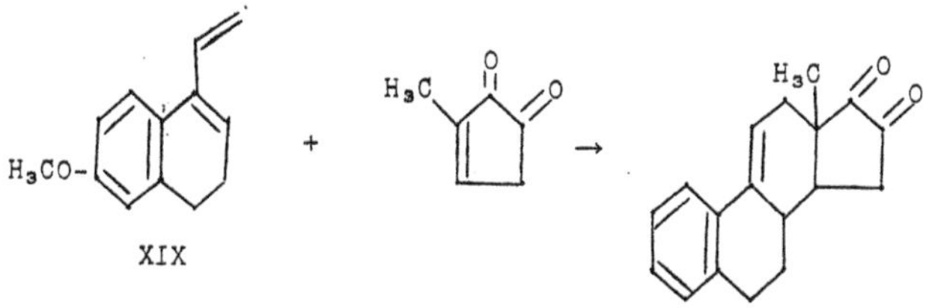

XIX

Terpenes and Terpenoid Bodies: Many terpenes react as dienes, even some which do not contain conjugated double bonds. But in such cases, isomerization probably occurs before the diene reaction.

Hultzch allowed maleic anhydride to react with dl-limonene, α-pinene, 3-carene and terpinolene, obtaining different products of indefinite composition for each. But when the same terpenes were treated with maleic acid, each yielded the same adduct. This adduct was found to be identical with the one obtained from 1,3-p-methadiene and maleic acid.

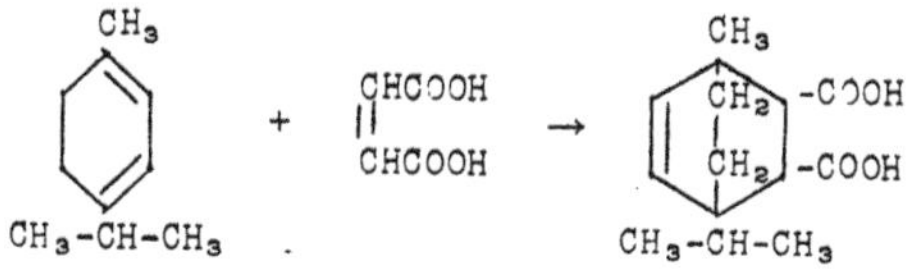

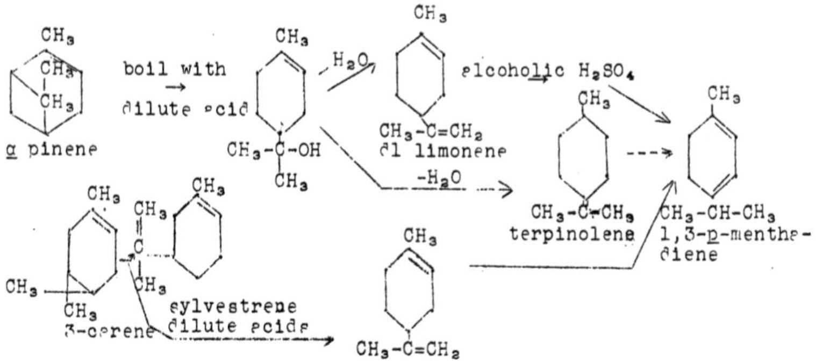
boil with
dilute acid
α pinene
$-H_2O$
alcoholic H_2SO_4
dl limonene
$-H_2O$
terpinolene
1,3-p-menthadiene
sylvestrene
3-carene
dilute acids

Thus it is reasonable to assume that each of the above terpenes will isomerize to 1,3-p-methadiene under the influence of maleic acid, whereas maleic anhydride is not sufficiently acidic to cause the same isomerization.

Dimerization: In 1895 Wallach prepared dl-limonene by the dimerization of isoprene at elevated temperature in the absence of oxygen and peroxides (reaction A.). Later Wagner - Jauregg found that if the reaction is run at room temperature, it gives rise to diprene (reaction B).

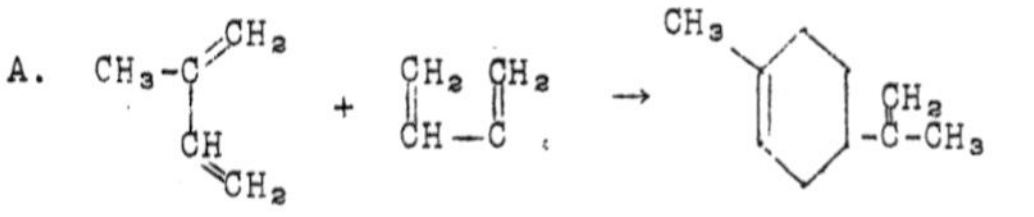

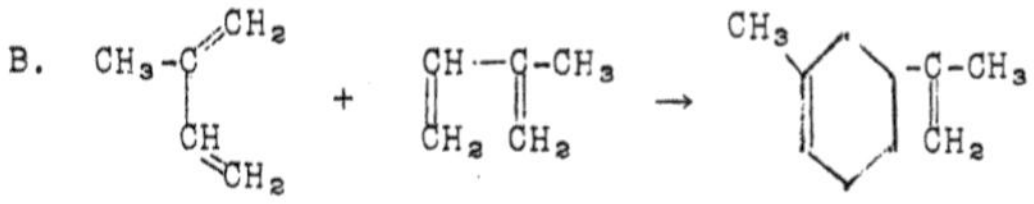

Many other examples have been found in which a compounds reacts both as diene and dienophile. Since such reactions often lead to trimers and tetramers, they are generally run in the presence of hydroquinone to avoid the formation of high molecular weight polymers.

In extension of the diene synthesis to the dimerization of other hydrocarbons, it was found that butadiene gave vinyl-3-cyclohexene when heated at 180°C. with hydroquinone:

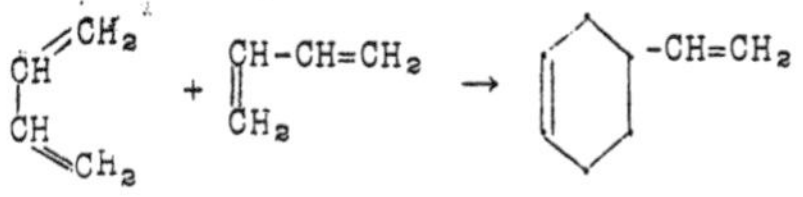

The product formed does not add dienophiles, thus rendering unlikely any proposed structure containing a conjugated system. Dehydrogenation gives ethylbenzene and styrene. Vinylcyclohexene-3 can further add butadiene at elevated temperature to give butadiene trimer.

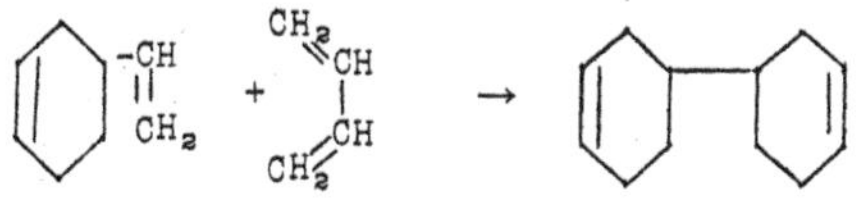
-CH
CH$_2$
CH$_2$
CH
CH
CH$_2$

<u>Dienic Acids and Derivatives</u>: Dienic acids in which the double bonds are conjugated can act as dienes in the diene synthesis. Esters of such acids, as well as their lactones and lactams, may react in the same fashion.

Allen and coworkers (5) allowed ethyl sorbate to react with benzoylethylene. No attempt was made to isolate the eight isomers from the oily ester formed, but rather it was hydrolyzed directly to the solid acids. The addition might have occurred so that the benzoyl group would appear either <u>ortho</u> or <u>meta</u> to the methyl group. Structure XX was shown to be correct by dehydrogenation and hydrolysis of some of the adduct (XX) to compound XXI which in turn was dehydrated to methanthraquinone (XXIII). The cyclization could only have occurred if the keto and acid groups were ortho to each other, thus confirming structure XX.

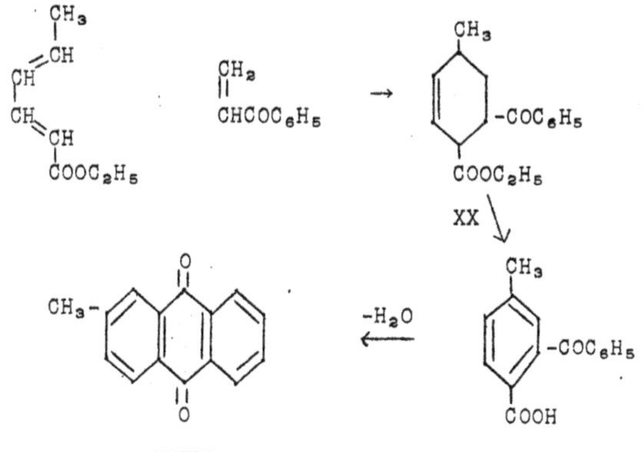
CH$_3$
CH
CH
CH
CH
COOC$_2$H$_5$
CH$_2$
CHCOC$_6$H$_5$
CH$_3$
-COC$_6$H$_5$
COOC$_2$H$_5$
XX
CH$_3$
-COC$_6$H$_5$
COOH
-H$_2$O
CH$_3$-
O
O

XXIII XXI

The addition of dienophiles to the dienic lactones which contain a six-membered lactone ring engenders products containing a lactone bridge. From such compounds carbon dioxide is lost readily. The general statement of the reaction depends upon whether olefinic or acetylenic dienophiles are used.

A. 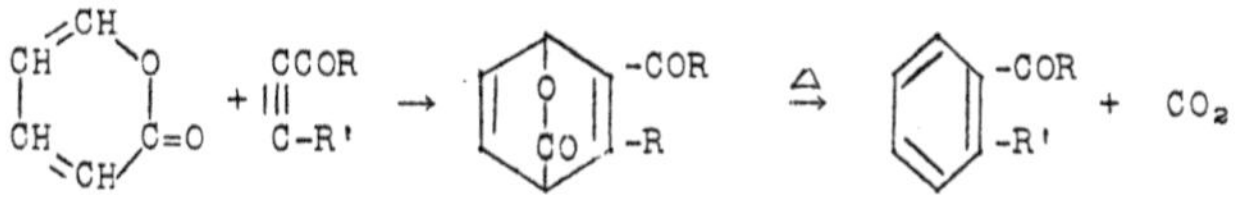

B. 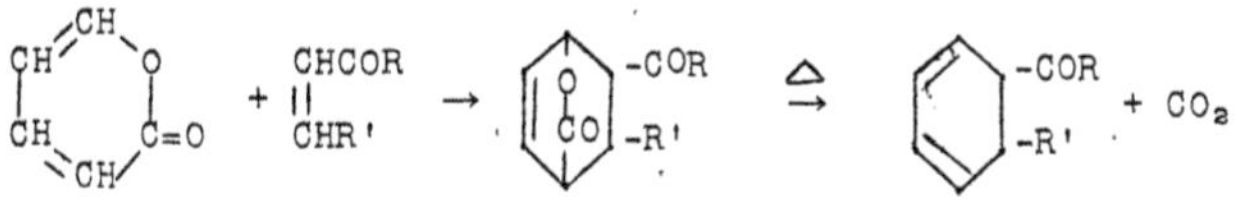

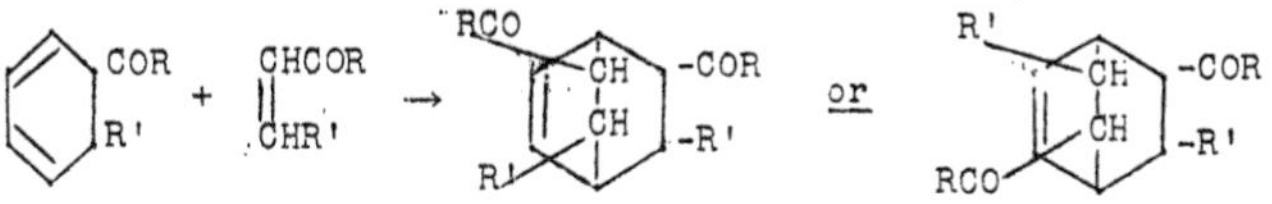

Aromatic Polynuclear Hydrocarbons: As a general rule, the polynuclear aromatic hydrocarbons which add dienophiles are those for which complete Kekulé structures cannot be drawn. One of the most familiar reactions of this type is the addition of dienophiles to anthracene according to the following general equation:

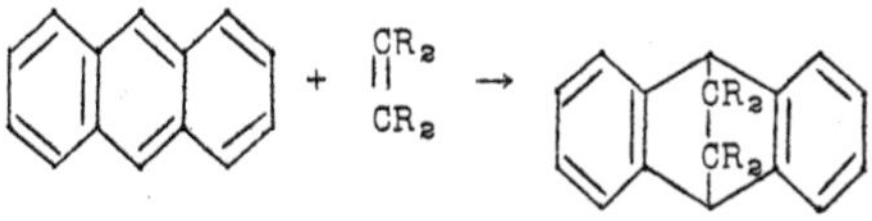

Reversal of the reaction upon heating may be expected however, with the regeneration of anthracene and an olefin. The extent of reversal is partially dependent upon the substituents in the diene and their position in the three rings. Electronegative substituents and alkyl radicals have little effect in the terminal rings, but if the former are situated in the meso position they strongly decrease the tendency for adduct formation. With alkyl groups in the 9,10 positions the addition generally is more rapid, whereas similarly located aryl groups tend to decrease the rate of addition.

Aromatic Hydrocarbons with Unsaturated Side Chains:

A. Vinyl Aromatics-- Another type of Polynuclear aromatic to be considered is that which contains an α,β-unsaturated side chain. Such compounds are homologues or benzologues of styrene and its derivatives. The conjugated system consisting of the unsaturated linkage in the side chain and one of the double bonds in the benzene ring will react with dienophiles to form a semicyclic double bond with respect to the former benzene ring, which is converted to a 1,3-cyclohexadiene ring.

Although equimolecular quantities of maleic anhydride on styrene itself gives a polymeric adduct, α-vinylnaphthylene reacts as shown below:

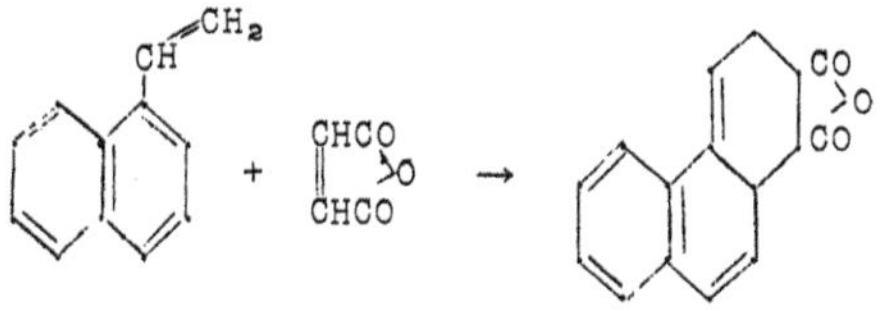

β-vinylnaphthalene adds maleic anhydride to give either compound XXIV or, should migration of the double bond occur, compound XXV.

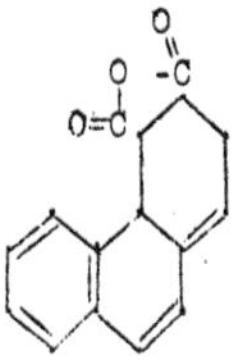

XXIV

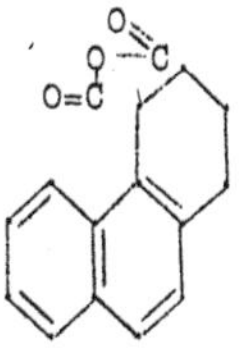

XXV

B. 9-Methyleneanthrone and Derivatives-- Partially hydrogenated benzanthrones are formed when 9-methyleneanthrone reacts with dienophiles. Since these adducts are readily oxidized to give benzanthrone, the addition reaction is often run in an oxidizing solvent such as nitrobenzene to give a completely aromatic product. For such reactions the necessary temperature range is 200-250°C.

The addition of cinnamic acid to 9-methyleneanthrone in boiling nitrobenzene gives Bz-1-phenylbenzanthrone, decarboxylation occurring during the synthesis.

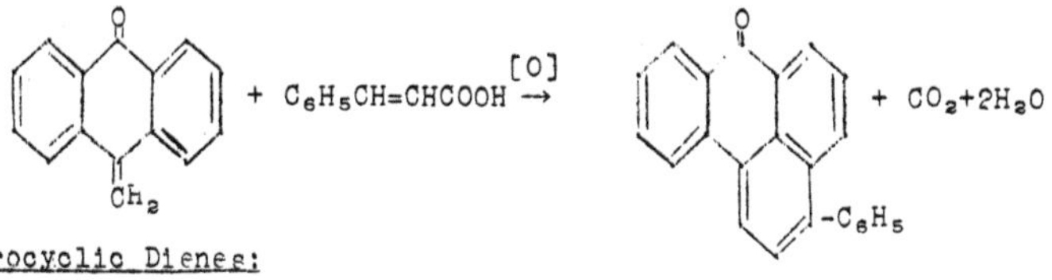

Heterocyclic Dienes:

A. Furan and Derivatives:-- The addition of dienophiles to furan and its derivatives generally results in an oxygen bridge compound. The partially hydrogenated adduct of sylvan (α methyl furan) and acetylene-dicarboxylic ester loses ethylene upon heating to give 3-methyl-3,4-furandicarboxylic ester.

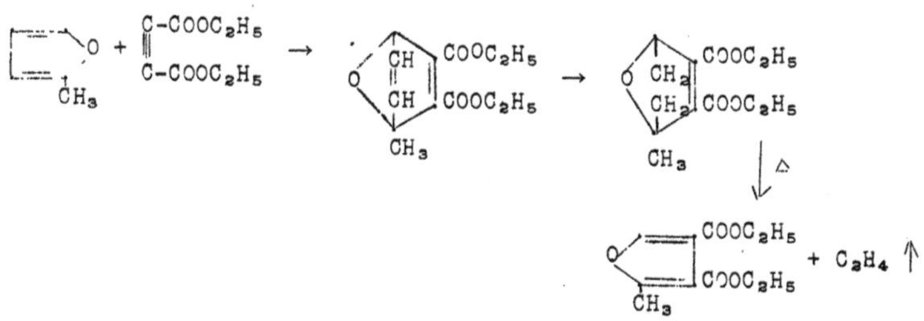

The partially or fully hydrogenated adducts of this type are attacked by hydrogen chloride or bromide to open the oxygen bridge. The dichloro compound thus formed then dehydrohalogenates, with slight heating if there are no substituents on the bridge carbon atoms, and spontaneously if alkyl or aryl groups are present.

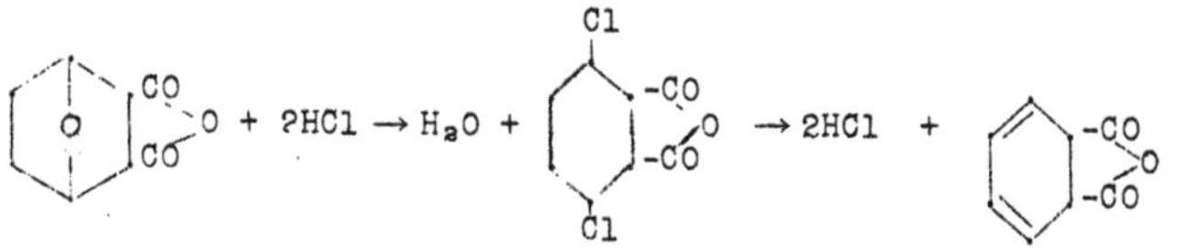

B. Isobenzofurans-- The isobenzofurans are conveniently prepared by a diene synthesis as shown below:

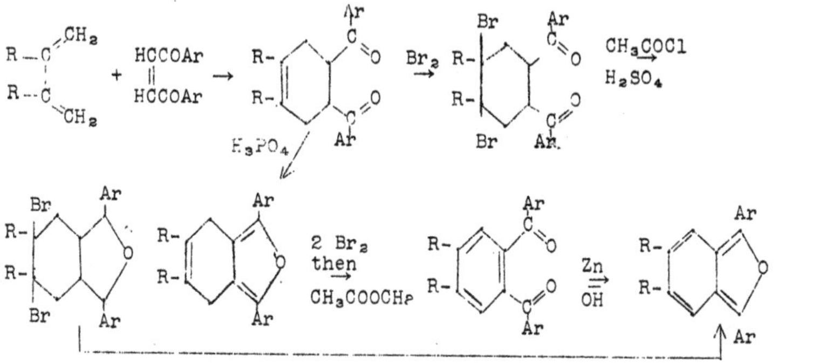

The adducts of isobenzofurans with various dienophiles are used in various syntheses. For instance 1,4-diphenylnaphthalene has been prepared from diphenylisobenzofuran by the following series of reactions;

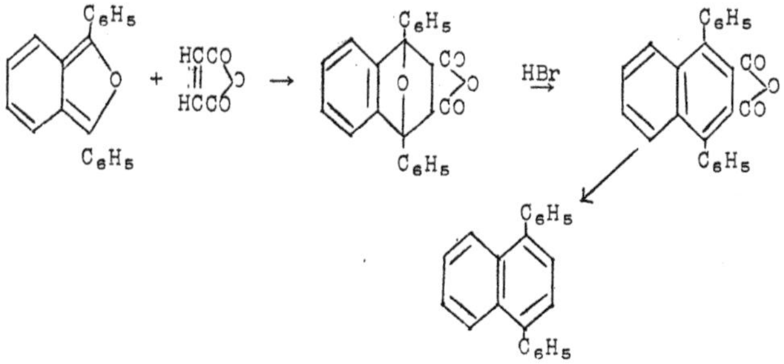

Miscellaneous Diene Syntheses

A. Dimerizations of Carbonylenic Dienophiles-- Alder and his coworkers have recently studied the dimerization of unsaturated aldehydes and ketones, which are unique since the function of the diene is played by the group C=C-C=O instead of C=C-C=C. Addition always occurs so that the RCO- group is ortho to the ring oxygen atom.

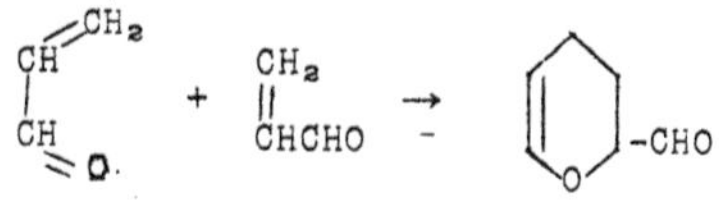

B. Reverse Diene Syntheses-- Certain compounds decompose under strong heating into a diene and an olefinic compound. The most familiar application of this type of reaction is the Kistiakowsky hot-wire method of producing butadiene from cyclohexene. The advantage of this method over other pyrolytic methods is that no polymeric material is formed.

Sterochemistry of the Diene Synthesis:

A. The Cis Principle-- The diene synthesis affords one of the few examples of pure cis-addition. A cis dienophile will never give rise to a trans adduct, and vice versa. Accordingly butadiene and maleic acid give cis-1,2,3,6,-tetrahydrophthalic acid, whereas fumaric acid leads to the corresponding trans adduct.

B. The General Orientation Scheme-- Another stereochemical

question to be considered is whether the diene adds cis on one side or the other of the dienophilic double bond. This presents no problem for butadiene, but cyclopentadiene may add to maleic anhydride in the following two ways:

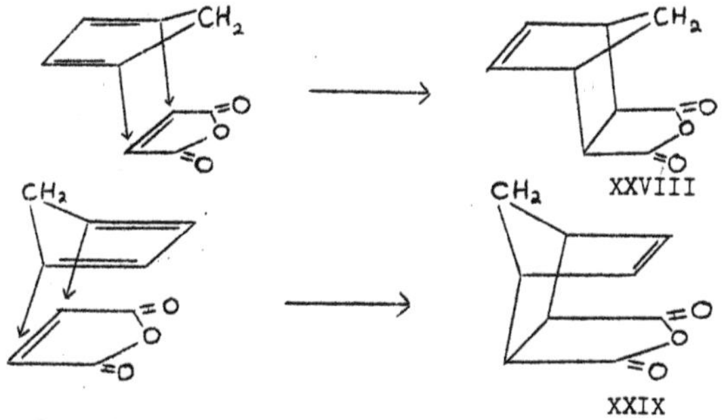
CH2
CH2
CH2
CH2

XXVIII

XXIX

Actually only adduct XXIX is formed in this case. This is designated as an endo configuration in which the endomethylene group is on the opposite side of the molecule from the carboxyl groups. The other configuration, in which the carboxyl groups and the endomethylene group are on the same side of the molecule, is called the exo form. But adducts with an endo configuration are not always formed. In general they appear to add in such a way that there is a maximum accumulation of double bonds just prior to the addition. For instance dephenyl fulvene leads exclusively to an exo adduct.

The next stereochemical question to be considered is whether the diene adds "above" or "below" the dienophilic double bond. The problem does not arise in most diene syntheses, but it is a factor in the addition of dienes to the adducts of cyclopentadiene and an acetylenic dienophile. In such cases the addition generally occurs exo with respect to the methylene bridge.

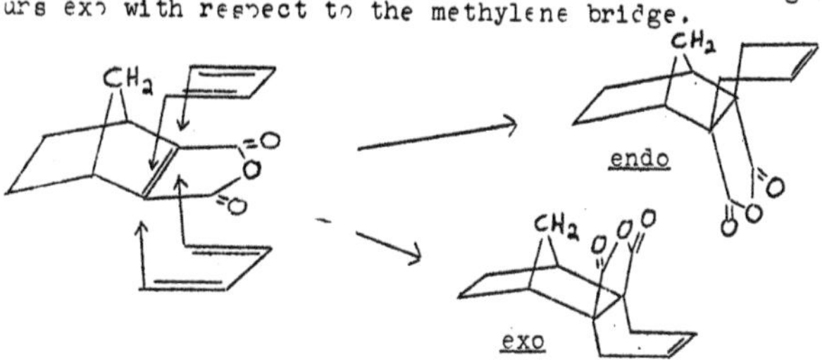
CH2
CH2
endo
CH2
exo

Miscellaneous Applications of the Diene Synthesis:

The Diels-Alder addition has been used for determining the amount of diene present in a sample, or the number of conjugated systems in a large molecule. Generally an excess of maleic anhydride is allowed to react with the sample; followed by titration of the unreacted anhydride removed by aqueous extraction.

Polyakova prepared 95-97% anthracene by heating the crude mixture with an excess of maleic anhydride followed by digestion with alkali and water extraction. The adduct was obtained by acidification and then sublimed to restore anthracene and maleic anhydride.

BIBLIOGRAPHY

1. Norton, Chem. Rev., 31, [2] 319 (1942)
2. Joshel and Butz, J. Am. Chem. Soc., 63, 3350 (1941)
3. Wasserman, J. Chem. Soc., 1942 618 and 623
4. Bergmann, Haskelberg and Bergmann, J. Org. Chem., 7, 303 (1942)
5. Allen, J. Am. Chem. Soc., 62, 658 (1940)
6. Adams, McPhee, Carlin and Wicks, J. Am. Chem. Soc., 65, 356 (1943)

D. H. Chadwick and J. B. McPherson, Jr.
April 21, 1943

SUBSTITUTION AT A SATURATED CARBON ATOM: CYCLIC INTERMEDIATES INVOLVING A NEIGHBORING GROUP 11

Lucas, Winstein et al.

Robinson (1932) and later Ingold (1934) have suggested an intermediate of the type -C(X)- $\overset{+}{C}$- in the halogenation of olefins. Roberts and Kimball in 1937 pointed out that this intermediate might be formulated as >C—C< (bridged by X+) since the difference between the ionization potentials of carbon and halogen is small (11.22 electron volts for carbon, 11.80 e. v. for bromine and 13.0 e.v. for chlorine) and consequently the positive charge may exist on the halogen atom as easily as on the carbon. They also pointed out that the contribution of the cyclic form would tend to prevent rotation of the molecule and would lead to the observed "trans" addition by the rearward attack of a halide ion on one of the carbon atoms in a manner analagous to the ruptune of oxide rings.

Lucas, Winstein and coworkers have recently shown that similar cyclic intermediates may be involved in substitution at a saturated carbon atom in a molecule in which a "neighboring" groups is present. By a neighboring group is meant a group attached to a carbon atom adjacent to the carbon atom in question. If As represents the neighboring group, Y the leaving group and Z the entering group, the general picture of the mechanism is:

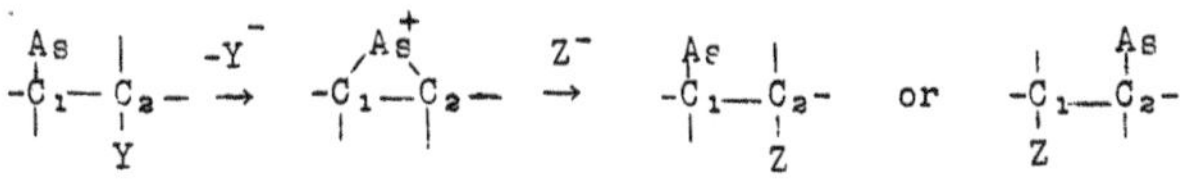

The first step is the simultaneous or nearly simultaneous removal of Y^- ion and the rearward approach of the neighboring group As to C_2. This step and consequently reaction by the above mechanism is favored by conditions which cause Y^- to ionize off. One Walden inversion occurs here. The next step is the attachment of Z^- by rearward approach to either C_1 or C_2. Another Walden inversion occurs; the net steric result is two inversions or retention of configuration.

The correct representation of the cyclic intermediate is a composite of the following resonating forms:

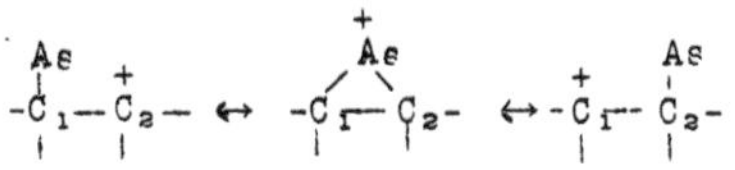

For convenience the cyclic form will be used as an abbreviation for this resonance mixture.

It would appear that the neighboring group As could be -OCOR, -NH, NR, -OR, -OH, -SR or halogen. Reactions which apparently proceed by this mechanism are known where As is -OCOR, -OH, -Cl and -Br.

If As is -OCOR, the cyclic intermediate is considered to be

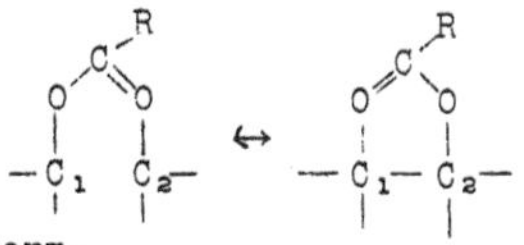

instead of the usual form.

The more important pieces of evidence which have necessitated the formulation of this mechanism in certain cases are:

1. The cis- and trans-2,3-dibromobutanes, the erythro- and threo-3-bromo-2-acetoxybutanes, trans-1-acetoxy-2-bromo-cyclohexane and trans-1,2-dibromocyclohexane react with silver acetate in anhydrous acetic acid to give the corresponding diacetates with essentially complete retention of configuration; whereas the reaction under the same conditions of a typical bromide which did not contain a neighboring group, 2-bromoöctane, proceeded with 86% inversion of configuration, which is a typical Sn1 steric result. Hence it seems that the presence of a neighboring group causes overall retention of configuration, a fact which is explained by the cyclic mechanism. It will be noted that in the case of the above dibromides the acetoxy bromide is first formed with essentially complete retention of configuration and then reacts again to form the diacetate, also with predominant retention.

2. Both d- and l- threo-2-butanol give inactive dl-2,3-dibromobutane with fuming aqueous HBr under conditions which do not isomerize the dl-dibromobutane to the meso-dibromobutane and consequently are not likely to racemize the d- and l- forms. The following mechanisms which might seem possible are inoperative because they would lead to active products:

(a) 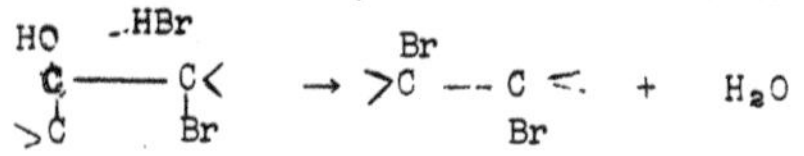

(b) 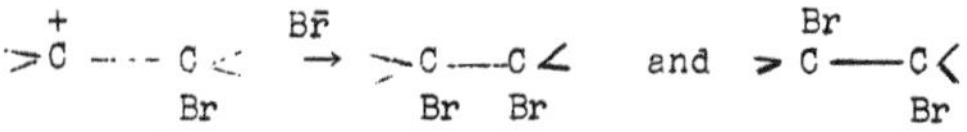

The mechanism involving the cyclic intermediate leads to the observed inactive products because the intermediate 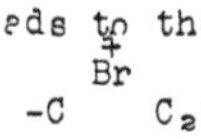

is symmetrical and therefore can lead only to inactive products, or in other words, because the attack of the Br^- ion at C_1 to give one optical isomer or at C_2 to give the mirror image of the first is equally likely; hence equal amounts of the d- and l- forms are produced. Several other examples of loss of activity on substitution apparently due to the effect of a neighboring group have been reported by Winstein, Lucas et al. Among these are the reaction of active trans

1-acetoxy-2-bromocyclohexane with silver acetate in dry acetic acid to give predominantly the inactive trans dl-diacetate and the reaction of active threo-2,3-butane chlorohydrin with thionyl chloride to give predominantly inactive dl-2,3-dichlorobutane. In all examples given experimental evidence indicated that racemization either did not occur or was highly improbable.

3. All of the bromides and acetoxybromides which are listed in (1) above as reacting with silver acetate in dry acetic acid to give the corresponding diacetates with essentially complete retention of configuration, namely the cis- and trans- 2,3-dibromobutanes, the erythro- and threo-3-bromo-2-acetoxybutanes, trans-1-acetoxy-2-bromocyclohexane and trans-1,2-dibromocyclohexane, react with silver acetate in acetic acid containing one mol of water to give the monoacetates of the corresponding glycols with essentially complete inversion of configuration. The monoacetates are slowly converted to diacetates under reaction conditions; hence the monoacetates are the first products of the reaction. These apparently anomalous results can be explained as follows using threo-2-acetoxy-3-bromobutane as an example:

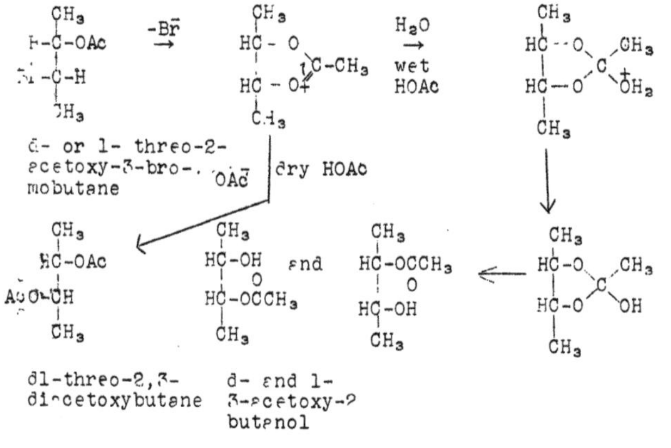
-Br̄
H₂O
wet
HOAc
d- or l- threo-2-acetoxy-3-bromobutane
OAc̄
dry HOAc
and
dl-threo-2,3-diacetoxybutane
d- and l- 3-acetoxy-2 butanol

Similar mechanisms account for the reaction of trans-2-acetoxycyclohexyl p-toluenesulfonate in absolute alcohol in which calcium carbonate is suspended or in dry acetic acid without silver acetate to give products which may be hydrolyzed to quite pure cis cyclohexene glycol. Here the EtOH or AcOH take the place of the H_2O.

4. If carbon atoms C_1 and C_2 in the picture of the mechanism are a part of a small ring, then reaction by this mechanism can take place only if the substituent groups As and Y are trans to one another, in which case the product will also be trans. If As and Y are cis, reaction by the ordinary Sn1 or Sn2 processes will result in inversion of configuration, that is, also a trans product. Hence there will be a tendency for either the cis- or trans-reactant to give trans- product. Winstein pointed out this effect was observed in a number of cases, e. g. the reaction of acetohalogen

sugars to give only acetates with acetate groups 1 and 2 *trans* to one another and the formation of *trans*- 1,2-dibromocyclohexane from the *cis*- and *trans*- diacetate. It is true that the same result would be predicted from the fact that the *trans* compound is more stable, but the stability of the *trans* compound may not be the only effect involved.

In conclusion, it may be pointed out that the mechanism involving the neighboring group is not operative in certain cases where it might be, presumably because some other mechanism is faster. This is particularly true of compounds in which chlorine must play the part of the neighboring group. For example, the erythro- and threo-2-butenechlorohydrins react with phosphorus trichloride or with thionyl chloride and pyridine with essentially complete inversion of configuration, corresponding to Sn1 or Sn2 mechanism. With thionyl chloride in the absence of pyridine, however, reaction proceeds through the cyclic intermediate.

BIBLIOGRAPHY

Winstein et al. J.A.C.S. 64, 2780 (1942)
Lucas, Schlatter and Jones, ibid., 63, 22 (1941)
Winstein and Lucas, ibid., 61, 1576 (1939)
Winstein and Lucas, ibid., 61, 2845 (1939)
Lucas and Gould, ibid., 63, 2541 (1941)
Roberts and Kimball, ibid., 59, 947 (1937)

Reported by Harry F. Kauffman, Jr.
April 28, 1943

PECHMANN CONDENSATION FOR THE SYNTHESIS OF SUBSTITUTED COUMARINS

In 1883, Pechmann and Duisberg found that phenol reacted with acetoacetic ester at ordinary temperatures in the presence of dehydrating agents to form 4-methylcoumarin. They tried this same reaction using simple polyhydroxy benzenes and substituted phenols with benzoylacetic ester and α-methylacetoacetic ester as well as acetoacetic ester itself, and found the reaction to be quite general.

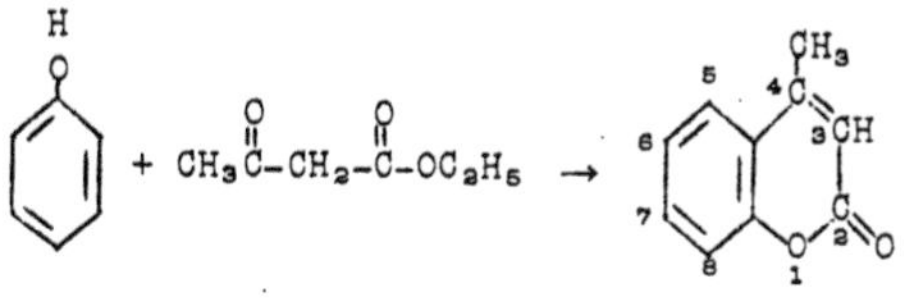

In 1884, Pechmann discovered that when malic acid was heated with phenol in the presence of sulfuric acid, zinc chloride, or a similar reagent, coumarin was formed and carbon monoxide and water were eliminated. Pechmann explained his results by the following series of reactions.

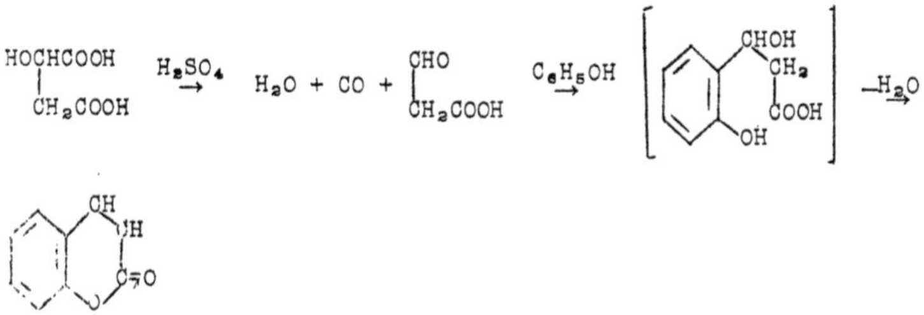

Pechmann used this reaction with various phenols to form substituted coumarins in good yields.

Since then, considerable work has been done to determine the limitations of the reactions and to determine the effect of various substituents on the ease of the reaction and the nature of the products formed. Considering the hydroxy benzenes themselves, reactivity decreases in the following order: resorcinol, pyrogallol, phloroglucinol, hydroquinone and phenol. α-Naphthol has about the same reactivity as phloroglucinol, and β-naphthol has about the same reactivity as phenol.

Substituents in the ring also affect the reactivity, negative groups and halogens decreasing the reactivity and alkyl groups increasing the reactivity. The order of the de-activating effect is:

Carboxyl, bromine, acetyl, carbethoxy, chlorine, hydrogen, methyl, ethyl. The position of the groups also affects the reactivity. The size of alkyl groups does not affect the reactivity appreciably.

The orientation of the coumarin is hard to predict. In the case of certain phenols, e.g., phenol, hydroquinone, phloroglucinol, pyrogallol, and pyrogallol derivatives, only one coumarin is possible, but in most cases, two or more configurations are possible, and it is difficult to make a choice. In most of the reactions, a single product is isolated. With orcinol and acetoacetic ester, it is theoretically possible to prepare two substituted coumarins, but only one, 4,7-dimethyl-5-hydroxycoumarin is obtained.

In acyl substituted resorcinols, the hydroxyl group next to the acyl group does not become part of the pyrone ring, so that the number of possible structures is limited. Thus resacetophenone will be expected to form either 4-methyl-6-aceto-7-hydroxycoumarin, or 4-methyl-5-hydroxy-6-acetocoumarin, but not 4-methyl-5-hydroxy-8-acetocoumarin. Desai and Vakil found that in certain instances the acyl group was eliminated during the course of the reaction, especially if the group is <u>ortho</u> or <u>para</u> to an alkyl group. Desai and Ekhlas found this to be true also of 4-acyl-1-naphthols.

With hydroquinones, the problem of orientation is much simpler; the substituent on the hydroquinone always appears in the 7-position on the coumarin.

In addition to acetoacetic ester, a number of other β-ketonic esters have been used: benzoylacetic ester, butyroacetic ester, ethyl acetonedicarboxylate, γ-phenylacetoacetic ester, α-methyl-, α-ethyl-, and α-propylacetoacetic esters, and also the cyclic β-ketonic esters, 2-carbethoxycyclohexanone, and 4- and 5-ethyl-2-carbethoxycyclohexanone. In addition to malic acid, diacetonedicarboxylic acid has been used. In all cases, except the cyclic esters, the reactivity is decreased.

A number of reagents have been used for effecting the condensation. Concentrated sulfuric acid and $POCl_3$ are the most common, but 85% phosphoric acid, P_2O_5, $AlCl_3$, and $ZnCl_2$ have also been used. When $AlCl_3$ is used as a catalyst with acyl-resorcinols, the products are usually 5-hydroxycoumarins, whereas most other catalysts give 7-hydroxycoumarins.

The conditions of the reaction vary considerably and depend upon the particular coumarin. Solvents, especially benzene, are sometimes used. In some cases, the reaction mixture is refluxed for a few hours, while with others, better yields are obtained by allowing the reaction mixture to stand at room temperature for twenty to sixty hours.

The chief advantage of the Pechmann reaction is that a complicated molecule may be synthesized from relatively simple reagents in a single step. Also it is one of the few methods for introducing a group into the 4-position in coumarin.

BIBLIOGRAPHY

Pechmann and Duisberg, Ber., 16, 2119 (1883).
Pechmann, Ber., 17, 929 (1884).
Pechmann and Cohen, Ber., 17, 2187 (1884).
Pechmann and Welsh, Ber., 17, 1646 (1884).
Desai and Mavani, Proc. Ind. Acad. Sci., 15A, 14 (1942).
Desai and Mavani, Proc. Ind. Acad. Sci., 15A, 1 (1942).
Dey, J. Chem. Soc., 107, 1606 (1915).
Sastry and Seshadri, Proc. Ind. Acad. Sci., 12A, 449 (1940).
Rao and Seshadri, Proc. Ind. Acad. Sci., 13A, 255 (1941).
Desai and Ekhlas, Proc. Ind. Acad. Sci., 8A, 569 (1938).
Sethna and Shah, J. Ind. Chem. Soc., 15, 383 (1938).
Desai and Vakil, Proc. Ind. Acad. Sci., 12A, 357 (1940).
Desai and Mavani, Proc. Ind. Acad. Sci., 14A, 100 (1941).
Chowdhry and Desai, Proc. Ind. Acad. Sci., 8A, 1 (1938).
Kotwani, Sethna and Advani, Proc. Ind. Acad. Sci., 15A, 441 (1942).
Kotwani, Sethna and Advani, J. Univ. Bombay, 10, 143 (1942).
Deliwala and Shah, Proc. Ind. Acad. Sci., 13A, 352 (1941).
Adams, Cain and Loewe, J. Am. Chem. Soc., 63, 1977 (1941).

Reported by R. S. Ludington
April 28, 1943

THE MECHANISM OF THE REACTIONS OF SECONDARY AMINES WITH

α,β- UNSATURATED KETONES

Norman H. Cromwell, University of Nebraska

The reaction of secondary amines with 1,2-dibromoketones and with the corresponding α-bromo-α,β-unsaturated ketones (II) has been studied extensively by Dufraisse and Moureu. Cromwell has reinvestigated these reactions using a variety of secondary amines with particular regard to proof of structure of the products and their mechanism of formation.

The general structures which have been found to occur are illustrated below.

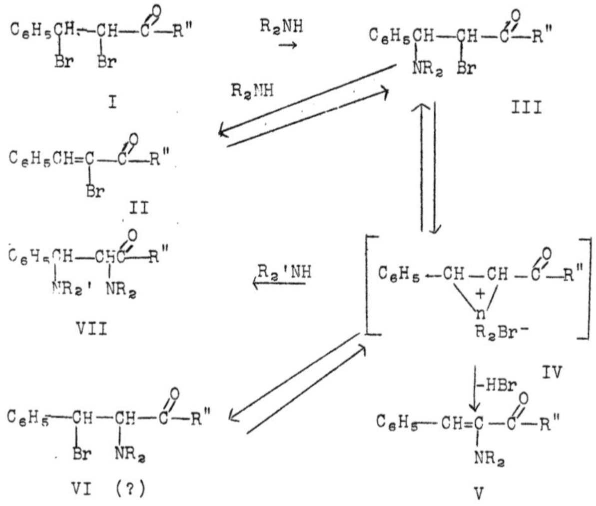

The rapid reaction of benzalacetophenone dibromide (I) with morpholine gave mostly α,β-dimorpholinobenzylacetophenone (VII) with small amounts of α-morpholinobenzalacetophenone (V). Experiments failed to show any addition of morpholine to the isolated unsaturated amino ketone (V). α-Bromobenzalacetophenone (II) reacted with morpholine in the cold to give α-bromo-β-morpholinobenzylacetophenone.(III).

Cromwell has found that α-bromo-β-piperidinobenzylacetophenone (III), prepared by Dufraisse, reacted with excess morpholine to give two products, α-piperidino-β-morpholino-benzylacetophenone (VII) and α-piperidinobenzalacetophenone (V). Also the strong heterocyclic base, pyrrolidine, has been found to resemble piperidine in these reactions.

Tetrahydroisoquinoline was found to add rapidly and completely

to α-bromobenzalacetone to give a bromo-amino ketone whose structure is now assigned as α-bromo-β-tetrahydroisoquinolinobenzylacetone (III). The product reacted with sodium ethoxide to give α-tetrahydroisoquinolinobenzalacetone (V). With excess tetrahydroisoquinoline the bromo-amino ketone (III) gave a 75% yield of α,β-di-tetrahydrooisoquinolinobenzylacetone (VII), which was also prepared directly from α,β-dibromobenzylacetone in 63% yield.

In either absolute alcohol or absolute ether the bromo-amino ketone (III) from tetrahydroisoquinoline reacted readily with a much weaker base, tetrahydroquinoline, to give good yields of α-tetrahydroisoquinolino-β-tetrahydroquinolinobenzylacetone (VII). The structure of this diamino ketone was established by hydrolysis to give the expected α-tetrahydroisoquinolinoacetone. When a base of almost equal or greater strength than tetrahydroisoquinoline was used, a mixture of products resulted due to the reversibility of reaction (II) to (III).

In an attempt to prove the structure of the primary products from the reaction of α-bromo-α,β-unsaturated ketones with secondary amines, one of them, α-bromo-β-piperidinobenzylacetophenone, was reduced with hydrogen over platinum. The reaction proceeded to give benzylacetophenone and piperidine hydrobromide. It had been hoped that only the bromine atom would be replaced. However, this was evidence for the structure A that has been assigned to these bromo-amino ketones and evidence against structures B or C. It has been shown by Cromwell, Wiles and Schroeder that β-amino ketones are unstable to catalytic reduction, as compared with α-amino ketones.

$C_6H_5-CH(NR_2)-CHBr-COR''$ (A) $\quad C_6H_5-CHBr-CH(NR_2)-COR''$ (B) $\quad C_6H_5-CH_2-CBr(NR_2)-COR''$ (C)

Another attempt at proof of structure of these addition products was carried out by reducing the known α-bromo-β-piperidinobenzalaceto-phenone (VIII). Unfortunately the bromine is labile even with catalytic hydrogen. It was hoped that (VIII) could be reduced to give A. However, the bromine was replaced by hydrogen and the resulting β-piperidinobenzalacetophenone apparently hydrolyzed to give the product, dibenzoylmethane, in good yields. It was also found that it was not possible to isolate a bromo-amino ketone of structure B or C when α-piperidinobenzalacetophenone was treated with dry hydrogen bromide.

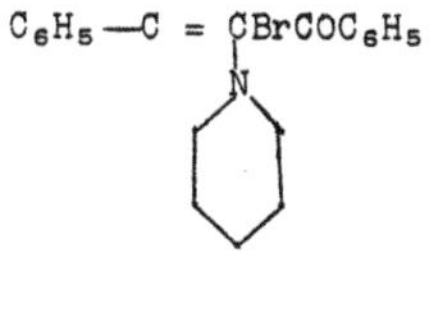
C6H5—C = CBrCOC6H5
N

VIII

In view of the difficulties encountered in an attempt to prove the structure of the addition product (III) the possibility of equilibrium with the isomer (VI) cannot be excluded.

The above mechanism has been extended by Cromwell to the reactions of piperidine with bromine derivatives of benzalacetone and with benzoylacetone. Also N-methylbenzylamine, an open chain type of secondary amine has been found to react with bromine derivatives of benzalacetone and benzalacetophenone in a manner similar to that of heterocyclic amines.

BIBLIOGRAPHY

Dufraisse and Moureu, Bull. Soc. Chim., (4e) 41, 457-472,850, 1370 (1927)
Dufraisse and Netter, ibid., (4e) 51, 550-562 (1932)
Moureu, Ann. Chim., (10) 14, 314 (1930)
Rubin and Day, J. Org. Chem., 5, 54 (1940)
Cromwell, J. Am. Chem. Soc., 62, 1672, 2897, 3740 (1940); 63, 837, 2984 (1941)
Cromwell, Wiles and Schroeder, ibid, 64, 2432 (1942)
Cromwell and Cram, ibid, 65, 301 (1943)
Cromwell and Witt, ibid., 65, 308 (1943)

Reported by H. W. Johnston
May 5, 1943

THE SYNTHESIS OF ISOQUINOLINES

The synthesis of isoquinolines may be divided into five types depending upon the point at which ring closure is effected, as shown by the following formulas:

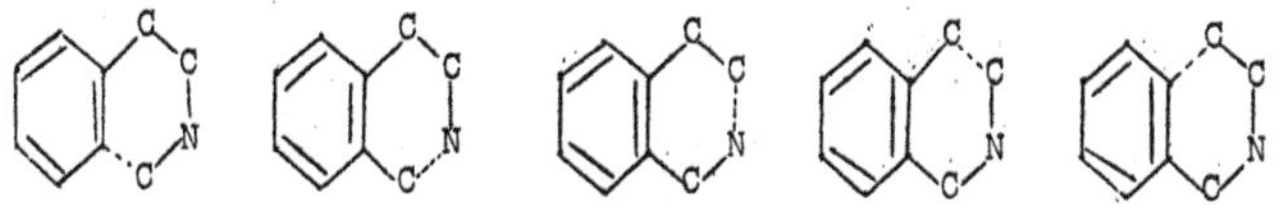

I II III IV V

Type I Syntheses.

Bischler and Napieralski prepared 1-alkyl or 1-aryl-3,4-dihydroisoquinolines by heating the acetyl or benzoyl derivative of β-phenylethylamine with phosphorus pentoxide or with zinc chloride at about 200°C. Later improvements using boiling toluene or xylene in the presence of phosphorus pentoxide or phosphorous oxychloride have made this reaction a practical one for the synthesis of benzylisoquinoline alkaloids. When β-aryl-β-hydroxyethylamines are used, an additional molecule of water is eliminated and isoquinolines are produced.

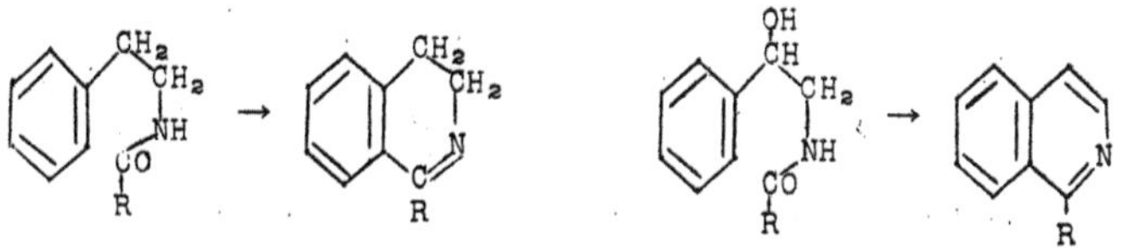

An important variation of the Bischler-Napieralski reaction is the condensation of aldehydes with β-phenylethylamines and ring closure by means of hot hydrochloric acid.

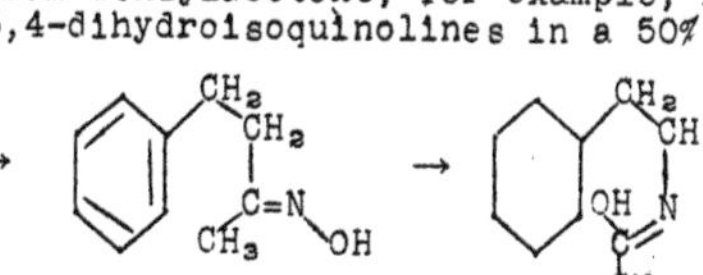

The Beckmann rearrangement of oximes leads to acyl or aroyl amines which may then be condensed to form isoquinolines or dihydroisoquinolines. From benzylacetone, for example, Pictet and Kay obtained 1-methyl-3,4-dihydroisoquinolines in a 50% yield.

Isocyanates, produced as an intermediate in the Curtius synthesis of amines, have undergone ring closure to give dihydrocarbostyrils, such as corydaldine.

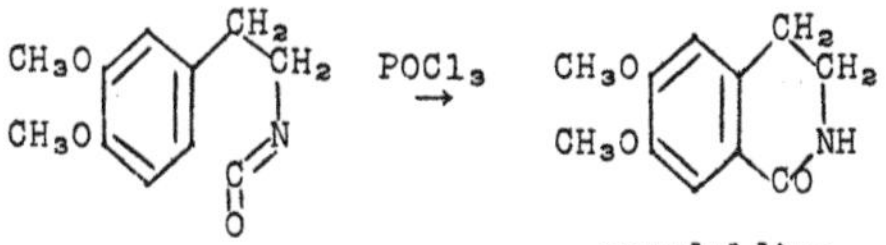
CH₃O
CH₃O
CH₂
CH₂
N
C
O
POCl₃
CH₃O
CH₃O
CH₂
CH₂
NH
CO

corydaldine

The acid chlorides of a series of β-arylethylglycines prepared by v. Braun and Wirz gave tetrahydroisoquinolines and carbon monoxide when treated with aluminium chloride. When the nitrogen was secondary it was necessary to make the p-toluenesulfonyl derivative first.

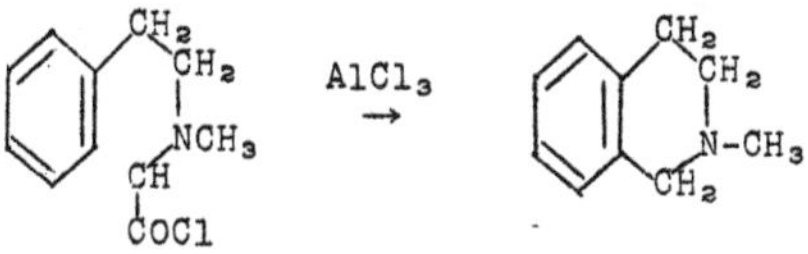
CH₂
CH₂
NCH₃
CH
COCl
AlCl₃
CH₂
CH₂
N-CH₃
CH₂

Type II Syntheses.

Treatment of homophthalic acid with ammonia will give homophthalimide, or 1,3-diketo-1,2.3.4-tetrahydroisoquinoline. Heating with zinc dust, or successively with phosphorus oxychloride and hydriodic acid yields isoquinoline. Homophthalimide may also be obtained by hydrolysis of o-cyanobenzyl cyanide with sulfuric acid.

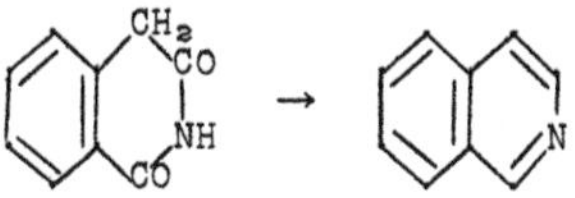
CH₂
CO
NH
CO
N

Helfer distilled the dihydrochloride of β-(o-aminomethylphenyl)-ethylamine and obtained tetrahydroisoquinoline in 60% yield. Treatment of homoxylene dibromide with aniline yields N-phenyltetrahydroisoquinoline.

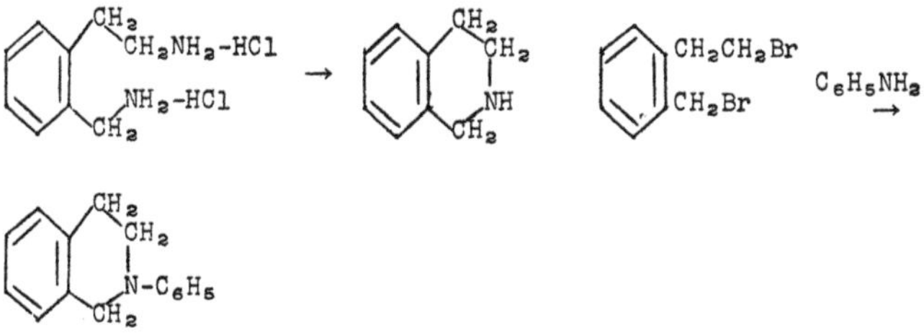
CH₂
CH₂NH₂-HCl
NH₂-HCl
CH₂
CH₂
CH₂
NH
CH₂
CH₂CH₂Br
CH₂Br
C₆H₅NH₂
CH₂
CH₂
N-C₆H₅
CH₂

Almost quantitative yields of isoquinoline derivatives may be obtained from the reaction of primary amines with isocoumarins. Isocoumarins, however, are rather difficultly accessible compounds. Isocoumarin itself may be obtained from the oxidation of β-naphthoquinone with sodium hypochlorite. The esters of isocoumaric acid may be prepared by the Claisen condensation of the esters of homophthalic acid with aliphatic and aromatic esters. Treatment of these esters with ammonia gives isocarbostyrils. The ketonic tautomers of isocoumaric acid, 1-phenacyl- and 1-acetonyl-benzoic acids, condense with primary amines to give isocarbostyrils.

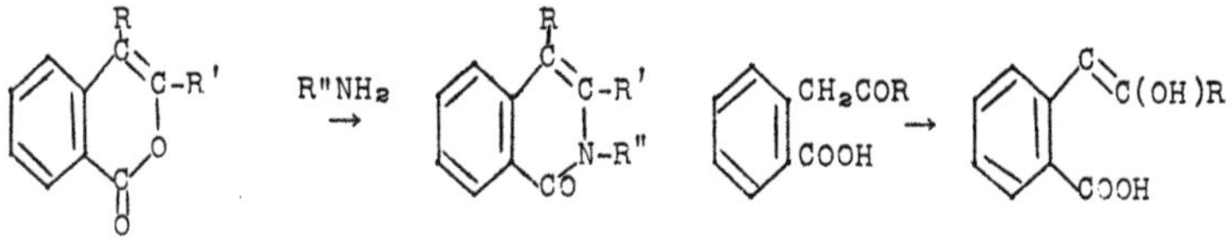
R
C
C-R'
O
C
O
R"NH2
R
C
C-R'
N-R"
CO
CH2COR
COOH
C
C(OH)R
COOH

Type III Syntheses.

No absolute examples of this type are known although some of the examples of type II may belong in this class.

Type IV Syntheses.

Reactions of this type all involve the condensation of a nuclear carboxyl with an active hydrogen.

Gabriel and Coleman treated ethyl phthalimidoacetate with sodium ethoxide and obtained 3-carbethoxy-4-hydroxyisocarbostyril by means of a modified Dieckmann reaction. This synthesis is limited to phthalimido derivatives which may undergo the Claisen condensation. Fischer and Krollpfeiffer condensed 2,3,4-trisubstituted pyrroles with phthalic anhydride in the presence of acetic anhydride to form isoquinoline derivatives of type (A).

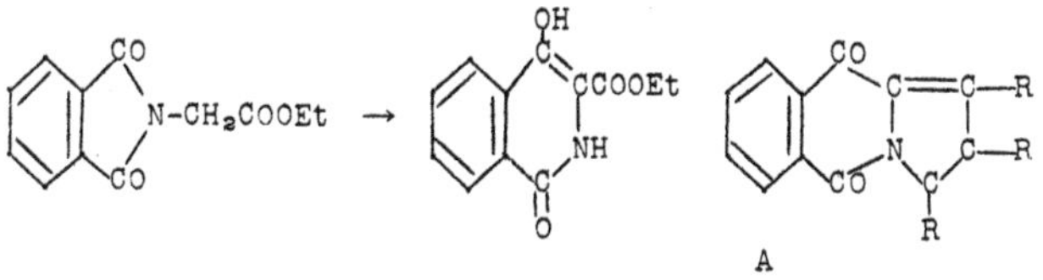
CO
N-CH2COOEt
CO
OH
C
CCOOEt
NH
C
O
CO
C
C—R
N
C—R
CO
C
R

A

Type V Syntheses.

Some of the reactions of this type closely resemble modifications of the Skraup synthesis. Ring closure can occur only in compounds of formula (B) and (C), where the double bonds may be only potentially present.

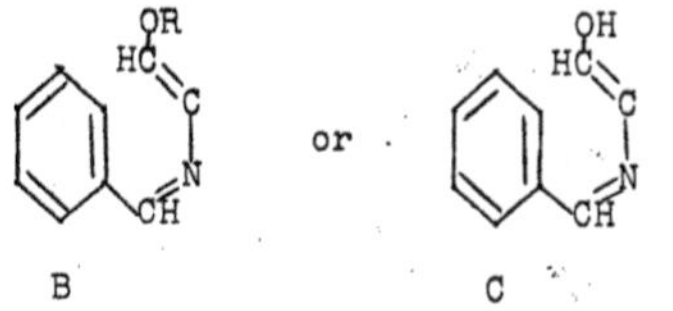

OR
HC
C
N
CH
or
OH
HC
C
N
CH

B C

Benzalaminoacetal, for example, gives isoquinoline in 50% yields when treated with sulfuric acid or phosphorus oxychloride, giving the required double bond on elimination of alcohol. Bensylaminoacetaldehyde on treatment with oleum is oxidized to benzalaminoacetaldehyde, the enol form of which has the formula of (C) and thus may be condensed to give isoquinoline. Arsenic pentoxide has also been used as the oxidizing agent in similar reactions. The enolic form of benzoylaminoacetal has one actual and one potential double bond and yields isocarbostyril on treatment with sulfuric acid.

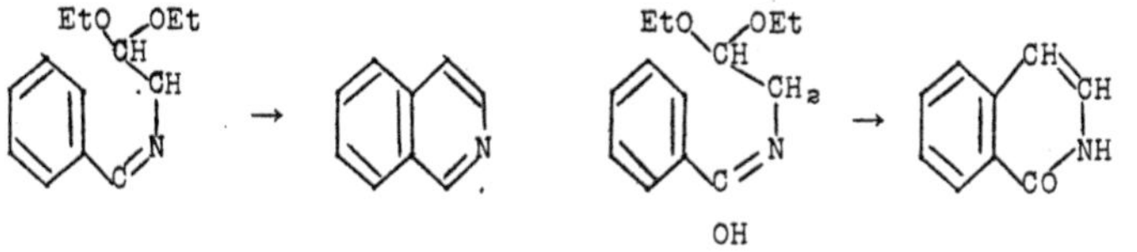

EtO OEt
CH
CH
C
N
N
EtO OEt
CH
CH₂
N
C
OH
CH
CH
NH
CO

Bz-hexahydroisoquinolines have been synthesized by Basu by means of a modification of the Guareschi synthesis of pyridines. Ethyl cyclohexanone carboxylate was condensed with cyanoacetamide to yield compound (E). Oxymethylenecyclohexanone was treated with ammonia to give an aminomethylene compound which was then condensed with acetoacetic ester in the presence of sodium to give compound (F). This compound, on hydrolysis, gave the acid which was easily decarboxylated and converted into isoquinoline by distillation with zinc dust.

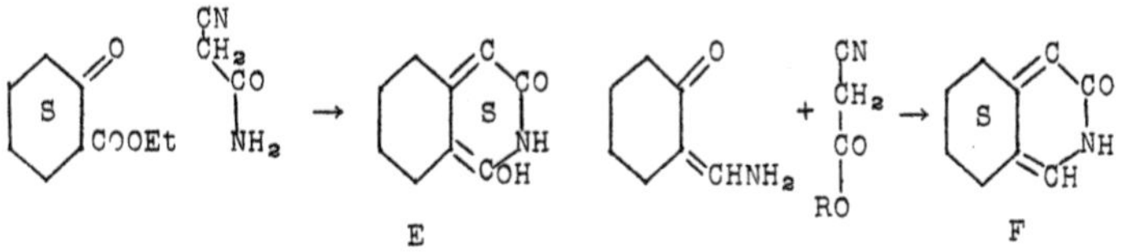

O
S
COOEt
CN
CH₂
CO
NH₂
C
CO
S
NH
COH
O
CHNH₂
+
CN
CH₂
CO
RO
S
C
CO
NH
CH

E F

BIBLIOGRAPHY

Menske, Chem. Rev., 30, 145 (1942)

Reported by W. R. Hatchard
May 5, 1943

CLEAVAGE OF KETONES WITH ALKALI

<u>Cleavage of Acetophenone Derivatives.</u>[2,3] Methyl phenyl ketones may undergo cleavage in two manners:

$$C_6H_5COCH_3 \xrightarrow{1} C_6H_5COOH + CH_4$$
$$C_6H_5COCH_3 \xrightarrow{2} C_6H_6 + CH_3COOH.$$

A smooth cleavage by alkali of acetophenone, itself, has not been successful but that of derivatives varies from one extreme to the other. Halogen substitution in the methyl group facilitates cleavage in direction 1. Mono, di, tri, and penta substitutions in the benzene nucleus range from poor yields in direction 1 to good yields in direction 2.

Experimental work on acetophenones shows that under the applied conditions and with at least one unsubstituted *ortho* position, cleavage results, if at all, on the aliphatic side (direction 1); likewise, if substitution is on the methyl group, only cleavage in direction 1 is observed. But if both *ortho* positions are occupied, smooth cleavage occurs in direction 2 except when the methyl group is also substituted.

Table I - Acetophenones

Method of cleavage (MC); cleavage direction (CD); percent cleavage (%C).

Compound	MC*	CD	%C
Acetophenone	1	1	5.5
Acetophenone	2	1	31
3-chloro-	1	1	14.5
4-chloro-	1	1	3.9
2,6-dichloro-	1	2	80
3,5-dichloro-	1	1	19
2,5-dichloro-	1	1	2.4
2,3,6-trichloro-	1	2	89
2,4,6-trichloro-	1	2	80
2,3,4,5,6-pentachloro-	1	2	78
3,4,5-tribromo-	1	1	30
2,4,6-tribromo-	1	2	93
2,3,4,5-tetrabromo-	1	-	--

Compound	MC	CD	%C
α,α,α,2,4,6-hexabromo-	1	1	---
4-methyl-	2	1	40.5
2-methyl-	2	1	27.5
3-methyl-	2	1	39
2,4-dimethyl-	2	1	23
2,5-dimethyl-	2	1	13
2,4,6-trimethyl-	2	-	---
2-nitro-	3	1	50
α,α,α-tribromo-2,4,6-trimethoxy-	1	1	---
α,α,α-trichloro-2,4,6-trimethoxy-	1	1	---
2,4-dichloro-	1	1	6

*Method of cleavage (1): 0.01-0.02 mole of ketone and 25-50 g. of 50% KOH are placed in a nickel tube fitted with reflux condenser and heated on an oil bath at 150° for 24 hours. Method of cleavage (2): For methyl-substituted acetophenones the ketone and a mixture of KOH and NaOH are heated for 3 hours in a nickel tube at 200°. Method of cleavage (3): For nitroacetopheones the ketone and 10% KOH are heated for 12 hours on a water bath.

Naphthalene aliphatic ketones[9] behave similarly to acetophenones.

Cleavage of Benzophenone and Derivatives. In the decomposition of benzophenones by fused alkali, the substituents do not play as decisive a role in directional cleavage as in the acetophenone series. However, Kozlov[4,5] says that p-alkylated benzophenones cleave so readily into p-alkylbenzoic acids that he recommends it as a method of preparing the corresponding benzoic acids. With a second substitution the cleavage goes chiefly in the opposite direction. Therefore, cleavage of benzophenones may give either one or both of the possible acids plus one or both of the possible aromatic hydrocarbons.

Lock's work on benzophenones is tabulated as follows:

Table II - Benzophenones[6]

Compound	% BzOH	%Sub. BzOH	Compound	% BzOH	%Sub. BzOH
2-methyl-	66	12	2,4-dichloro-	94	0
3-methyl-	41	42	2,5-dichloro-	91	0
4-methyl-	33	48	2,6-dichloro-	96	0
2-chloro-	89	0	3,5-dichloro-	97	0
3-chloro-	81	5	2,3,4,5,6-		
4-chloro-	66	18	pentachloro-	85	0
2-nitro-	57	--	3-nitro-	53	-

Cleavage of other benzophenones appear in Table III.

Table III - Benzophenones[4,7,8]

p-ethyl → p-ethylbenzoic acid, benzene (81%)
p-isopropyl → p-isopropylbenzoic acid, benzene
2,4-diisopropyl → m-diisopropylbenzene, benzoic acid
p,p'-diisopropyl → p-isopropylbenzoic acid, isopropylbenzene (100%)
p,p'-diethyl → p-ethylbenzoic acid, ethylbenzene (90%)
2-methyl-5-isopropyl → 2 acids, 2 hydrocarbons
2,4'-dimethyl-5-isopropyl → 2 acids, 2 hydrocarbons
2,5-diethyl → benzoic acid, p-diethylbenzene
2-methyl-5-ethyl → benzoic acid, p-methylethylbenzene
2,4'-dimethyl-5-ethyl → 2 acids, 2 hydrocarbons
4'-methyl-2,4,6-triphenyl → triphenylbenzene, p-toluic acid
2,4,6-triphenyl → triphenylbenzene, benzoic acid (100%)

Naphthyl phenyl ketones [9] give a mixture of acids and hydrocarbons. Beta isomers tend to yield more of the mixed product than do alpha isomers. Dinaphthyl ketones have not been studied to any great extent.

Ring-Cleavage of Fluorenones[10] Huntress has recently been investigating the effects of strong alkali on fluorenones. In cleaving the fluorenones 0.5 gm. of ketone, 2 gms. of KOH and 15-40 ml. of diphenyl ether are heated on an oil bath at 160-200° with vigorous stirring.

Fluorenone gave an almost quantitative yield of o-phenylbenzoic

acid; 2-chlorofluorenone gave 4'-chlorobiphenyl-2-carboxylic acid; 2-hydroxyfluorenone gave 4-hydroxybiphenyl-2-carboxylic acid in 50% yield; 2-aminofluorenone gave a gummy mass; fluorene-2-sulfonic acid gave 4'-hydroxybiphenyl-2-carboxylic acid; fluorenone-2-potassium sulfonate gave after a short time and in a good yield a mixture of biphenyl-2-carboxylic acid-4-potassium sulfonate and biphenyl-2-carboxylic acid-4'-potassium sulfonate; but with five hours of treatment the same two biphenyls resulted along with some 4'-hydroxybiphenyl-2-carboxylic acid. It becomes evident that the opening of the ketonic ring of 2-sulfofluorenone occurs with greater ease as compared with the replacement of sulfonic acid group by hydroxyl.

1,8-Dichlorofluorenone gave 50-60% yield of 3,3'-dichlorobiphenyl-2-carboxylic acid plus some lactone of 2'-hydroxy-3-chlorobiphenyl-2-carboxylic acid. 1,6-Dichlorofluorenone gave a 50% yield of the corresponding biphenyl plus some lactone. 3,6-Dichlorofluorenone gave 90-92% yield of 3',5-dichlorobiphenyl-2-carboxylic acid and no lactone. Huntress explains the lactone formation as replacement of halogen by hydroxyl and rearrangement along with the ring opening. These independent reactions occurred only with halogen atoms ortho to the carbonyl group of the dichlorofluorenones.

Sodium Amide as a Cleaving Agent.[11,12] Aromatic ketones with sodium amide in the molten state undergo cleavage of the carbonyl group according to the following equation:

$$RCOR' + 2NaNH_2 \rightarrow NaHCN_2 + NaOH + RH + R'H$$

along with considerable carbon, as well as small amounts of HCN, NaCN, $H_2NC(NH)NHCN$, and NH_3. The reaction which is strongly exothermic proceeds in all cases at temperatures below 150° and for diketones is often very violent.

Benzophenone, benzil, and benzoin with sodium amide at 95-110° give benzene; fluorenone and phenanthrenequinone give biphenyl. Benzopinacolone at 200° gives triphenylmethane and benzene; Michler's ketone at 180° gives dimethylaniline.

In the aliphatic and terpene series, the solid ketones were fused with sodium amide while the volatile ones were passed over sodium amide in an electric furnace. $(Me_3C)_2CO$ gave 33% yield of Me_3CH; fenchone gave 36% yield of 1-methyl-3-isopropyl-pentane.

BIBLIOGRAPHY

1. Fuson and Walker, J. Am. Chem. Soc., 52, 3269 (1930).
2. Lock and Bock, Ber., 70B, 916 (1937).
3. Lock and Schreckeneder, Ber., 72B, 511 (1939).
4. Kozlov, Fedoseev, and Drabkin, J. Gen. Chem. (U.S.S.R.), 6, 1686 (1936); C.A., 31, 2591.
5. Fedoseev, C.A., 36, 5471.

6. Lock and Rodiger, Ber., 72B, 861 (1939)
7. Kozlov, Fedoseev, and Olifson, J. Gen. Chem. (U.S.S.R.), 6, 259 (1936); C.A. 30, 4845
8. Fedoseev, J. Gen. Chem. (U.S.S.R.), 7, 1364 (1937); C.A., 31, 8527.
9. Olifson, J. Gen. Chem. (U.S.S.R.), 9, 36 (1939); C.A., 33, 6291.
10. Huntress and Seikel, J. Am. Chem. Soc., 61, 816, 1066 (1939).
11. Freidlin and Bulanova, J. Gen. Chem. (U.S.S.R.), 9, 299 (1939); C.A., 33, 9299.
12. Freidlin, Lebedeva, and Kuznetsova, J. Gen. Chem., (U.S.S.R.), 9, 1589 (1939); C.A., 34, 2788.

Reported by David B. Guthrie
May 12, 1943

EFFECT OF METAL HALIDES ON CERTAIN GRIGNARD REACTIONS

Kharasch et al., University of Chicago

Small amounts of cobalt chloride and certain other metallic halides have been found to materially affect the course of certain Grignard reactions. The postulation has been made that the reactions proceed through a chain mechanism in which cobalt subhalide (CoCl) is the active chain carrier. Thus the cobalt chloride is simply an oxidation - reduction catalyst in most cases. The following reactions will illustrate the mechanisms proposed.

Reaction of Aryl Grignard Reagents and Aryl or Alkyl Halides.--Aryl Grignard reagents and alkyl or arly halides do not react under ordinary conditions. In the presence of a small amount of cobaltous chloride a vigorous reaction takes place for which the following reactions have been postulated.

$$C_6H_5MgBr + CoCl_2 \rightarrow C_6H_5CoCl + MgBrCl$$

$$2C_6H_5CoCl \rightarrow C_6H_5-C_6H_5 + 2\cdot CoCl$$

$$.CoCl + BrC_6H_5 \rightarrow CoClBr + C_6H_5\cdot$$

$$XC_6H_5\cdot \rightarrow C_6H_6 + C_6H_5-C_6H_5 + \text{poly phenyls}$$

An alkyl halide such as ethyl bromide can be substituted for the bromobenzene. In this case ethylene and ethane are produced from the free radicals formed by the reaction of cobalt subhalide with ethyl bromide, and no polyphenyls are formed. This proves that the polyphenyls come from the bromobenzene rather than from the Grignard reagent.

An application of this reaction has been made in the preparation of hexestrol. Cobalt subchloride, formed by the reaction of phenyl- or methylmagnesium bromide and cobalt chloride, is allowed to react with anethole hydrobromide to produce the free radical. The free radical formed dimerizes to hexestrol dimethyl ether. Yields of 42 per cent have been obtained in this reaction.

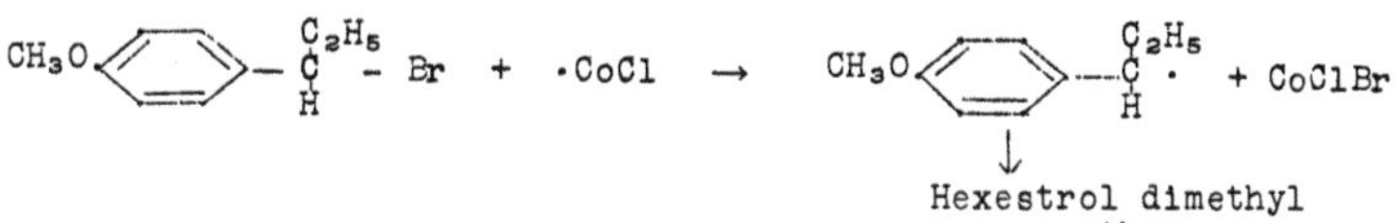

CH3O
C2H5
C
H
- Br
+ ·CoCl
→
CH3O
C2H5
C ·
H
+ CoClBr

Hexestrol dimethyl ether

The Reaction of Grignard Reagents and Aromatic Acyl Halides.--When phenylmagnesium bromide is added to an ethereal solution of benzoyl chloride, the normal reaction products are benzophenone and triphenylcarbinol. In the presence of a small amount of cobalt chloride ethyl benzoate, biphenyl, benzophenone, benzoic acid, phenylbenzoin, tetraphenylethylene oxide and stilbene dibenzoate are obtained. These products can be accounted for by a chain mechanism

in which cobalt subhalide is the active chain carrier. In such a reaction free benzoyl radicals would be formed by the reaction of cobalt subhalide with benzoyl chloride. The benzoyl radicals could react with the molecules present in solution, or dimerize to benzil which would then undergo subsequent reactions. The great reactivity of benzil would preclude the possibility of its being isolated in any great amount from such a reaction mixture.

$$C_6H_5MgBr + CoCl_2 \rightarrow C_6H_5CoCl + MgBrCl$$

$$2C_6H_5CoCl \rightarrow C_6H_5-C_6H_5 + 2\cdot CoCl$$

$$\cdot CoCl + C_6H_5COCl \rightarrow CoCl_2 + C_6H_5CO\cdot$$

$$C_6H_5CO\cdot + C_2H_5OC_2H_5 \rightarrow C_6H_5CO_2C_2H_5 + C_2H_5\cdot$$

$$2C_2H_5\cdot \rightarrow C_2H_4 + C_2H_6$$

$$2C_6H_5CO\cdot \rightarrow C_6H_5COCOC_6H_5$$

$$C_6H_5COCOC_6H_5 + C_6H_5MgBr \rightarrow (C_6H_5)_2COHCOC_6H_5$$

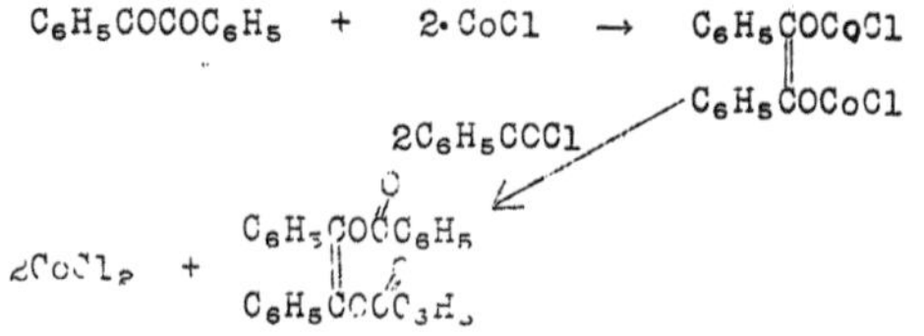

The Reaction of Grignard Reagents with Vinyl Halides.--Under ordinary conditions vinyl halides cannot be coupled with Grignard reagents, but in the presence of small amounts of cobaltous chloride this reaction takes place.

$$C_6H_5MgBr + H_2C{=}CHCl \xrightarrow{CoCl_2} \underset{50\text{-}75\%}{C_6H_5-CH{=}CH_2} + MgBrCl$$

The condensation is general with aryl or aryl-aliphatic Grignard reagents and vinyl halides which have a hydrogen on the unsaturated carbon holding the halogen.

The Oxidation of Grignard Reagents.--Aliphatic Grignard reagents are readily oxidized to give high yields of the corresponding alcohol while aromatic Grignard reagents give only small yields of the corresponding phenol. Cobaltous chloride has very little effect on the oxidation of alkyl Grignard reagents; with aryl Grignard reagents biaryls are formed and oxidation is suppressed. These differences are explained by the fact that the oxidation reaction is faster with alkyl Grignard reagents than the reaction involving cobaltous chloride and hence is the predominate reaction. With aryl Grignard reagents, the opposite is true.

Bibliography

Kharasch and coworkers, J. Am. Chem. Soc., 63, 2305, 2308, 2315, 2316, 3239 (1941); 65, 491, 493, 495, 498, 501, 504 (1943).

Reported by J. R. Elliott
May 12, 1943

NATURAL STILBENES

The first indication that stilbene derivatives occur in nature was due to Asahina and Asano[1] who obtained a small amount of phenolic acid in their investigation of hydrangenol, a phenolic substance isolated from hydrangea. They assigned to this acid structure II, which they believed is interconvertible with strucqture I, representing hydrangenol:

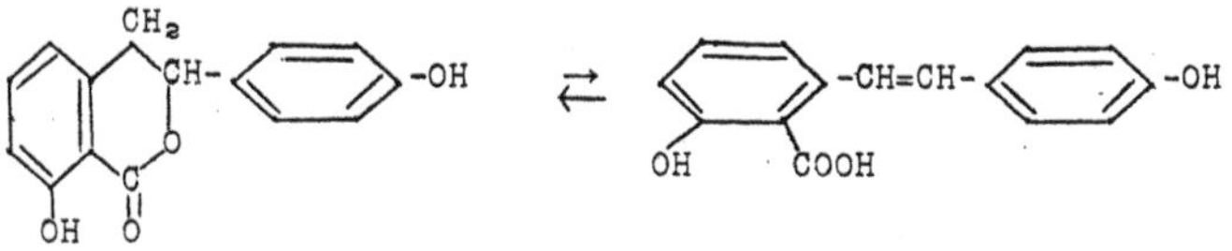

I II

The isolation of rhapontin from Turkish rhubarb roots by Kawamura,[3] and the proof that it is the glucoside of 3,5,3'-trihydroxy-4'-methoxystilbene by the same author seems to be the first evidence that stilbene derivatives can be isolated as such from natural products.

Shortly afterwards, Takaoka (10-15,17) established the structures of resveratrol and hydroxyresveratrol, both isolated previously by Saito and Suginome,[4] from the dried roots of white hellebore; and Erdtman[2] isolated from pine heart wood pinosylvin and its monomethyl ether and assigned the structures 3,5-dihydroxystilbene and 3-hydroxy-5-methoxystilbene respectively. Späth added pterostilbene, 3,5-dimethoxy-4'-hydroxystilbene, to the list.

From suitable solvents these substances can be obtained in colorless crystals with definite melting points. Their reactions are characteristic of the functional groups. They dissolve in concentrated sulfuric acid with yellow to red coloration, take up one molecule of hydrogen for each molecule of substance, decolorize potassium permanganate solution, and yield upon oxidation phenolic acids (together with benzoic acid in the case of pinosylvin and its monomethylether). They easily can be methylated, acetylated, or benzoylated. Their structures have been well established both by analysis and by synthesis.

The method used by Takaoka[11] in elucidating the structure of hydroxyresveratrol may serve as an illustration of the general method of attack. The formation of tetraacetyl- and tetrabenzoylhydroxyresveratrol shows the presence of four hydroxyl groups in the molecule. The formation of the former was achieved by boiling hydroxyresveratrol in anhydrous acetic acid for one hour and that of the latter by allowing the substance to react with benzoyl chloride in pyridine at 0°C.

Distillation with zinc dust yielded resorcinol. Oxidation with chromic oxide in glacial acetic acid converted the tetraacetate to β-diacetylresorcylic acid. The same treatment converted the tetrabenzoate to two products, α-dibenzoyl- and β-dibenzoylresorcyclic acid.

first indication that stilbene derivatives occur
o Asahina and Asano[1] who obtained a small amount
heir investigation of hydrangenol, a phenolic
from hydrangea. They assigned to this acid st
[illegible] interconvertible with structure
[illegible]

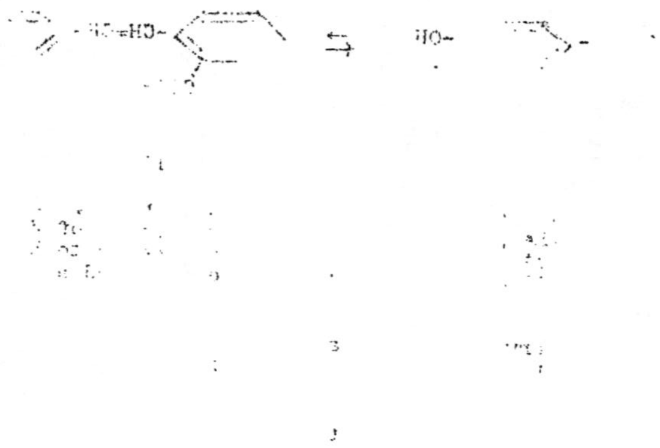

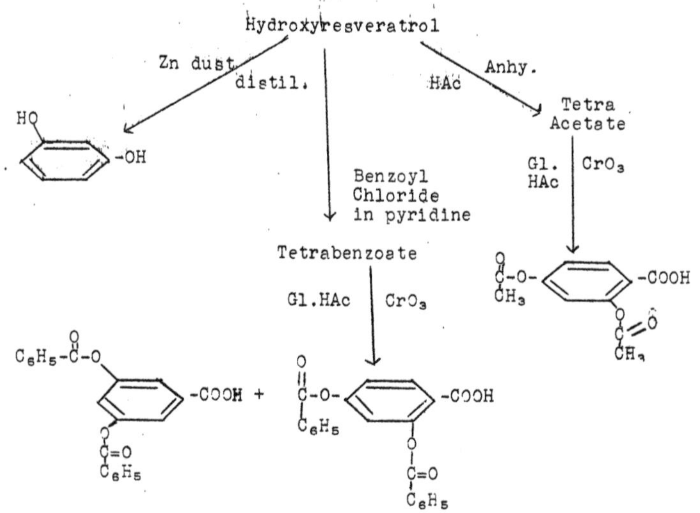
Hydroxyresveratrol
Zn dust distil.
Anhy. HAc
Tetra Acetate
HO
-OH
Benzoyl Chloride in pyridine
Gl. HAc
CrO3
Tetrabenzoate
Gl.HAc
CrO3
-COOH
CH3
C6H5-C-O
-COOH +
C-O-
C6H5
-COOH
C=O
C6H5

Absorption study showed that the spectrum is similar to that of stilbene. Therefore, hydroxyresveratrol must be 3,5,2', 4'-tetrahydroxystilbene.

The synthesis of these compounds consisted in subjecting a phenolic aldehyde and a phenylacetic acid (substituted or unsubstituted as the case may require) to a Perkin's condensation conducted in acetic anhydride, decarboxylating the acetoxystilbene-α-carboxylic acid thus obtained, and finally hydrolizing the acetyl groups.

The synthesis of resveratrol is interesting because it was achieved both by Takaoka and by Späth, employing the same underlying principle but different starting materials.

Takaoka[14,17] started with 3,5-dihydroxyphenylacetic acid and p-hydroxybenzaldehyde:

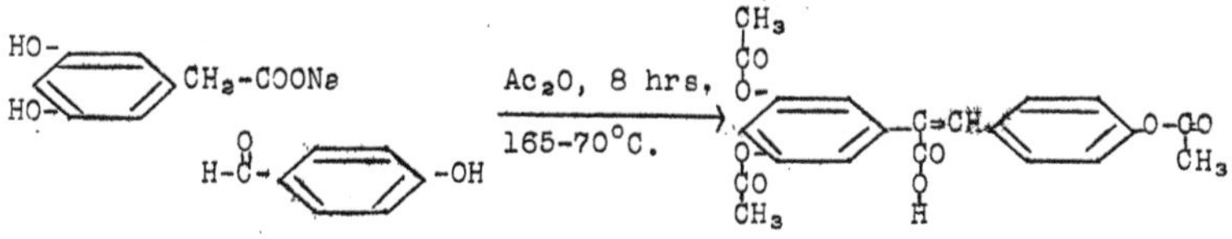
HO-
HO-
CH2-COONa
H-C-
-OH
Ac2O, 8 hrs,
165-70°C.
CH3
CO
O-
C=CH
CO
O
H
O-CO
CH3

$\xrightarrow[\text{200-10°C., 40 min.}]{-CO_2 \text{ / Naturkupfer C in quinoline}}$ HCl → triacetoxystilbene (identical with triacetate of resveratrol.)

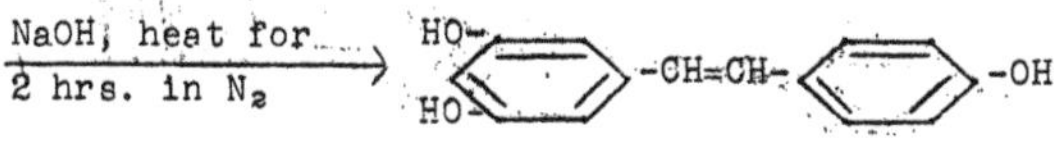
NaOH; heat for
2 hrs. in N_2
HO-
HO-
-CH=CH-
-OH

Spath[7] started with p-hydroxyphenylacetic acid and 3,5-dihydroxybenzaldehyde:

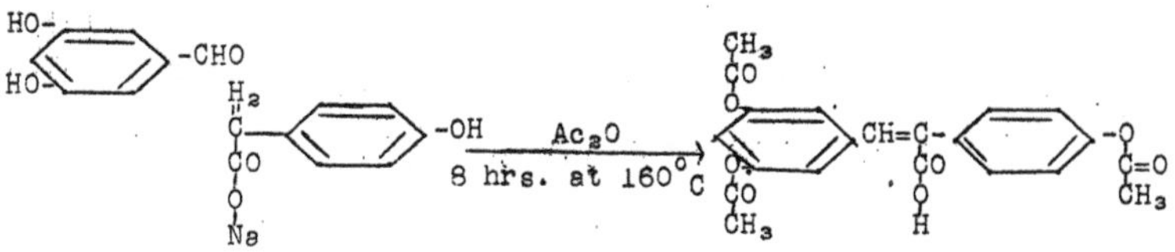
HO-
HO-
-CHO
H_2
C
CO
O
Na
-OH
Ac_2O
8 hrs. at 160°C
CH_3
CO
O
CH=C-
CO
O
H
-O
C=O
CH_3

10% NaOH, 25° for 20 hrs. → HCl → 3,5,4'-trihydroxy-α-stilbene carboxylic acid. → Naturkupfer C, quinoline, 220°C.

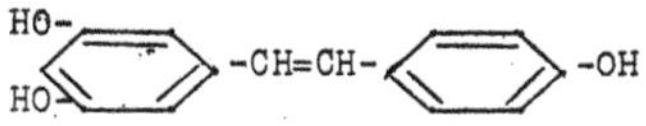
HO-
HO-
-CH=CH-
-OH

Indications of *cis*-*trans* isomerizm were noticed by Späth and Kromp[6] when they discovered that after decarboxylation, synthetic pterostilbene refused to crystallize even when seeded with the natural product. They ascribed this to the production of a *cis* or a mixture of *cis* and *trans* forms during the decarboxylation, while the natural product is assumed to be the *trans* form. By allowing it to stand for a long time with hydrogen chloride in a methyl alcohol-water mixture, a product was finally obtained which was found to be identical with the natural product.

This phenomenon again was noticed when they synthesized pinosylvin monomethyl ether.[8] In this case, the isomerization was e effected by heating the product in an evacuated bomb for 2.5 min. at a temperature of 350°C.

It is remarkable that all of the six natural stilbenes so far identified contain the 3,5-dihydroxyphenyl group.

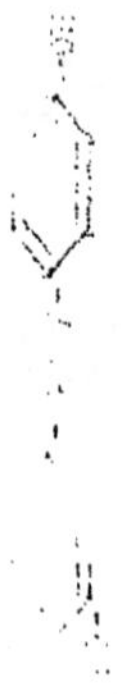

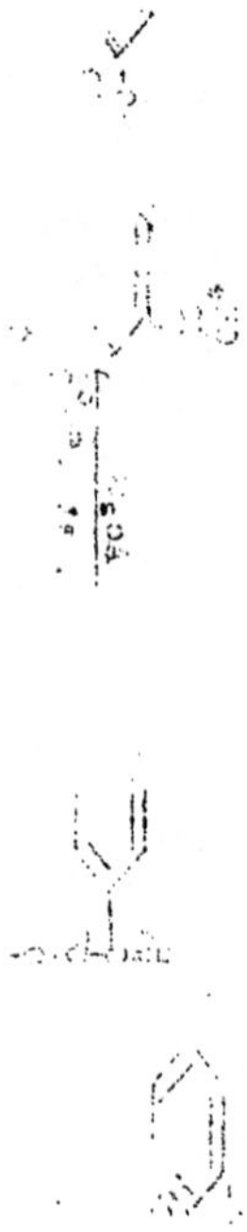

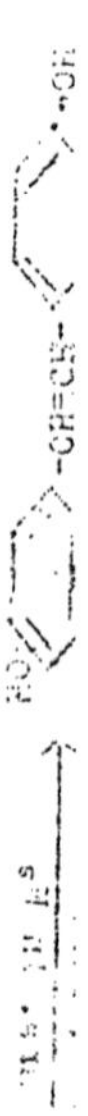

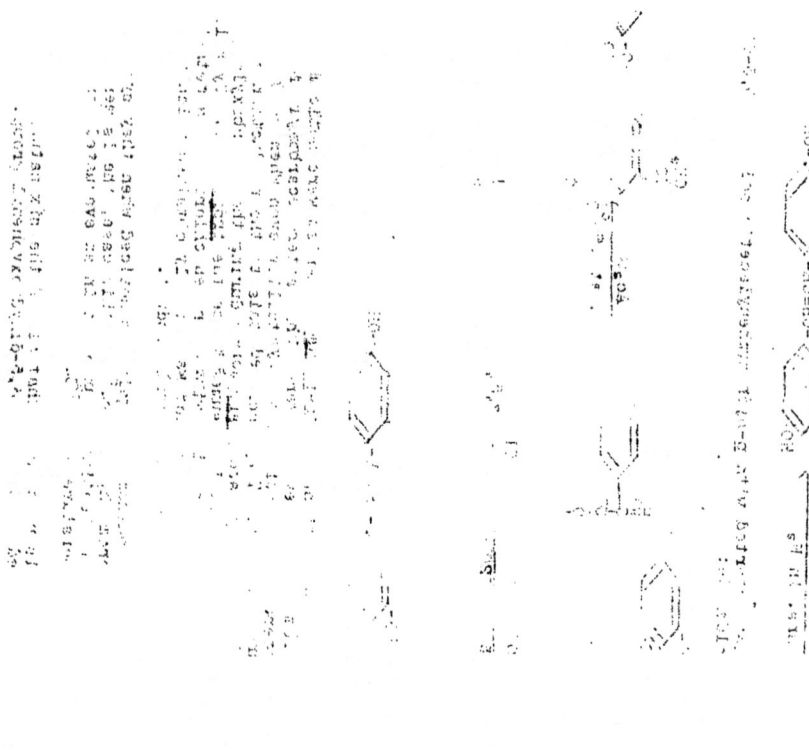

BIBLIOGRAPHY

1. Asahina and Asano, J. Pharmac. Soc. Japan, 50, 573 (1930); Ber., 63, 429 (1930)
2. Erdtman, Ann., 539, 116 (1939)
3. Kawamura, J. Pharmac. Soc. Japan, 58, 83 (1938)
4. Saito and Suginome, Bull. Chem. Soc. Japan, 11, 168 (1936)
5. Späth and Schläger, Ber., 73, 881 (1940); C. A., 35, 740 (1941)
6. Späth and Kromp, Ber., 74, 189 (1941); C. A., 35, 4002 (1941)
7. Späth and Kromp, Ber., 74, 867 (1941); C. A., 35, 6948 (1941)
8. Späth and Liebherr, Ber., 74, 869 (1941); C. A., 35, 6949 (1941)
9. Späth and Kromp, Ber., 74, 1424 (1941); C. A., 36, 5472 (1942)
10. Takaoka, J. Chem. Soc. Japan, 60, 1091 (1939)
11. Takaoka, ibid., 1261 (1939); C. A., 36, 5162 (1942)
12. Takaoka, J. Faculty Sci. Hokkaido Imp. Univ., Ser. III, 3, 1-16 (1940); C. A., 34, 7887 (1940)
13. Takaoka, Proc. Imp. Acad. (Tokyo), 16, 405 (1940); C. A. 35, 1398 (1941)
14. Takaoka, J. Chem. Soc. Japan, 61, 30 (1940)
15. Takaoka, ibid., 61, 96 (1940)
16. Takaoka, ibid., 61, 374 (1940); C. A., 35, 1399 (1941)
17. Takaoka, ibid., 61, 1067 (1940)

Reported by K. H. Chen
May 19, 1943.

Substance	Structure	Source	Synthesized from	Ref.
Resveratrol m.p. 261°	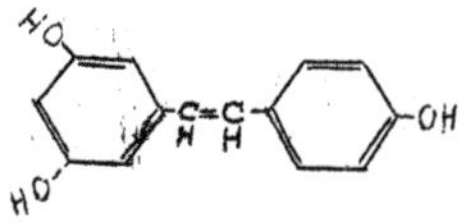	White Hellebore roots	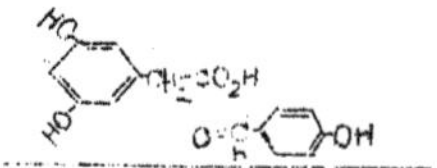 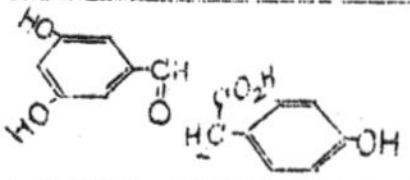	4,10, 13,14, 17
Hydroxy-resveratrol m.p. 199.5°	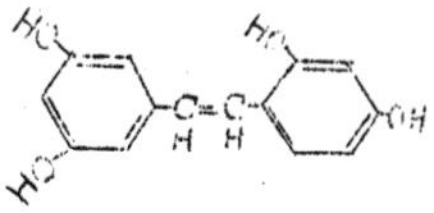	"	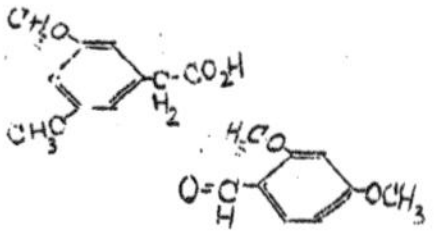	4,11, 15
Pterostilbene m.p. 85-6°	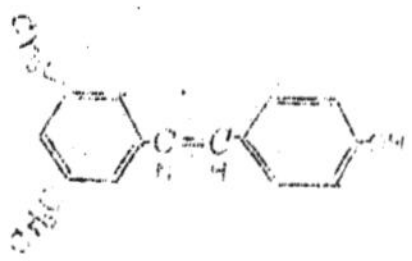	Red Sandal wood	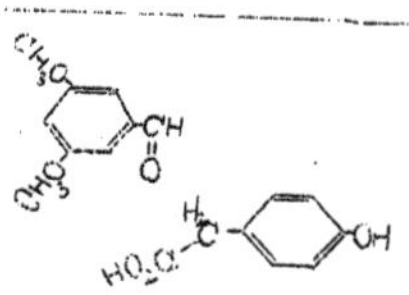	5,6
Pinosylvin m.p. 156°	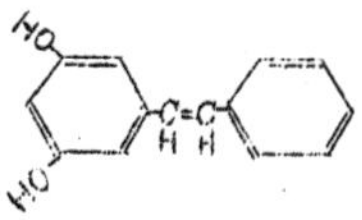	Pine heart wood	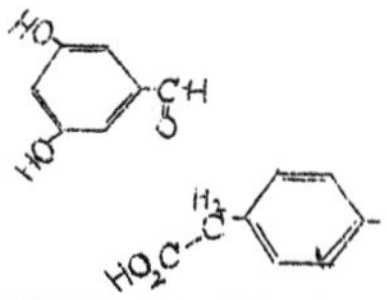	2,8
Pinosylvin monomethyl ether m.p. 122-3°	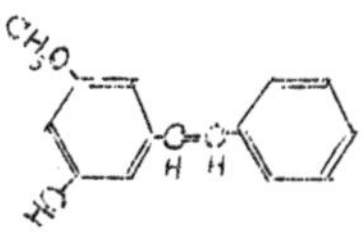		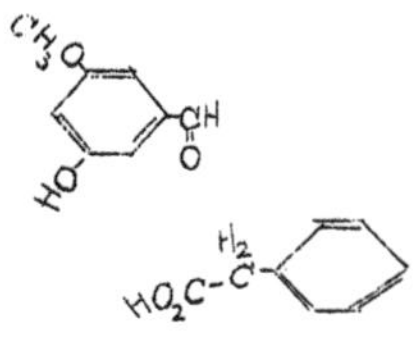	2,9
Rhapontigenin m.p. 186-7°	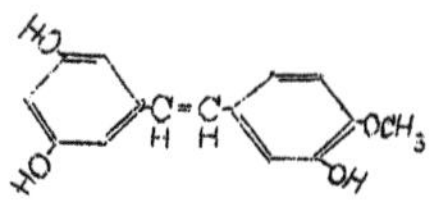	Turkish rhubarb roots	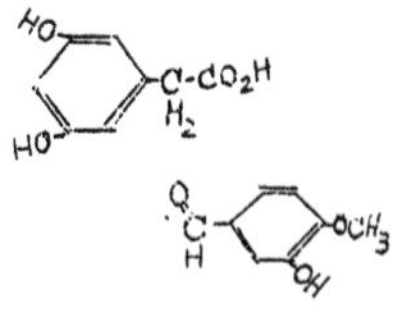	3,16

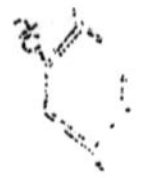

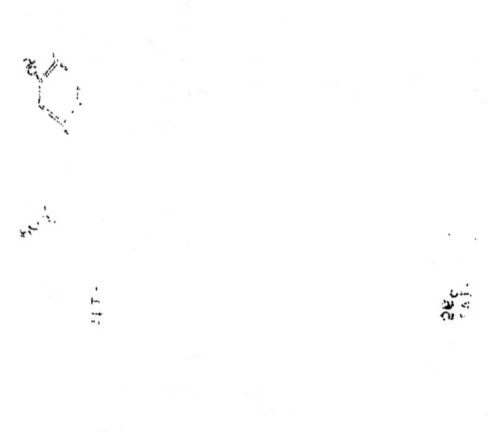

INDUSTRIAL FERMENTATIONS

Industrial catalytic and enzymatic syntheses differ in that the former usually involves the use of a simple type of chemical made outside the reacting mixture which is then added to the reactants under controlled conditions. In the latter processes, the catalyst, the enzymes, are manufactured during the course of the reaction. Different organisms bring about different chemical changes in the same medium, and the chemism of a given organism varies in the same substrate if the nutrients and physical or chemical environements are altered. For example, molds commonly produce citric acid from carbohydrates, while bacteria do not. The butanol-acetone fermentation industry depends for maximum yields upon the use of the appropriate organism under anaerobic conditions. Yeast produces only small amounts of glycerol relative to ethanol in the normal or slightly acid medium, while in alkaline medium or in the presence of sodium bisulfite, the relative yields are reversed.

The catalyst in purely chemical syntheses brings about a practically complete final reaction without a complexity of intermediates— the action is drastic. In fermentations you are dealing with a series of graded products or stages in oxidation and reduction. The formation of the various fermentation products is the process whereby the organism secures energy for growth and maintenance.

Emphasis will be placed predominately on the fermentation of carbohydrates and their dissimilation products since they furnish the important sources of energy for the organisms and for the enzymatic production of chemicals on a large scale. The following chart shows some of the many products obtained by fermentation of carbohydrates employing the proper mold, yeast or bacteria.

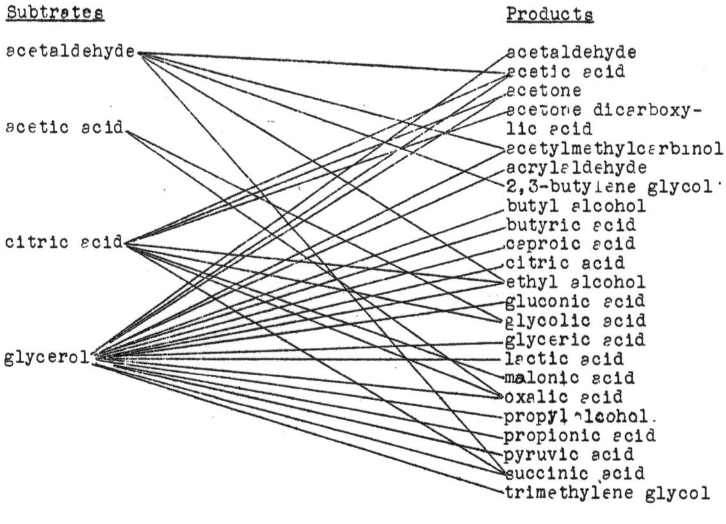
Subtrates
Products
acetaldehyde
acetic acid
citric acid
glycerol
acetaldehyde
acetic acid
acetone
acetone dicarboxylic acid
acetylmethylcarbinol
acrylaldehyde
2,3-butylene glycol
butyl alcohol
butyric acid
caproic acid
citric acid
ethyl alcohol
gluconic acid
glycolic acid
glyceric acid
lactic acid
malonic acid
oxalic acid
propyl alcohol
propionic acid
pyruvic acid
succinic acid
trimethylene glycol

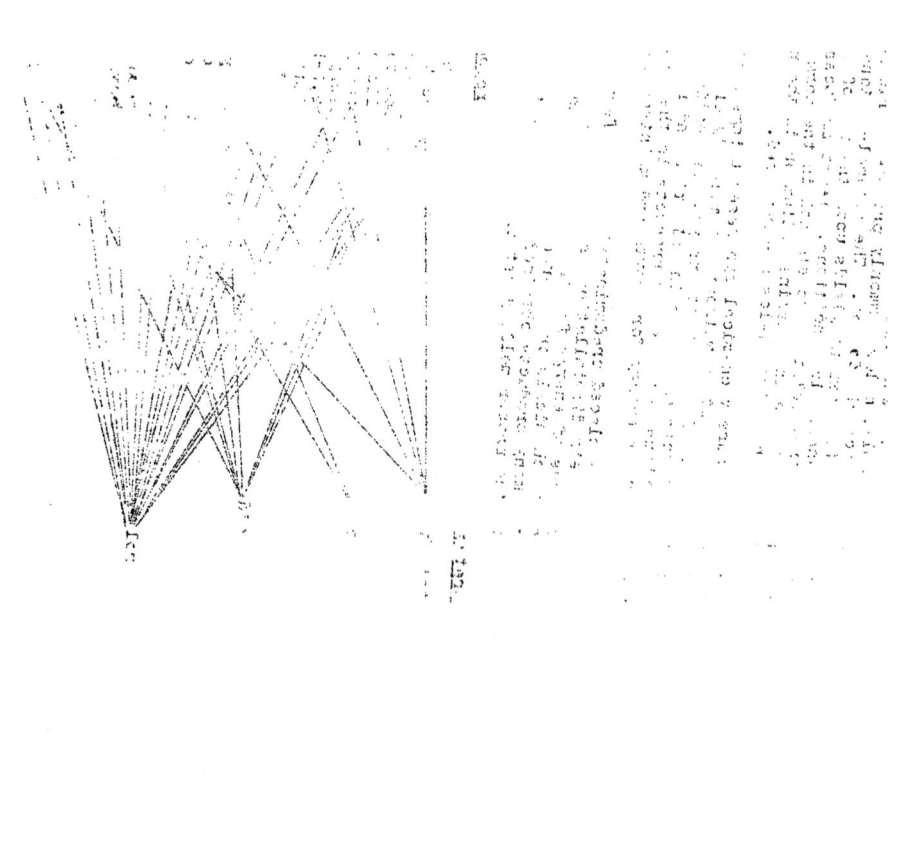

It is interesting to note that there is evidence of synthesis (glycerol $\rightarrow$ acetylmethylcarbinol or citric acid) as well as degradation.

<u>Ethyl alcohol</u>

A suitable mechanism scheme for alcoholic fermentation has long been sought ever since Gay-Lussac proposed the empirical equation:

$$C_6H_{12}O_6 \rightarrow 2CO_2 + 2C_2H_5OH$$

After several decades of speculation as to the intermediates of this reaction, a scheme was suggested by Wohl, Neubauer, Neuberg and Kerb.

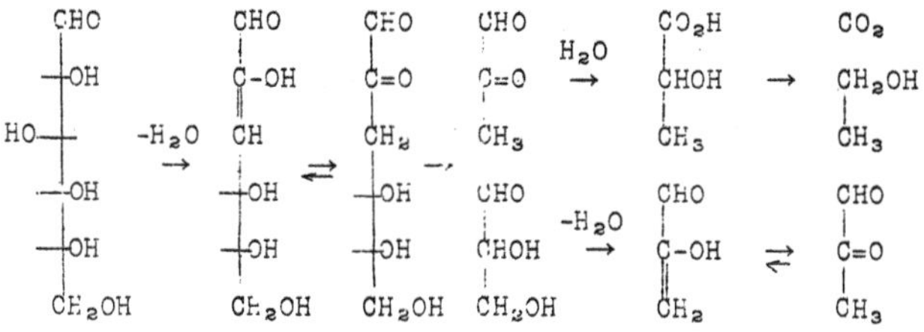

Later, the scheme was modified.

$$C_6H_{12}O_6 - 2H_2O \rightarrow \text{methylglyoxal aldol } (C_6H_8O_4)$$

$$C_6H_8O_4 \rightarrow 2CH_3CO\text{-}CHO \rightleftarrows CH_2{=}COH\text{-}CHO$$

$$CH_2{=}COH\text{-}CHO + H_2O \xrightarrow{H_2} C_3H_5(OH)_3$$
$$CH_2{=}COH\text{-}CHO + O \quad CH_3CO\text{-}CO_2H$$

$$CH_3CO\text{-}CO_2H \rightarrow CH_3CHO + CO_2$$

$$CH_3CHO + CH_3CO\text{-}CHO + H_2O \rightarrow CH_3CO\text{-}CO_2H + CH_3CH_2OH$$

Embden's study of phosphoglyceric acid seemed to bring nearer an understanding of the later phases of sugar dissimilation. The essential feature was that the methylglyoxal disappers from the scheme and pyruvic acid is formed from the phosphoglyceric acid.

$$\underset{OPO_3H_2}{CH_2}\text{-}\underset{OH}{CH}\text{-}CO_2H \rightarrow CH_3CO\text{-}CO_2H + H_3PO_4,$$

The pyruvic acid is then converted to acetaldehyde by the action of carboxylase. Fischer and Baer's work on the fermentation of phosphoglycericaldehyde showed the conversion of acetaldehyde to ethyl alcohol by the oxidation-reduction system:

$$\underset{OPO_3H_2}{CH_2}-\underset{OH}{CH}-CHO \xrightarrow{\text{oxidation}} \underset{OPO_3H_2}{CH_2}-\underset{OH}{CH}-CO_2H$$

$$CH_3CHO \xrightarrow{\text{reduction}} CH_3CH_2OH$$

Thus, 3-phosphoglyceric acid is generated so as to continue the cycle of glucose → ethyl alcohol.

The concept today of alcoholic fermentation intermediates involves enzyme systems concerned with phosphorylation-phosphorolysis, hydrogen-transport and decarboxylation:

$$\text{glucose} \rightleftarrows \text{hexosemonophosphate} \xrightleftharpoons{\text{3 steps}} \text{fructose-1,6-diphosphate}$$

$$\rightleftarrows \underset{\text{phosphodihydroxyacetone}}{\text{phosphoglyceraldehyde} \atop \uparrow\downarrow} \xrightleftharpoons[\text{A.D.P.}]{\text{cozymase}} \text{3-phosphoglyceric acid} \rightleftarrows \text{2-phospho glyceric acid.}$$

$$\rightleftarrows \underset{+ \text{ A.T.P.} + H_2\text{-cozymase}}{\text{phosphopyruvic acid}} \xrightleftharpoons{\text{adenosine}} \underset{\text{adenylic acid}}{\text{pyruvic acid} \atop +} \xrightleftharpoons{\text{carboxylase}} \text{acetaldehyde}$$

$$\xrightleftharpoons[H_2\text{-cozymase}]{} \text{ethyl alcohol} + \text{cozymase}$$

Molasses has been the most important raw material for alcoholic fermentation. However, other materials must be employed even though the yields are not as great or the cost as reasonable. Corn, wheat, sugar beets and the Jerusalem artichoke are being successfully fermented to boost the production of ethanol. Sulfite wastes are being fermented in Sweden and Germany to produce alcohol economically. A process involving the fermentation of wood sugars from sawdust has been successfully employed.

Butanol-acetone

$$\text{corn, rice, wheat} \xrightarrow{\text{2-3 days}} \underset{6}{\text{n-butanol}} + \underset{3}{\text{acetone}} + \underset{1}{\text{ethanol}} + CO_2 + H_2$$

96% utilization of corn

$$\text{dilute slop} \rightarrow \text{acetic acid} \xrightarrow{\text{n-butanol}} \text{butyl acetate}$$

Lactic acid

$$\text{cane and beet sugar, molasses, corn starch, corn sugar} \xrightarrow[CaCO_3]{55^\circ\text{-5-6 days}} \underset{(60\%)}{\text{Ca-lactate}} \xrightarrow{SO_2} \text{lactic acid}$$

$$\text{Jerusalem artichoke} \xrightarrow{H^+} \text{levulose} \rightarrow \text{d-lactic acid (94\%)}$$

Aeration is highly important to obtain high yields; however, if yeast extract is used instead of bacteria, aeration will decrease the yield.

Glycerol

$$\text{molasses} \xrightarrow[NaHSO_3]{Na_2CO_3 \text{ or}} \text{glycerol (20-25\% conversion)}$$

Normally, in acid or neutral media, only small amounts of glycerol are obtained along with good yields of ethyl alcohol.

Other important and interesting fermentation reactions are summarized in the table below.

Product	Substrate	Yield
acetic acid	molasses, cane-juice	-----
acetic acid	cellulose	40
acetol	1,2-propylene glycol	-----
butanol-isopropyl alcohol-acetone	glucose	-----
2,3-butylene glycol	corn, glucose	40
butyric acid	starch, molasses, pentoses	-----
citric acid	fructose or sucrose	90
diacetyl	acetylmethylcarbinol	-----
fumaric acid	glucose	60
gallic acid	tannin from gallnuts	70
gluconic acid	glucose	97
5-ketogluconic acid	glucose	90
2-ketogluconic acid	glucose	82
itaconic acid	glucose	20
propionic acid	any soluble carbohydrates	-----
kojic acid	glucose	45
oxalic acid	glucose	80
l-sorbose	d-sorbitol	92

BIBLIOGRAPHY

Fulmer, J. Ind. Eng. Chem., 22, 1148 (1930)
Ward, ibid., 30, 1233 (1938)
Boruff, ibid., 31, 1335 (1939)
Jacobs, ibid., 31, 162 (1939)
Wells, ibid., 31, 1518 (1939)
Stubbs, ibid., 32, 1626 (1940)
Fulmer and Wendland, ibid., 33, 1078 (1941)
Wiley and Peterson, ibid., 33, 606 (1941)
Anderson and Greaves, ibid., 34, 1522 (1942)
Bilford, ibid., 34, 1406 (1942)
Walker, J. Soc. Chem. Ind., 49, 946 (1930)
Butlin and Wince, ibid., 59, 41 (1940)
Chrzaszcz, ibid., 55, 884 (1936)
Campbell, ibid., 60, 47 (1941)
Bakonyi, Chem. Ztg., 58, 759 (1934)
Warburg and Christian, Biochem. Z., 303, 40 (1939)
Negelein and Broemel, ibid., 303, 231 (1939)
Nord, Chem. Rev., 26, 423 (1940)

Reported by Robert E. Allen
May 19, 1943.

α-KETO ETHERS

Henze, University of Texas

Although various methods have been used for the preparation of particular keto ethers, only two are of general application and only one of these is of practical importance. The ketonic cleavage of certain β-keto esters has been used to prepare a few keto ethers[3], but the yields are not good nor the reactions clean cut.

$$CH_3\overset{O}{C}\,\overset{OEt}{CHCO_2Et} \xrightarrow{HCl} CH_3\overset{O}{C}\,CH_2OEt + CO_2 + EtOH$$

However, various modifications of Sommelet's method have been used with good results by many workers--notably, Henze and coworkers[1], and Maruyama[5]. This involves the following series of reactions.

I. $ROH + RCHO \xrightarrow{HCl} ROCHClR$

II. $ROCHClR + MCN \rightarrow ROCHR(CN)$

III. $ROCHR(CN) + RMgX \rightarrow \xrightarrow{H_2O} ROCH(R)\overset{O}{C}R$

Preparation

I. To prepare the α-chloro ether, the usual procedure is to mix equimolar amounts of the alcohol and aldehyde and then pass dry HCl into the mixture until it is saturated, or in some cases until two layers separate. The chloro ether is then separated from the water layer, dried over calcium chloride and distilled. Formaldehyde and acetaldehyde are usually employed in the form of trioxymethylene and paraldehyde, respectively, although formalin has been used also. Chloro ethers have thus been obtained in yields of from 50-95%.

II. The corresponding nitriles have been prepared by adding the cloro ethers to either cuprous or silver cyanide. The former salt is usually satisfactory, but where it gives a poor yield, use of the silver salt may cause an improvement. The reaction may be carried out in the presence of a dry organic solvent such as ether or benzene, or merely by mixing the dry reactants. Under these conditions, no evidence of carbylamine formation was observed when silver cyanide was employed.

III. The ketones are then obtained by adding the α-cyano ethers to the appropriate Grignard reagent. The reaction mixture, from which a solid precipitates, is allowed to stand overnight, after which the complex is decomposed by the addition of hydrochloric acid.

The keto ethers which have been prepared by Sommelet's method are those numbered 1-11 in the table at the end of this report. In the preparation of the α-ethoxy γ-chloropropyl alkyl ketones (#11), acrolein was used as the aldehyde and addition of HCl to the double bond took place at the same time as ether formation.

$$CH_2{=}CHCHO + HCl + ROH \xrightarrow{HCl} ClCH_2CH_2CHClOR$$

α,α'-Dialkoxyketones have been prepared by Grimaux and Lefevre[2] using ethers of ethyl glycolate.

$$ROCH_2CO_2Et \xrightarrow{NaOEt} ROCH_2\overset{O}{C}\overset{OR}{C}HCO_2Et \xrightarrow{HCl} (ROCH_2)_2C{=}O$$

However, Henze and coworkers have employed a simpler synthesis which gives better yields. α,γ-Dichloroglycerol is treated with two moles of the appropriate sodium alcoholate. The resulting dialkoxymethyl carbinol is then oxidized with chromic acid to give the ketone.

An unusual type of keto ether has been prepared by Staudinger and Meyer.[4] They treated dimethyl ketene with ethyl vinyl ether and obtained a substituted cyclobutane to which they assigned the following structure.

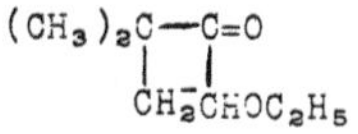

(CH3)2C—C=O
CH2CHOC2H5

Treibe[6] isolated 1-keto-2-methoxy-3-methylcyclohexene-2 from the reaction of 1-keto-3-methylcyclohexene-2 oxide with alkaline methanol.

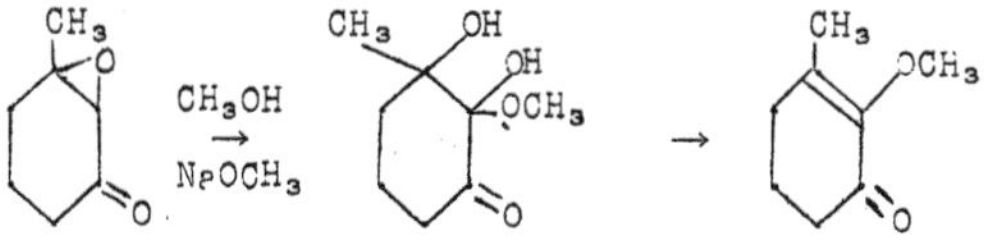

CH3
O
CH3OH
NaOCH3
O
CH3
OH
OH
OCH3
O
CH3
OCH3
O

Similar products were formed with other cyclohexenone oxides.

Killian and coworkers[8], and Rothrock[10] prepared 1-alkoxy-3-butanones by the addition of an alcohol to methyl vinyl ketone, which they obtained by the hydration of vinyl acetylene according to Conaway.[9]

$$CH_2{=}CHC{\equiv}CH \xrightarrow{H_2O} CH_2{=}CHC(=O)CH_3 \xrightarrow{ROH} ROCH_2CH_2C(=O)CH_3$$

<u>Hydantoins</u>

Nirvanol-(ethylphenyl hydantoin) and dilantin-(diphenyl hydantoin) are anticonvulsants used in the treatment, respectively, of St. Vitus dance and epileptic seizures. Although dilantin is effective in controlling these seizures, it is not a cure for epilepsy Henze and coworkers have converted many of their keto ethers to substituted hydantoins in the hope of finding a better anticonvulsant, which they have not yet succeeded in doing. They first used the following series of reactions:

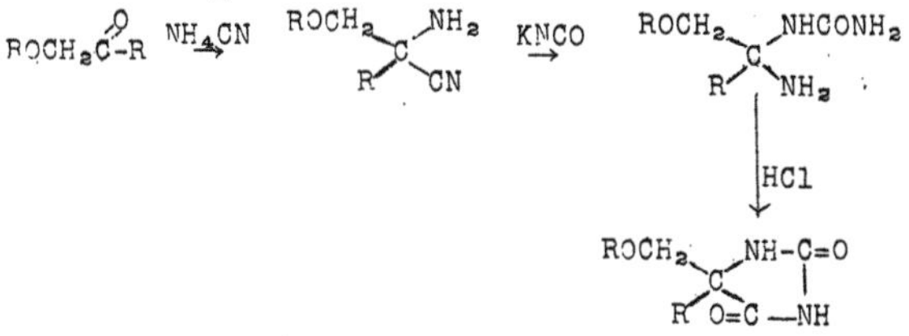

However, since 1934, the Bucherer reaction has been employed to obtain the desired hydantoin in one step by refluxing the keto ether with ammonium carbonate and potassium cyanide in alcohol solution. Those keto ethers from which hydantoins have been prepared are marked with an asterisk in the table.

Quinolines

Henze and coworkers have also prepared substituted quinolines from some keto ethers by means of the Pfitzinger reaction. 3-Alkoxy and 2-(1-alkoxyalkyl)-quinolines were thus made. None of those tested on canaries or mice showed any anti-malarial activity. 2-(1-methoxyethyl)-3- Methylcinchoninic acid, prepared by this method, proved to be an interesting compound. Under suitable conditions, it could be demethylated, demethoxylated, decarboxylated, reduced with or without demethoxylation and decarboxylation, or esterified.

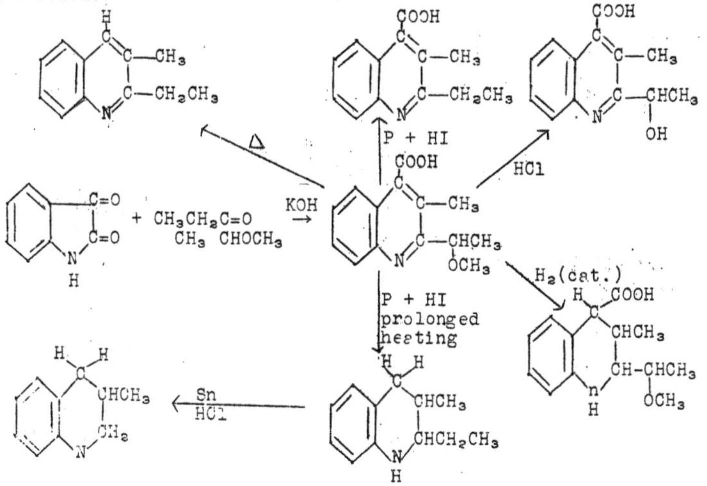

The following keto ethers, RCOR', have been prepared by Henze and coworkers. The numbers under R' refer to the following radicals: 1, methyl; 2, ethyl; 3, n-propyl; 4, i-propyl; 5, n-butyl; 6, i-butyl; 7, s-butyl; 8, t-butyl; 9, n-amyl; 10, i-amyl; 11, phenyl; 12, benzyl. Numbers in parentheses refer to the high and low percentage yields in the series.

R	R'
1.* CH_3OCH_2-	1,2,3,4,5,6,7,8(19),9,10(71).
2. $CH_3CH_2CH_2OCH_2$-	1,2,3,(64),4,(35),5,9,10,11.
3. $(CH_3)_2CHOCH_2$-	1(48),2,3,4,(8),5,6,7,9,10,11.
4.* $ClCH_2CH_2OCH_2$-	1,2(40),3,5,9,10,11(69).
5. $ClCH_2CHCH_3OCH_2$-	1,2,3,(67),5,9,11,12(31).
6. CH_3OCH- CH_3	1,2,3,4(13),5(63),6,7,9,10
7. $ClCH_2CH_2OCH$- CH_3	1(17),2,3(80).
8. $BrCH_2CH_2OCH$- CH_3	1,2(71),3,4,5,6,7,9,10(43).
9. $(ClCH_2)_2CHOCH$- CH_3	1,2(79),3,4,5,6,7(19),9,10.
10.* CH_3CH_2CHOCH- CH_3CH_3	1(68),2,3,4,5,6,7(35),9,10.
11. CH_3CH_2OCH- $ClCH_2CH_2$	1,2,3(75),4,5,6,7,8,9,10(40).
12.* $(ROCH_2)_2C=O$	1(60),2,3,4,5,6,7(16),9,10.

BIBLIOGRAPHY

1. Henze and coworkers, J. Am. Chem. Soc., 56, 1350; 58, 474; 59 540; 61, 433, 1226, 1355, 1574, 2730, 3376; 62, 1758; 63, 2112; 64, 1222, 1897, 2882, J. Org. Chem., 2, 508; 4, 234.
2. Grimaux and Lefevre. Bull. Soc. Chem., III, 1, 11
3. Erlenbach, Ann., 269, 22.
4. Staudinger and Meyer. Helv. Chim. Acta, 7, 19
5. Masuyama, Sci. Rep. Inst. phys. chem. Res. 20, 53.
6. Treibs, Ber. 63, 1493.
7. Lipscomb, Seminar report, "The Pfitzinger Reaction," 12/9/42.
8. Killian, Hennion, and Nieuwland, J. Am. Chem. Soc., 58, 982.
9. Conaway, U.S. Pat. 1,967,225.
10. Rothrock, U.S. Pat. 2,010,828

Reported by O.H. Bullitt, Jr. May 19, 1943

THE PHOTO-ADDITION OF HYDROGEN BROMIDE TO OLEFINIC BONDS

Kharasch and Mayo, University of Chicago
Vaughn and Rust, Shell Development Company

I. The Bromine Atom Theory of Abnormal Addition

Addition of Hydrogen Halides. Markownikoff, in 1870, suggested that in the addition of hydrogen halides to olefinic double bonds, the halogen atom becomes attached to the carbon atom bearing the least number of hydrogen atoms. While much of the early work designed to test this theory proved very confusing because of the contradictory results and conflicting hypotheses of different investigators, most of this confusion has been removed by the extensive liquid-phase studies of Kharasch, Mayo, and their co-workers during the years 1931-1940. They showed that the addition of hydrogen chloride or hydrogen iodide to olefins usually yields only one addition product, the one predicted by Markownikoff's original rule or by an extension of it; the addition of hydrogen bromide can, on the other hand, yield either the "normal" addition product predicted by the rule or the "abnormal" addition product contrary to the rule, or any mixture of the two isomers, depending upon the conditions used.

Mechanism of Normal Addition. The mechanism of normal addition is not well understood. It has been shown that several olefins which form 1:1 addition complexes with the hydrogen halides at low temperatures also react more readily at room temperature to form the normal addition products, while some olefins which do not form the low-temperature addition complexes react less readily at room temperature to form the normal addition products; this would seem to indicate that one step in the normal addition reaction is the formation of an olefin-hydrogen halide complex. The fact that excess hydrogen halide is more effective than excess olefin in accelerating the normal reaction lends support to this theory. Furthermore the negative temperature coefficient of the reaction between 45° and 70°C. might be ascribed to a dissociation of the complex. Kinetic studies have indicated that the normal addition reaction is largely, if not entirely, of an order higher than second, the rate of addition being greatly reduced by the presence of an inert solvent. Certain metallic halides and tetrasubstituted ammonium halides catalyze the normal reaction. These facts seem to indicate that the reaction proceeds by the formation of an olefin-hydrogen halide complex which then reacts in some way with a halide ion or proton to give the normal addition product.

Mechanism of Abnormal Addition. A satisfactory mechanism for the abnormal addition reaction must explain (1) the effect of traces of oxygen or peroxides, (2) the reason for reversal of normal addition, and (3) why hydrogen bromide is the only hydrogen halide capable of abnormal addition. The theory of Kharasch, Mayo, and their co-workers is the only one yet proposed which meets all these requirements.

Effect of Oxygen and Peroxides. Since the reaction is caused by relatively small quantities of oxygen or peroxides, and is inhibited by equally small quantities of antioxidants, it would appear that the abnormal addition is due to a chain reaction. Since hydro-

... ADDITION OF HYDROGEN BROMIDE TO OLEFINIC BONDS

... University of Chicago
... Shell Development Company

I. The Free Radical Atom Theory of Abnormal Addition

Addition of Hydrogen Halides. Markownikoff's rule ... the addition of hydrogen halides to olefinic ...

[illegible]

gen bromide is the only hydrogen halide which is capable of reacting in this way, the oxygen or peroxides must function through the hydrogen bromide rather than through the olefin. The action of oxygen or peroxides on hydrogen bromide should result in oxidation of the bromide ion, either to free bromine or to positive bromine ions. Since more energy is required to separate charged particles in non-polar solvents, the formation of free bromine is more likely. Since molecular bromine alone has no effect upon the reaction, Mayo and Walling suggest that the effect of oxygen or peroxides is probably the slow oxidation of hydrogen bromide to bromine atoms as indicated schematically by equation (1). The fate of the $HO_2\cdot$ radical is unimportant; the $Br\cdot$ atom formed is the particle necessary to carry on the addition chain reaction indicated in equations (2) and (3).

$$HBr + O_2 \xrightarrow{\text{alkene}} H\text{-}O\text{-}O\cdot + Br\cdot \qquad (1)$$

$$RCH{=}CH_2 + Br\cdot \rightarrow R\dot{C}HCH_2Br \qquad (2)$$

$$R\dot{C}HCH_2Br + HBr \rightarrow RCH_2CH_2Br + Br\cdot \qquad (3)$$

Reversal of Normal Addition--Radical Stability. There have been two explanations offered for the reversal of normal addition. Mayo and Walling suggest that the point of attack of the bromine atom depends upon the relative stability of the two bromoalkyl radicals which may be formed. If the directions of all additions by the chain mechanism are to be explained on this basis, the following orders of decreasing stabilities of free radicals are required.

Radicals from hydrocarbons: tertiary > secondary > primary.

Radicals from vinyl-type halides: $R\text{-}CHBr\text{-}\dot{C}H\text{-}X > R\text{-}\dot{C}H\text{-}CHBr\text{-}X$

Radicals from acids, esters: $R\text{-}CHBr\text{-}\dot{C}H\text{-}COOR > R\text{-}\dot{C}H\text{-}CHBr\text{-}COOR'$

If these relative stabilities can be proved correct, the hypothesis of radical stability explains the products of abnormal additions very nicely. Unfortunately definite information on the relative stability of these radicals is lacking and difficult to obtain.

Reversal of Normal Addition--Electron Density at Carbon Atom. Kharasch, Engelmann, and Mayo have offered another explanation, based upon somewhat sounder experimental evidence. They suggest that the free bromine atom, because of its oxidizing properties, will tend to approach the doubly-bonded carbon atom of higher electron density, where it would have the best opportunity of exerting its oxidizing effect by gaining an electron from the carbon. Pauling, Brockway, and Beach, by measurement of interatomic distances in the chloroethylenes, have shown that resonance increases the electron density of the unhalogenated doubly-bonded carbon, as indicated schematically by the following electronic formulas.

Alkyl groups probably cause reduced electron density, at the doubly-bonded carbon to which they are attached. In general, it is seen that the doubly-bonded carbon having the greater number of hydrogen atoms attached will have the greater electron density, and will be attacked preferentially by the free bromine atom, resulting in addition contrary to Markownikoff's Rule. Incidentally, this explanation also lends support to the theory that normal addition proceeds through the formation of an intermolecular addition complex between the olefin and hydrogen bromide, since the mutual effect of the hydrogen bromide dipole and the above-postulated polarity of the olefinic double bond would result in normal Markownikoff's Rule addition to the double bond, as indicated schematically by the following electronic formulas.

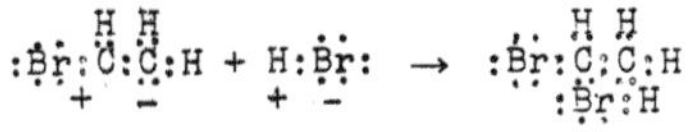

Abnormality of Hydrogen Bromide. The reason that hydrogen bromide is the only hydrogen halide capable of abnormal addition is best explained thermodynamically. The tendency for propagation of the abnormal chain reaction indicated in equations (2) and (3) will depend upon, or at least be measured by, the enthalpy change in each step.

Reaction	ΔH kcal/mol for the Halogens F	Cl	Br	I
(2)	+64	+27	+13	-1
(3)	-60	-15	0	+16

If it be assumed that a rapid chain reaction cannot occur when either step is appreciably endothermic, then atomic iodine would not add to the olefin readily enough, while hydrogen fluoride and hydrogen chloride would not react with the haloalkyl free radical readily enough, to propagate the chain reaction. Since atomic bromine reacts exothermically with olefins to form the bromoalkyl radicals, and these radicals react with hydrogen bromide without absorption of heat, hydrogen bromide is the only one of the hydrogen halides which could be expected, upon thermodynamic grounds, to undergo abnormal addition by the proposed mechanism.

II. The Work of Vaugn, Rust, and Evans

Vapor-Phase Photo-Addition. The only previous report of vapor-phase abnormal addition is found in Bauer's patents for hydrobromination of acetylene under the influence of light and oxygen; his method is not clear, as the transmission limits of the apparatus were not specified and a liquid phase complicated the process. It has now been shown that vapor-phase photo-addition of hydrogen bromide to olefins proceeds even more readily than the corresponding liquid-phase reaction, yielding almost entirely the abnormal addition product. In quartz vessels using light of less than 2900Å wavelength, the reaction proceeds too rapidly to permit measurement of the rate;

in thin-walled pyrex there is sufficient short wave length transmission to permit fairly rapid reaction. The reactions of ethylene, propylene, 1-butene, isobutene, and vinyl chloride have been studied, and each gave almost entirely the abnormal addition product. Thus the photolytic abnormal addition of hydrogen bromide to olefins has been extended from the liquid-phase to the vapor-phase with increased effectiveness.

Sensitizers. Many substances which undergo photolytic dissociation to free radicals at longer wave lengths than 2900 A.U. can act as sensitizers for the abnormal addition reaction; the free radicals formed react with hydrogen bromide to liberate free bromine atoms as a secondary reaction. It has been shown that acetone, which decomposes to yield methyl radicals at 3100 A.U., can effectively initiate the abnormal addition reaction; acetaldehyde has a similar effect. Tetraethyllead, which yields ethyl radicals readily below 3500 A.U., even when used in very small amounts, has a large sensitizing effect upon the reaction. Small amounts of molecular bromine, which are photo-dissociated to bromine atoms, can also serve as sensitizer, but dibromination will complicate the results.

Inhibitors. The action of inhibitors on the abnormal reaction has also been explained by this free radical chain mechanism. When peroxides were believed to be the essential catalysts for the abnormal addition, the inhibitors were believed to function because of their antioxidant properties. It has been shown, however, that methyl iodide and iodine, which certainly are not antioxidants, are powerful retardants of the abnormal photo-addition reaction. The mechanism of chain termination may be postulated as a reaction of

$$R\cdot + I_2 \rightarrow RI + I\cdot$$

$$R\cdot + I\cdot \rightarrow RI$$

$$2I\cdot \rightarrow I_2$$

iodine with free radicals, breaking the chain. As explained thermodynamically above, the reaction of iodine atom with the olefin is a relatively infrequent process, so that the iodine atoms can combine either with free radicals or with each other, but cannot propagate the chain.

Summary. The work of Vaugn, Rust, and Evans has thus made several important contributions to the bromine atom theory of abnormal addition. The identity of the wave length necessary to dissociate hydrogen bromide and the wave length necessary to cause the abnormal addition reaction indicates that the primary step in the process is the formation of free bromine atoms. The rapid rates of reaction and quantitative yields of pure abnormal addition products in vapor-phase photo-addition suggests the method for practical syntheses. The extension of the abnormal addition reaction to the vapor phase answers a question of long standing posed by many previous investigators, and the success of the reaction even in pyrex reaction vessels indicates that the reaction chains are very long. The use of other photo-dissociable materials than oxygen and peroxides to sensitize the reaction, and other materials than antioxidants to retard

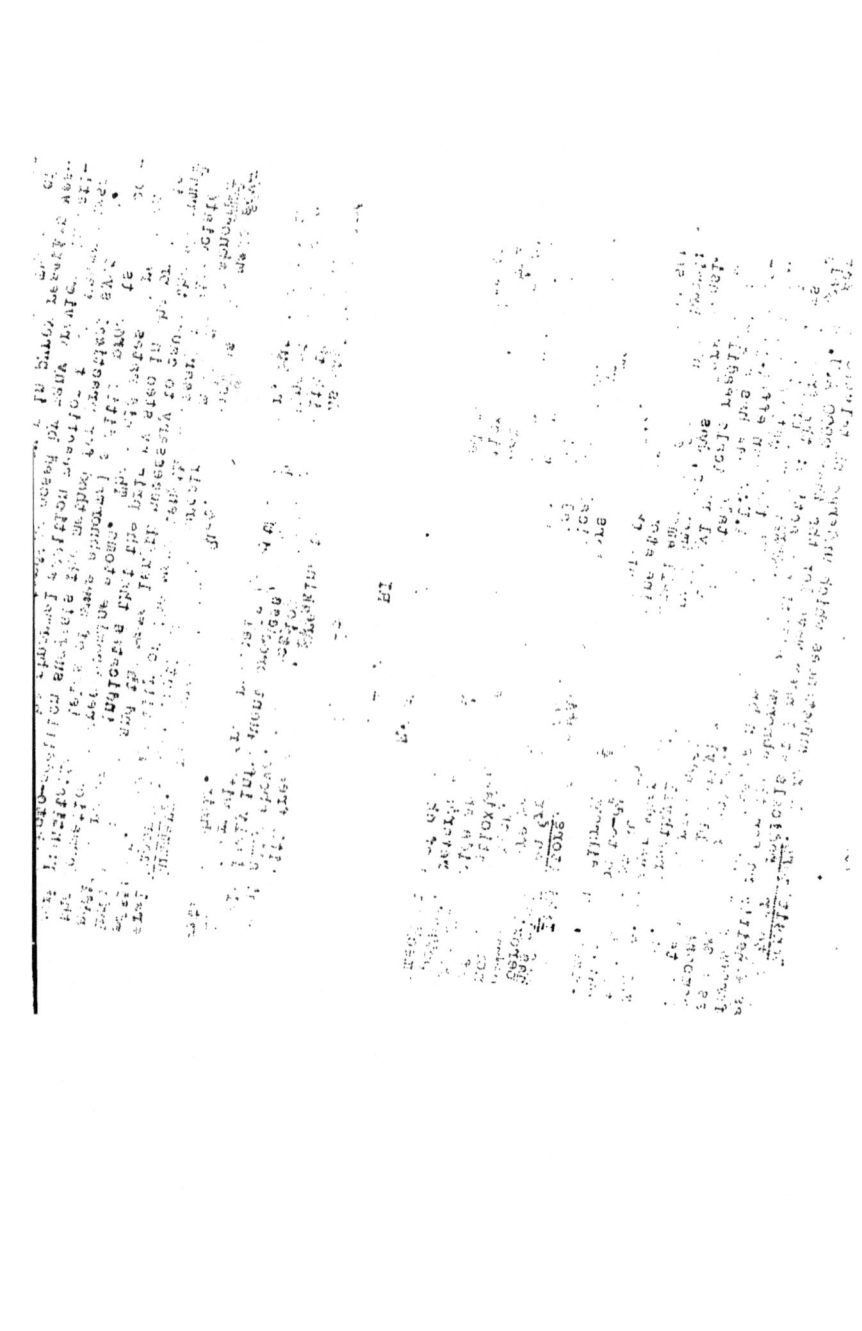

it, explains the catalytic effect of longer wave length and shows that oxygen, peroxides, and certain finely divided metals are not essential promoters of the reaction. All of the evidence substantiates the bromine atom chain mechanism of abnormal addition of hydrogen bromide to olefins.

Bibliography

Bauer, U. S. Patent 1,540,748 (1925).
Bauer, U. S. Patent 1,414,852 (1922).
Conant, Small, and Taylor, J. Am. Chem. Soc., 47, 1971 (1925).
Conn, Kistiakowsky, and Smith, J. Am. Chem. Soc., 60, 2764 (1938).
Kharasch, Englemann, and Mayo, J. Org. Chem., 2, 288 (1937).
Kharasch and Hinckley, J. Am. Chem. Soc., 56, 1212 (1934).
Kharasoh, McNab, and Mayo, J. Am. Chem. Soc., 55, 2531 (1933).
Leighton and Mortensen, J. Am. Chem. Soc., 58, 448 (1936).
Markownikoff, Ann., 153, 256 (1870).
Markownikoff, Compt. rend., 81, 670 (1875).
Mayo and Walling, Chem. Rev., 27, 351 (1940).
Michael, J. Org. Chem., 4, 519 (1939).
Michael and Weiner, J. Org. Chem., 4, 128 (1939).
Pauling, Brockway, and Beach, J. Am. Chem. Soc., 57, 2693 (1935).
Rollefson and Burton, "Photochemistry", Prentice-Hall, Inc., N.Y. 1939, pp. 189 ff, p. 234 ff.
Smith, Chem. and Ind., 15, 833 (1937).
Smith, Chem. and Ind., 16, 461 (1938).
Smith, Ann Repts. on Prog. of Chem. in 1939, Chem. Soc., 36, 219 (1940).
Smith, Chem. and Ind., 57, 461 (1939).
Urushibara and Takebayashi, Bull. Chem. Soc. Japan, 12, 173 (1937).
Vaugn, Rust, and Evans, J. Org. Chem., 7, 477 (1942).
Winstein and Lucas, J. Am. Chem. Soc., 60, 836 (1938).

Reported by Rudolph Deanin
May 19, 1943

THE AZULENES

I. Introduction

The blue coloring material occurring in many natural oils has been observed since the 15th century, and frequent descriptions of blue fractions obtained by distillation are scattered through the literature of the essential oils. The compounds responsible for this color are related to sesquiterpenes; they may be concentrated in the sesquiterpene-alcohol fractions by distillation, or they may be produced by dehydrogenation of these fractions.

Sherndal in 1915 first obtained the blue coloring component in pure form. He proved it to be a liquid hydrocarbon of empirical formula $C_{15}H_{18}$. It was inferred that the structure was of a new type because such intense color had never been observed before in molecules of so low molecular weight containing only carbon and hydrogen.

The structure remained unknown until 1936 when Pfau and Plattner, working in the laboratories of the firm of L. Givaudan and Co. in Vernier, Switzerland, and later Plattner, working at the Technischen Hochschule in Zurich, began an extensive investigation of these compounds. As a result of their work the structure typical of the azulenes has been elucidated, and a number of naturally occurring as well as purely synthetic azulenes have been prepared.

II. Proof of Structure

Proof of the skeleton of the azulenes necessitated first an isolation and structure determination of their sesquiterpene precursors. Plattner and Pfau have carried out this work for guajol and for β-vetivone, the precursors of guajazulene and vetivazulene, respectively. The more important points in the latter proof are outlined below:

Extraction of vetiver oil with Girard's reagent yielded substances which gave crystalline semicarbazones. By fractional crystallization of the semicarbazones a compound was isolated which, when hydrolyzed, gave an optically active ketone, β-vetivone, $C_{15}H_{22}O$. According to molecular refraction determinations this ketone (1) was bicyclic, and (2) had one of the two double bonds conjugated with the carbonyl.

$$\underset{\beta\text{-vetivone}}{C_{15}H_{22}O} \xrightarrow[\text{2. } PBr_3]{\text{1. Na + EtOH}} C_{15}H_{25}Br \xrightarrow[\text{aniline}]{\text{Heat with}} C_{15}H_{24} \xrightarrow[\text{275}^\circ \text{ for 22 hrs.}]{\text{Heat with 3S}}$$

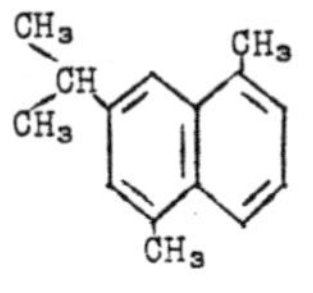
CH3
CH
CH3
CH3
CH3

+ $C_{15}H_{18}$

Vetivazulene

(identical with synthetic sample)

Formation of the naphthalene hydrocarbon indicated an ortho condensation of the two cycles in $C_{15}H_{24}$.

$$\underset{\beta\text{-vetivone}}{C_{15}H_{22}O} \xrightarrow{H + Ni} C_{15}H_{24}O \xrightarrow{H + Ni} C_{15}H_{26}O \xrightarrow{\text{ozone}} \begin{matrix} CH_3 \\ CH_3 \end{matrix}\!>\!CO +$$

$$C_{12}H_{20}O_2 \xrightarrow[\text{distil. with } NaHSO_4]{-H_2O} \underset{\text{(a ketone)}}{C_{12}H_{18}O}$$

All products after the catalytically produced dihydro-β-vetivone ($C_{15}H_{24}O$) were optically inactive. Racemization under the mild conditions of this reduction was unlikely so a plane of symmetry was attributed to dihydro-β-vetivone which was missing from β-vetivone. When dihydro-β-vetivol ($C_{15}H_{26}O$) was produced by a sodium-ethanol reduction it was optically active. On the basis of the isoprene rule the following formula was provisionally adopted for β-vetivone as a working hypothesis:

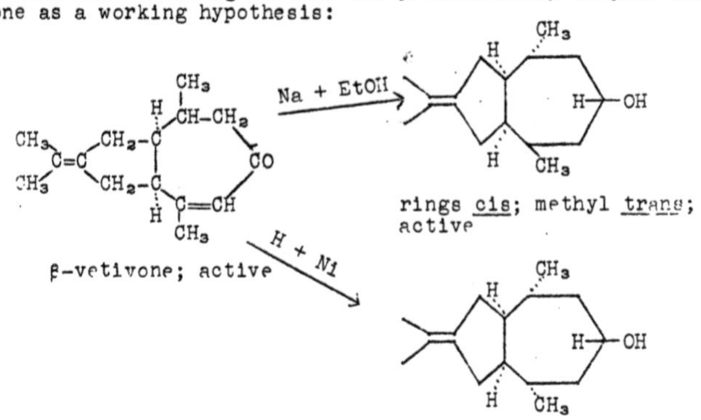
CH3
H
CH-CH2
CH3
C=C
CH2-C
CH2-C
CO
C=CH
CH3
Na + EtOH
H + Ni
H
OH
CH3

β-vetivone; active

rings cis; methyl trans; active

rings cis; methyls cis; inactive

Direct proof for this provisional formula for β-vetivone was obtained by oxidative degradation.

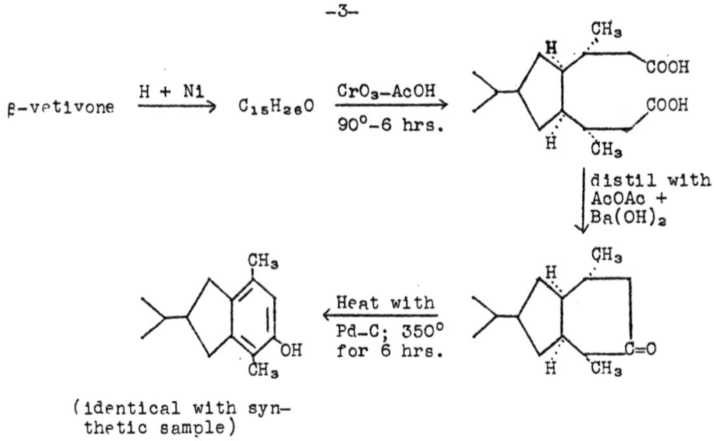
β-vetivone
H + Ni
$C_{15}H_{26}O$
CrO_3-AcOH
90°-6 hrs.
COOH
COOH
CH_3
H
distil with
AcOAc +
$Ba(OH)_2$
C=O
Heat with
Pd-C; 350°
for 6 hrs.
OH
(identical with synthetic sample)

Vetivazulene is then the normal dehydrogenation product of β-vetivone while 1,5-dimethyl-7-isopropyl naphthalene is a side-product formed by a retro-pinacolin rearrangement during dehydrogenation.

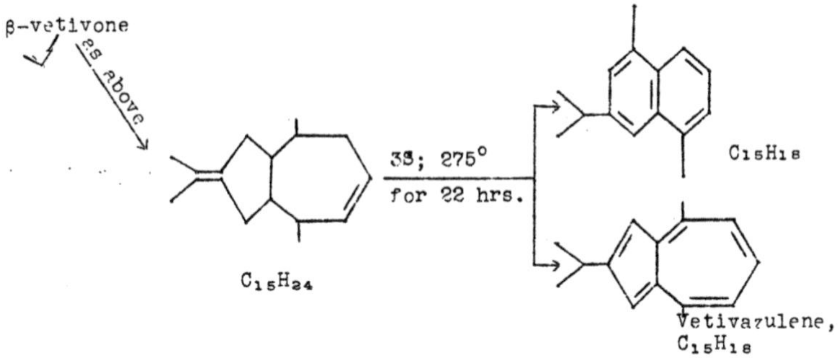
β-vetivone
as above
S; 275°
for 22 hrs.
$C_{15}H_{24}$
$C_{15}H_{18}$
Vetivazulene,
$C_{15}H_{18}$

Vetivazulene has been synthesized according to the method illustrated in the following section

III Synthesis

Azulenes have been synthesized by two general methods. Representative examples are outlined below.

1. Cyclopenteno-cycloheptanone Synthesis:

This method is useful for producing azulene itself or azulenes substituted in the 4-position.

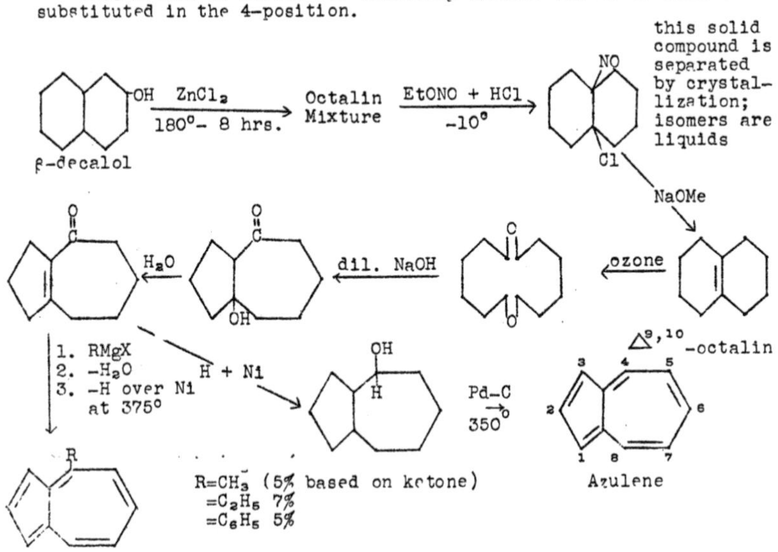
this solid compound is separated by crystallization; isomers are liquids
OH
ZnCl₂
180°- 8 hrs.
Octalin Mixture
EtONO + HCl
-10°
NO
Cl
β-decalol
NaOMe
ozone
dil. NaOH
H₂O
OH
Δ9,10-octalin
1. RMgX
2. -H₂O
3. -H over Ni at 375°
H + Ni
OH
H
Pd-C
350°
3 4 5
2 6
1 8 7
Azulene
R
R=CH₃ (5% based on ketone)
=C₂H₅ 7%
=C₆H₅ 5%

2 Diazoacetic Ester Synthesis:

Diazoacetic ester attacks the benzene ring preferably in the unsubstituted positions. The yields vary according to the constitution of the azulene but are always low e.g., 15% for 2-methylazulene and 1% for 1-methylazulene, based on the corresponding indan.

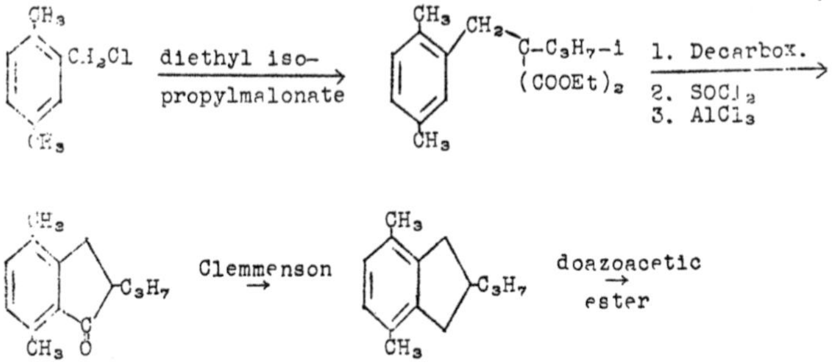
CH₃
CH₂Cl
CH₃
diethyl isopropylmalonate
CH₃
CH₂
C-C₃H₇-i
(COOEt)₂
CH₃
1. Decarbox.
2. SOCl₂
3. AlCl₃
CH₃
C₃H₇
CH₃ O
Clemmenson
CH₃
C₃H₇
CH₃
doazoacetic ester

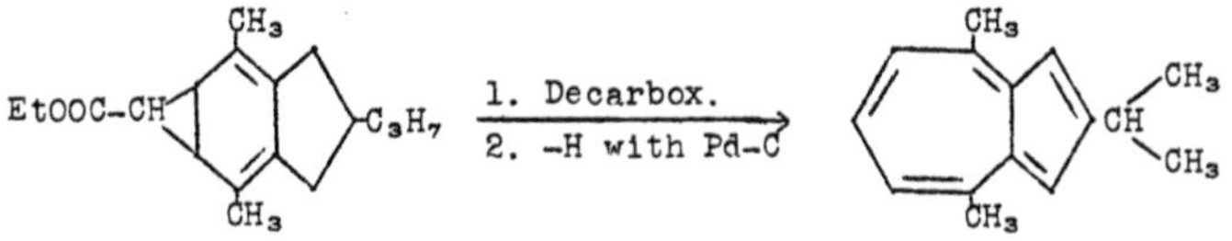

CH3
EtOOC-CH
C3H7
CH3
1. Decarbox.
2. -H with Pd-C
CH3
CH
CH3
CH3
CH3

Vetivazulene

In a similar manner the following azulenes have been synthesized from the appropriate indan: 1-methyl-, 2-methyl-, 1,2-dimethyl-, 4,8-dimethyl-, 1,4-dimethyl- and 5-methyl-.

The azulenes were isolated by extracting the dehydrogenation mixture with 95% phosphoric acid. They were removed from the acid by dilution with ice water and ether extraction. Purification was effected by recrystallization of the trinitrobenzolates. These derivatives were then decomposed chromatographically on Brockmann alumina from a cyclohexane solution.

IV. Absorption Spectra

Plattner made a study of the absorption spectra of the azulenes. He found that they fall into three characteristic groups. The spectrum of a given azulene shows

(1) a definite number of lines with relative intensities characteristic of the group, and
(2) a constant difference between the wave numbers of the bands, also a characteristic of the group.

Schechter and Haller have used these measurements as a basis for tentatively assigning positions to the methyl groups in a new trimethylazulene which they have obtained from pyrethrosin.

The blue to violet color of the azulenes is unique in such a low molecular weight compound. In compounds whose color is due exclusively to the presence of carbon-to-carbon double bonds, the combination occurring in azulene is by far the most active chromophoric system known.

Sklar has demonstrated, however, that the blue color of the azulenes is theoretically what is expected when color is attributed to the transitions between the levels which arise from resonance among the different possible structures. Using heat of hydrogenation data and the Heitler-London-Pauling-Slater quantum mechanical method the calculated maximum absorption band for azulene was 6014 Å in good agreement with the observed value of 7000 Å.

Bibliography

Semmler, Die Ätherische Öle, Vol. III, 260ff.
Sherndal, J. Am. Chem. Soc., 37, 167 (1915).
Hückel and Schnitzpahn, Ann., 505, 274 (1933)
Pfau and Plattner, Helv. Chim. Acta, 19, 858 (1936).
Sklar, J. Chem. Phys., 5, 669 (1937).
Susz, Pfau and Plattner, Helv. Chim. Acta., 20, 469 (1937).
Plattner and Pfau, ibid., 20, 224 (1937).
22, 202, 640 (1939).
23, 768 (1940).
Plattner and Lemay, ibid., 23, 897 (1940).
Plattner and Wyss, ibid., 23, 907 (1940).
Plattner and Magyar, ibid., 24, 191, 1163 (1941).
Plattner and Wyss, ibid., 24, 483 (1941).
Plattner, ibid., 24E, 283 (1941).
Schlecter and Heller, J. Am. Chem. Soc., 63, 3507 (1941).
Plattner and Magyar, Helv. Chim. Acta., 25, 581 (1942).
Plattner and Roniger, ibid., 25, 590 (1942).

Reported by Robert Gander
May, 1943

Lightning Source UK Ltd.
Milton Keynes UK
UKHW011001021118
331648UK00007B/319/P

9 780265 875933